A PHILOSOPHY OF AIR POWER

James Trapier Lowe

UNIVERSITY PRESS OF AMERICA

LANHAM • NEW YORK • LONDON

4720 Boston Way
Lanham, MD 20706

3 Henrietta Street
London WC2E 8LU England

Printed in the United States of America

Library of Congress Cataloging in Publication Data

Lowe, James Trapier.
A philosophy of air power.

Includes index.
1. Air power. 2. Aeronautics, Military. 3. World War, 1939-1945—Aerial operations. I. Title.
UG630.L74 1984 358.4'03 84-5254
ISBN 0-8191-3953-X (alk. paper)
ISBN 0-8191-3954-8 (pbk.: alk. paper)

All University Press of America books are produced on acid-free paper which exceeds the minimum standards set by the National Historical Publications and Records Commission.

Respectfully dedicated to -

"Possum" Hansell

and

Malcolm W. Moss

the acknowledgements

The management team that made the preparation of this manuscript possible consisted of Mary Jane, Jim, Rick, Alan, Kaye, William, Matt, Stephanie, Karen, Laura, Jessica Anne, James Matthew, Al Carey, and Sheila. Other important members were Mary Gill, who did all the typing, and Flo Carey and Bob Lamme, who did the art work. However, the author must take the responsibility for any mistakes, errors of judgement, or wrong conclusions.

CONTENTS

CHAPTER I

THE IDEA OF AIR POWER

Man has always, even in earliest antiquity, felt the closest association between control of the air space and domination of the land and sea areas below. This is why the ancients seem to have endowed only their greatest deities with the power of flying. Considered as an attribute of power, the ability to fly was second in importance only to Jupiter's ability to flash the lightning and hurl the thunderbolt. The feeling seemed to be universal among the ancients that he who controlled the air space above dominated, as a logical corollary thereof, the ground and sea spaces below.

The idea that man could learn the secret of the gods, that is that he could learn to fly, is one of the oldest and most persistent idea in history. Thus air power is not an exception to the general rule that "everything comes by antecedents and meditations, and these may always be traced farther and farther back without the mind ever coming to rest."[1]

The concept of air power is so deeply embedded in the human experience that to get any idea of its origin it is necessary for the mind to penetrate the veil of pre-history, to go back to the world without remembrance. A good case can be made for the fact that without the pre-history of aviation there would not have been any history.

"Our modern progress in avaition," Professor Berthold Laufer said in 1928, "is not solely due to the efforts of the present generation, stupendous and admirable as they may be, but presents the process of a gradual evolution of ideas which have grown out of the imagination, endeavors, experiments, triumphs and failures of many past ages."[2]

He continued by saying, "stress must be laid on the word 'imagination,' for there is no field of human exertions in which imagination and romantic dreams have played a greater role and have proved more fertile than in the development of aviation. Intuition, romance, and adventure are its leading motives; for man, from the very moment he had grown into a full-fledged human being, never lived on bread and love alone."[3]

Professor Laufer concluded by saying:

> We have conquered the air in this age of science and unprecedented progress of mechanics, but in the last instance this conquest goes back to the trend of man's mind toward the romantic and adventurous. Describing merely the gradual perfection of mechanical devices does not make a complete history of aviation. It is the spirit and the idea behind the devices that count, the idea itself means everything. The will to fly is the will to conquer, and this will has pervaded the hearts of men in the earliest stages of the great civilizations of Asia."[4]

There are many records of flying in the Bible. The first example is in the Book of Kings (II King, 2:11), and describes the ascension of Elijah before the eyes of his successor, Elisha: "And, behold, there appeared a chariot of fire, and horses of fire, and parted them both asunder; and Elijah went up by a whirlwind into Heaven." The Psalms of David refer constantly to flying, both human and divine, and to the delights of flying, as in the case of the Fifty-fifth Psalm: "Oh that I had wings like a dove! Then would I fly away, and be at rest." Most of the great religious writings of the world, including many of the philosophical essays and nearly all the classical legends, contain abundant references to the art of flying.

One of the earliest flyers in history was the Chinese Emperor Shun, who lived in the third millennium before Christ, and he learned the art at the hands of the two daughters of the Emperor Yao, who at the instance of Yao instructed Shun in the "art of flying like a bird."[5] What the legend probably means is that the two daughters taught Shun the art of parachuting. For the story continued that Shun, who was disliked by his parents, was forced to build a granary and ascend it, at which point the parents set the structure on fire, and Shun, by spreading out two umbrella-shaped reed hats, which he used as a parachute, jumped to safety.

This feat preceded by nearly five thousand years the theory of parachuting advanced by the great scientist and scholar, Leonardo da Vinci, and it was similarly far in advance of the first successful jump in modern times, by a Frenchman from the top of a house at Lyons, who believed, curiously enough, that the importance of his experiment was that it demonstrated a practical means of escaping from fire.

A very early legend concerning air fighting is found in the Hindu epic poem, *The Mahabharata of Krishna-Dwaipayana-Vyasa*, the traditional date of which is somewhere between the fifth century B.C. and the year A.D. 200.[6] The story concerns a warlike aerial city, complete with inhabitants, that ranged about at will, sometimes on the earth, on other occasions submerged in water, and sometimes in the upper air. The city was attacked, and apparently destroyed by demons who possessed the ability to fly and fight in the air.

The demons evidently built an aerial chariot, complete with armor of iron and steel, and by means of this chariot maneuvered themselves into a position above the aerial city. The demons then destroyed the city by aerial bombardment, using missiles that demolished everything on which they fell. Coming some two thousand years before Christ, this is believed to be one of the earliest records of the practice of the modern theory of area or saturation bombardment.

In the year 1500 B.C. a very spectacular aerial journey for purposes of world conquest is recorded in the ancient legends of Persia. The journey was made by the Flying King of Persia. It seems that King Kai Kaoos had conquered the known world and yearned for new worlds to conquer. Consequently, the Demons whispered to Kai Kaoos:

> Thou art great as King can be
>   Boundless in thy majesty;
> What is all this earth to thee,
>   All beneath the sky?
> Peris, mortals, demons hear
>   Thy commanding voice with fear;
> Thou art lord of all things here,
>   But, thou canst not fly![7]

The preparations for Kai Kaoos' aerial journey went on apace. First, his aerial chariot was built. This consisted of light wood, and its corner posts were four javelins. Then some young male eagles were captured, and tied to the posts. On top of the javelins, just out of reach of the eagles, were impaled huge chunks of flesh. Kai Kaoos then had the ship provisioned with ample food, not to forget quantities of rare old wine, and hopped in prepared to take off.

The eagles, when released, strained upward for the meat that was just beyond their reach, and Kai Kaoos was airborne. The aerial journey was fun and exciting until the eagles, by common consent, seemed to tire of the futile game and descended to the earth, of course bringing Kai Kaoos and his chariot to earth with them.

Kai Kaoos, with all four engines dead, got out of his aerial chariot, and very chargined, faced his royal councilors, who had traced him to his inglorious landing. One of the councilors, more outspoken than the others, told Kai Kaoos that he was more fit for a madhouse than for a conqueror of the air. Kai Kaoos felt so mortified that he retired to his palace and did not come out again for forty days.

There are practically no important differences between this legend and the celebrated aerial exploits of Alexander the Great, something over a century later. According to the Alexander tale, the great warrior, having exhausted the worlds to conquer, and being bored for lack of military activity, himself dreamed up the idea of conquering all the inhabitants of the air space, and when that was accomplished to try his hand at conquering the inhabitants of the world's oceans. Accordingly, he had a chariot built in the form of a glass cage of wide dimensions, surrounded himself with magicians from the east, and embarked upon his conquest of the inhabitants of the air. At first, all went well. His magicians knew the language of the birds and developed appropriate intelligence concerning their laws and customs, while the great conqueror received their voluntary submissions to his empire. Eventually, he came upon a bird with a human countenance, and this bird, so the story ends, advised Alexader to return to the earth, which he did.

The folklore of India is replete with legends concerning the exploitation of the air for military purposes. One of the best of these concerns the god Vishnu, who normally rode about the air on a huge bird, Baruda, and a humble weaver who sought to emulate the god. The weaver had a wooden airship constructed in the image of the huge bird, Garuda, and he then dressed himself up to look like Vishnu. Thus equipped the weaver flew to the King's palace and landed on the seventh floor, where the princess lived. The princess, feeling extremely favored at the attention of a powerful god, married the weaver instantly. The King, when he learned that his daughter had been favored with the attention of a god, took steps immediately to take advantage of the happy circumstance by declaring war on his neighbors. The reasoning of the King was perfectly logical, that with the assistance of the powerful god Vishnu, his son-in-law, he could conquer the world. When the day of the battle drew near, the King, through his daughter, called on Vishnu to scatter his enemies. The little weaver, feeling that the game was up, donned his god-like raiment, mounted his flying machine and appeared over the battlefield prepared to die in a blaze of glory--the first of the Kamikazes. But the god Vishnu, realizing that his own prestige was at stake, entered the weaver's body and routed the enemy forces. The weaver then confessed everything to the King. But the King, after recovering from his disappointment when he realized his plans for world domination would have to be laid aside, honored the little weaver greatly, instead of ordering his death, and in recognition of his bold and successful prank, gave him his daughter in marriage, in addition to settling him on a large landed estate.

Greek history and mythology are full of winged gods who possessed the power of flight and of extraordinary individuals who acquired the art from a god to meet some particular emergency. One of the most celebrated of the Greek legends is the story of Daedalus and Icarus.

Daedalus was an Athenian who fled to Crete with his son Icarus, after he had killed his nephew, Talus. Daedalus incurred the enmity of Minos, King of Crete, and consequently was put in prison. Daedalus then made wings for himself and son and fastened them to their bodies with wax. He cautioned his son to straighten up and "fly right"; neither too low, so that the sea brine would soften the wax, nor too high, so that the heat of the sun would melt the wax. Icarus became excited, however, and flew too high during the escape from Crete. The result was that the sun melted the wax and Icarus plunged to his death. The Icarian Sea is still named for the fallen flyer. Daedalus continued on safely and landed on the island of Sicily.

The story of Daedalus and Icarus is perhaps better known in history than any of the other ancient legends concerning flying. What is the importance of the legend? In the words of Professor Laufer: "of all flying stories of classical antiquity it is this one which has left a lasting impression on future generations and fired the ambition of many imitators; and it is on this point, its moral effect, that the importance of the story rests. Daedalus was an historical personage, a many-sided artisan who surely made some attempts to fly. Like many others of his type he was not understood or even was misunderstood by his contemporaries, and his story has been handed down in the form of poetic romance and exaggerated legend."[8]

Laufer went on to say, "and what, after all, is the difference whether the Daedalus story is true or not? It is not the gray, cold naked objective truth that counts in the history of mankind and will advance the cause of civilization, but it is the flight of human imagination, the impulses and visions of a genius, very often his errors and miscalculations, which have stimulated inventions and progress. Ever since Daedalus' alleged or real flight, men in Europe have tried and died until finally success was insured."[9]

Another Greek legend that illustrates in what esteem the art of flying was held by the ancients is the story of Phaeton and Apollo. Apollo, in a moment of weakness, consented to allow his son, Phaeton, to drive the god's "Horses of the Sun" for a single day. As the winged horses were harnessed, Phaeton sprang into the great chariot and "darted forward to cleave the clouds. They rushed headlong from the travelled road--and soon Earth and Sky were ablaze. His knees shook with terror. Clouds began to smoke and planets took fire. Mountains and forests burned. Mighty oceans and broad rivers dried up, great cities perished and whole nations were consumed."[10]

The account of this air catastrophe continues: "Amidst intolerable heat, Phaeton was hurled on-- he knew not where. The people of Ethiopia became black through blood forced suddenly to the surface. The Nile fled away and hid its head in the desert, where it remains concealed to this day. Earth, surrounded by waters but with head and shoulders bare, with husky voice called on Jove to loose his thunderbolts and strike down the reckless charioteer. And Phaeton, his hair afire, fell like a shooting star into the river Eridanus."[11]

Winged horses, mechanical airships, chariots drawn through the air by giant birds--these were **not** the only devices whereby the ancients exploited the air space for military advantage. Kites were also used, particularly by the Chinese.

In the fifth century before Christ, according to Chinese legend, kites were used for aerial reconnaissance. Kungshu, a clever mechanician and contemporary of Confucius, constructed a wooden kite, which he used to spy on a city that he had marked for capture. Also, kites were used for military communications; for military engineering; and in a way to put modern strategists to shame, for purposes of psychological warfare.

An unnamed general, in the reign of Liu Pang, was such a brilliant strategist, greatly superior to any general of the East or the West who succeeded him, that he took an army on the verge of defeat, and by an imaginative strategem with kites alone, turned defeat into victory.

Though the name of the general, in the reign of Liu Pang, who won the decisive military victory against great odds by means of kites alone is not known, his exploits should go down in military history in a chapter ahead of those of any of the great captains of war, including the most gifted and renowned.

This general, so the story goes, besieged, outnumbered and, in a hopeless military situation, "resolved to make a last vigorous effort to drive Liu Pang from the throne he had recently usurped. A battle, however, resulted in the army of the general being hemmed in and threatened with annihilation. At his wits' end to devise a method of escape, he at last conceived the ingenious idea of frightening the enemy by flying kites, fitted with aeolian strings, over their camp in the dead of night. The wind was favorable and, when all was wrapped in darkness and silence, the forces of Liu Pang heard sounds in the air resembling *Fu Han*, Beware of Han! "It was their guardian angels," they declared, "who were warning them of impending danger." So great was their consternation at this aerial visitation that, the story concludes, "they precipitately fled, hotly pursued by the general and his army."[12]

Musical kites were known to the Chinese. The incident of their use in battle, therefore, is entirely within the realm of possibility. The general who executed the maneuver was not only an imaginative genius, but a great soldier, and a sound tactician and strategist. He did not gamble on the kites until the last battle; only when his alternatives were annihilation or surrender. Then he prepared his coup, but more than that, he foresaw the possibility of success, and seeing ahead, made his dispositions and followed through on the situation created by his musical kites.

This is the true test of military genius: the ability to create conditions favorable for battle, which is the realm of strategy--and actual conduct on the field of battle, in this case the "hot pursuit" of a beaten foe, which is the field of tactics. The unnamed Chinese general, therefore, was both a brilliant strategist and a sound tactician.

It is precisely to the realm of strategy and tactics that the Chinese story of musical kites has its present-day significance. All air operations, from the beginning, have depended on two things for their success. One is the actual impact of such operations on the capacity of the enemy, in military terminology the "capability impact," and the other is the psychological effect of such operations on the morale of the enemy, again in the language of the soldier, the effect of the operations on the enemy's "will to resist." Of course, will is a part of capability, because where there is no will all the capability in the world is meaningless. Perhaps that is why Napoleon said that "the moral is to the material as three is to one."[13]

In any case, it has always been incumbent on Air Power to take cognizance of the effect of air operations on the will of the enemy to resist.

Instinctively (as witness the legends in every land) man has associated control of the air space with domination of the earth below. Air warfare produces a feeling of helplessness in man wholly out of proportion to the actual firepower to which he is exposed.

One thunderbolt in the hands of Jove makes him a god for man to worship, offer sacrifices to, and obey. It was this fear to which Hitler appealed so successfully when he attached the screaming devices to the Stuka dive-bombers in World War II. They were his aeolian strings, and like the good Chinese general, Hitler sent his tanks and infantry crashing through the Low Countries and France to exploit the advantage opened up mainly by his terrifying air weapons.

Fear of the unknown and fear of being exposed to air action are two of man's most deeply rooted instincts. They go hand in hand and are augmented today by another great fear, the fear of being atomized. As Quincy Wright said in a monumental understatement, "only a few writers, such as Hermann Kahn, have given serious attention to winning a nuclear war, and his suggestion that losses of life in the United States of more than twenty million would be tolerable has not gained great support."[14]

But coming back to the antecedents of air power, to man's instinctive fears of the unknown, and to the uses of kites in ancient China for psychological warfare, kites were also used for military communications, such as signalling, and there is even an example in history of the use of kites for military engineering. Han Sin was desirous of digging a tunnel into the palace of his enemy and is said to have used a kite to measure the distance to the palace. But with this use of kites to exploit the air space for military advantage there logically developed the counterweapon, anti-aircraft artillery or flak, in the form of bows and arrows that were used, not too successfully, to bring down the offending vehicles of the air.

A Chinese city was being besieged in the year 549, and the defending general dispatched a kite-message calling for reinforcements. The kite was shot at, but the flak evidently failed of its purpose, for, according to the legend, the kite was transformed into a bird, which disappeared into the clouds. On another occasion, a loyal general in a Chinese civil war, being bottled up in a city, sent a call for help by means of a kite. The kite was shot at by the besiegers, but again the kite got through, and the defending general held out successfully until a loyal army arrived and dealt the rebel general a crushing blow.

In ancient Japan, kite flying as an adjunct to military operations was fairly common. Most of the Japanese stories are about men riding on kites for the purpose of spying on enemies, but there is one story in particular of a very imaginative brigand, who used a kite in order to steal gold from the tower of the Castle of Nagoya. It is said that the construction of large kites from that time on was strictly forbidden.

The British Navy used kites in military operations well into the nineteenth century. In 1806, to cite one example, military kites were used as a means of waging a very modern type of psychological warfare. In the words of the British commander who used the kites:

> A devise practised by us, when at various times running close in to the French shore, must not be omitted. A number of printed proclamations, addressed to the French people, had been put on board, with instructions to embrace every opportunity of getting them distributed. The opportunities for this were, of course, few, being chiefly confined to the crews of boats or small fishing craft; who would scarcely have ventured on the distribution had the proclamation been entrusted to them. The device resorted to was the construction of small kites to which a number of proclamations were attached. To the string which held the kite a match was appended in such a way that when the kite was flown over the land, the retaining string became burnt through and dispersed the proclamations which thus became widely distributed over the country to the great annoyance of the French government.[15]

Another example occurs in one of the ancient books of India. There it is written how information on the movement of hostile troops was kept current by means of carrier pigeons. Sometimes, though, this use of pigeons backfired with disastrous results. In the year 1262 during the Mongol siege of the city of Mosul, for example, the Mongols caught a carrier pigeon resting on one of their catapults, and discovered from the message he was carrying that an army to relieve the city was on its way. The captured message revealed enough other details about the approaching army so that the Mongols could ambush and destroy their enemy. Caesar is known to have used the pigeon-post to obtain information on incipient revolts throughout the Roman Empire.

Onc oΓ the most interesting uses of pigeons was the occasion in the tenth century when six hundred pigeons, each carrying a small silk bag containing a cherry, were dispatched from Balbek to Cairo, so that the Caliph Aziz could satisfy his desire for a dish of cherries from Balbek.

European nations of the north seem to have learned of the use of pigeons only during the Crusade. However, their use for military purposes persisted, and they were employed on a large scale in the Franco-Prussian War of 1870, particularly during the siege of Paris. Pigeons were also used in World War I, and to a lesser extent in World War II.

During the period of the Middle Ages man continued to think about flying. His thoughts, however, took a curious twist, and he began to identify the art with the evil one. The hallmark of the witch was her broomstick which she used to transport herself through the air with the greatest of ease. Fakers of all kinds sprang up, along with persons of true inventive genius.

The Franciscan monk, Roger Bacon, is a good example of the type of inventive genius that appeared during this period of history. There was no end to the number and kind of inventions that occurred to his active mind. Among them was a recipe for gunpowder written before the year 1249. In addition, Friar Bacon had ideas for all kinds of marvelous machines such as steamships, automobiles, submarines, and even flying machines.

"These machines were made in our times," Bacon said, "except possibly a flying machine which I have not seen nor did I know anyone who has, but I know an expert who has thought out the way to make one. And such things can be made almost without limit, for instance, bridges across rivers without piers or other supports, and mechanisms, and unheard of engines."[16]

A full two centures after Roger Bacon made his splash with the discovery of gunpowder and the invention of all those other marvelous machines, another inventive genius appeared on the scene, Leonardo da Vinci, the natural son of a Florentine nobleman and a peasant girl. Da Vinci was primarily an architect and military engineer who boasted that he could do painting and sculpture as well as any other man. He was also a close friend and admirer of Niccolo Machiavelli. In addition, da Vinci was a town planner, and a student of inventions, with a special interest in flying machines, military tanks, and submarines. Lewis Mumford said that da Vinci "made the first scientific observations of the flight of birds, designed and built a flying machine, and designed the first parchute."[17]

In 1507, just before da Vinci's death in 1519, an Italian in the service of King James IV of Scotland made an actual attempt at flying. The name of the Italian aeronaut of the early sixteenth century was John Damian. He made his attempt at flying for the purpose of overtaking a diplomatic mission that had set out from the Scottish court for France.

Preliminary to his feat, Damian had a pair of wings constructed for himself out of bird feathers. He fastened the wings onto his body, and became airborne by jumping off the top of Stirling Castle. The result was disastrous. Damian fell to the ground with a resounding thud. He also broke both legs. Later he explained his failure by pointing out that some hen feathers had gotten mixed in with the other feathers that had constituted his wings, and the natural inclination of the hen feathers for the barnyard had pulled him earthward instead of skyward, thus ruining what otherwise would have been a successful flight. There is no record, however, that Damian ever tried to correct this simple error by having a new set of wings made minus the offending hen feathers. Perhaps in his book of values there was never another occasion when the urgency of travel was great enough to justify flying.

The idea of flying like a bird by means of wings dominated the thinking of man from the beginning. Friar Bacon thought this way. "Flying machines can be constructed," he said, "so that a man sits in the midst revolving some engine by which artificial wings are made to beat the air like a flying bird."[18] Even today, the inscription on the Wright memorial at Kitty Hawk is a quotation from da Vinci: "There Shall Be Wings."

The idea of whirling wings and of artificial wings to be attached to the human frame or to a machine to enable man to fly was one that died very hard. However, Baron Ebner of the Austrian Corps of Engineers questioned the idea in 1861 by pointing out that bird flight was not the answer to military aviation. He said that just as the locomotive did not resemble the horse, and that the wheel was not made in imitation of the leg, so the practical airship of the world would probably not resemble the bird.

Two hundred years before Baron Ebner suggested his new approach to human flight the old idea of an "aerial ocean" came to the fore in connection with aviation. Man visualized his boats sailing on the "aerial ocean" just as his ships sailed on the surface oceans, and the only problem seemed to be the one of getting up through the thin layer of heavy air that covered the earth's surface.

One of the solutions to the problem of the heavy air that suggested itself was the morning dew. Clearly, this substance was of heavenly origin, coming as it did from the stars at night, and returning as it also did to the sun in the morning. Why not capture some of this stuff and use its ascensive power to lift man skyward so he could sail about on the ocean of the air?

While man was in this frame of mind about the easy sailing on the ocean of the air, a Jesuit priest, Francesco de Lana, published a very learned trestise on the subject of air travel. His book was published at Brescia in 1670 and dealt, not only with the technical problem of human flight, but more importantly with the social and military consequences of man's conquest of the air.

De Lana did not consider human flight to be a problem from the technical point of view. In fact, he was quite confident that he had solved the problem himself and had perfected the design of an airship that would fly, one that would enable man to float about on the ocean of the air with the same ease and facility that he then navigated the oceans of the earth.

The key to de Lana's invention consisted of four perfectly round copper balls from which the heavy surface air could be exhausted, and which would then be lighter than the surface air, and so would rise and could carry up a boat for passengers and cargo, as well as a sail for purposes of aerial navigation. He did not actually construct and try out the airplane he believed he had invented because, as he said in his book, his "vows of poverty prevented my expending 100 ducats, which sum at least would be required to satisfy so laudable a curiosity."[19]

De Lana said that he was not unaware that the idea of an airship would strike some people "like unto a fable no less strange or incredible than those mad fantasies that issued from the fertile brain of Lucian." But he was convinced that his invention was sound, for he went on to add, "yet, on the other side, I clearly discern that I have not erred in any of my proofs, especially as I have conferred on those matters with many sage and well-instructed persons who have not been able to find any errors in my discourse, having expressed a desire to have a visible proof by one of these spheres which would, of itself, ascend in the air."[20]

De Lana's machine (the word in his), which was "fashioned like unto a ship," and "will float on the air," was examined by him from all possible angles to determine whether he had made any miscalculations. On the important subject of air safety, he said, "much greater is the difficulty of preventing the ship from being driven at too great a speed by a strong wind, whereby it runs the risk of hurling itself against the mountains, which are the rocks of the aerial oceans, and be wrecked."[21] To this, he replied:

> I confess that our aerial ship might run great perils, but not more than the ships on the seas are liable, because, as in their case, so would we be furnished with anchors, which would

> easily grapple to the trees. Besides, although our aerial ocean has no shores, it has this advantage, that there is no necessity for ports of refuge for the airships, they being able every time they are in danger to descend from the high air and land on the ground.[22]

De Lana saw no "other difficulties," to use his language, "that could prevail against this invention, save one only, which to me seems the greatest of them all, and that is that God surely would never allow such a machine to be successful since it would create many disturbances in the civil and political governments of mankind."[23]

De Lana did not want to leave the world in doubt as to what he considered these disturbances would be, however, for in his final paragraph he narrates them as follows:

> Where is the man who can fail to see that no city would be proof against surprise, as the ship could at any time be steered over its squares, or even over the courtyards of dwelling-houses and brought to earth for the landing of its crew? And in the case of ships that sail the seas, by allowing the aerial ship to descend from the high air to the level of their sails their cordage could be cut; or even without descending so low iron weights could be hurled to wreck the ships and kill their crews, or they could be set on fire by fireballs and bombs; not ships alone, but houses, fortresses and cities could be thus destroyed, with the certainty that the airship could come to no harm as the missiles could be hurled from a vast height.[24]

De Lana's invention did not turn out to be an invention at all, even though all the principles were sound. The globes that were to be the key to the ascent were impossible of construction with the materials available at that time, and as man was to find out centuries later, all de Lana's ideas on the "easy sailing" on the oceans of the air had to come in for very considerable revisions. Meanwhile, however, Galien, a Dominican professor of philosophy at the papal University at Avignon, conceived of a bold scheme of exploiting the air for military advantage. He published his ideas in a small volume entitled *L'Art de Naviger dans les Airs.*

Very briefly, Galien's ideas of the technical problems involved were much the same as de Lana's, that the surface air could be exhausted from the globes with tubes fitted with stopcocks, and that the globes would then be lighter than air and rise, but his military ideas were much in advance of de Lana's. Not only were Galien's ideas ahead of de Lana's, but ahead of any developments that have taken place up to the present time. Galien conceived of an airship of huge proportions, extending more than a mile in every direction, possible of transporting fifty-four times as much cargo as Noah's Ark and "capable of transporting a large army to the middle of Africa."[25]

Thus between them these two Catholic priests--de Lana, the Jesuit, and Galien, the Dominican--anticipated not only the future of military Air Power but possibly the hereafter as well.

De Lana clearly foresaw the high-altitude, long-range bombers that were to lay waste the cities of Europe and Asia; he visualized clearly the decisive naval engagements of World War II, fought from the air without the surface ships exchanging a single shot; and he even visualized and foresaw the aerial task forces that the Germans were to employ so decisively, before an astonished world, as they flew across the water barrier to occupy Norway, to say nothing of their Mediterranean hop to conquer the island of Crete.

Also clearly within de Lana's vision was the German blitzkrieg, the airplane-tank team that knifed its way into Poland and won the victory in three weeks; that slashed into the Low Countries and overran Holland in five days, Belgium in eighteen; that conquered France in thirty-five days, Yugoslavia in twelve, and Greece in eighteen.

Galien's vision of whole armies being transported by air may belong to yet another war. If the United

States were on the receiving end of an atomic attack, and its losses were twenty million or more on the first day, the problem of occupation could be solved by Galien's airborne army. It has been estimated that a force of three hundred thousand well trained troops would be enough to take over the remains of this once proud and prosperous country.[26] Future historians may say, in retrospect, that de Lana and Galien, between them, may well have outlined and foreseen, not only World War II, but possibly, as well, the strategy and tactics of a World War III.

The concept of Air Power is thus not the product of the twentieth century, it is the fulfillment of an evolution of ideas that have their origin in remotest antiquity, going back to the beginning of the human experience in a world without remembrance. Professor Laufer was quite right, therefore, when he said that "our modern progress in aviation is not solely due to efforts of the present generation, stupendous and admirable as they may be, but presents the process of a gradual evolution of ideas which have grown out of the imagination, endeavors, experiments, triumphs and failures of many past ages."[27]

FOOTNOTES--CHAPTER I

[1]Herbert Butterfield, *The Origins of Modern Science* (New York: The Macmillan Company, 1951), 141.

[2]Berthold Laufer, "The Pre-history of Aviation" (Chicago, Field Museum of Natural History, 1928, Publication 253, *Antropological Series;* v. XVIII, no. 1), 10-11.

[3]*Ibid.*

[4]*Ibid.*

[5]*Ibid.*, 14.

[6]C.F. Snowden Gamble, *The Air Weapon* (London: Oxford University Press, 1932), 2.

[7]Francis Trevelyan Miller, *The World in the Air* (New York: G.P. Putnam's Sons, 1930), 74.

[8]Berthold Laufer, "The Pre-history of Aviation" (Chicago: Field Museum of Natural History, 1928, Publication 253, *Antropological Series*, v. XVIII, no. 1), 63.

[9]*Ibid.*

[10]Francis Trevelyan Miller, *The World in the Air* (New York: G.P. Putnam's Sons, 1930), 79.

[11]*Ibid.*

[12]Berthold Laufer, "The Pre-History of Aviation" (Chicago: Field Museum of Natural History, 1928, Publication 253, *Antropological Series*, v. XVIII, No. 1), 34.

[13]Bernard Brodie, *Strategy in the Missile Age* (Princeton: Princeton University Press, 1965), 42.

[14]Quincy Wright, *A Study of War* (Chicago: The University of Chicago PRESS, 1965), 34.

[15]C.F. Snowden Gamble, *The Air Weapon* (London:' Oxford University PRESS, 1931), 34.

[16]Lewis Mumford, *The Pentagon of Power* (New York: Harcourt, Brace, Jovanovich, Inc., 1970), 116.

[17]Lewis Mumford, *Technics and Civilization* (New York: Harcourt, Brace & World, Inc., 1963), 140.

[18]Lewis Mumford, *The Pentagon of Power* (New York: Harcourt, Brace, Jovanovich, Inc., 1970), 116.

[19]Francesco de Lana, *The Aerial Ship* (London: Sell & Olding Ltd., printed and published for the Aeronautical Society of Great Britain, first printed in Brescia, 1670), 21-27.

[20]*Ibid.*

[21]*Ibid.*

[22]*Ibid.*

[23]*Ibid.*

[24]*Ibid.*

[25]J.e. Hodgson, *The History of Aeronautics in Great Britain* (London: Oxford University PRESS, 1924), 14.

[26]Edmund A. Walsh, *Total Power* (New York: Doubleday & Company, Inc., 1948), 252.

[27]Berthold Laufer, "The Pre-History of Aviation" (Chicago: Field Museum of Natural History, 1928, Publication 253, *antropological Series*, V. XVIII, no. 1), 10.

CHAPTER II

18TH AND 19TH CENTURIES

The use of military air power, so confidently anticipated in all past ages, particularly by such sages as de Lana and Galien, respectively, the Jesuit and Dominican prophets, was revived in the final quarter of the 19th Century. The occasion was the Spanish siege of Gibraltar, which pinned the British to the Rock and denied them ingress or egress by land or by sea. And it was to Joseph Michael Montgolfier, the son of a wealthy paper manufacturer in France, that the idea came in 1782: "why not by air?"[1]

The thought was a springboard to action. Montgolfier increased his efforts to solve the riddle of human flight, but the solution came almost by accident. Watching smoke rise from a fire, nothing new, he queried why could not some of the smoke be captured, say in a paper bag, and then would not the smoke, if it were imprisoned in the bag by a tightly drawn string, pull the bag skyward? If so, man would have discovered the secret of human flight.

All that would remain would be the construction of a larger bag, its inflation with more smoke, or whatever the substance was that filled the bag when held over an open fire, and then the attachment to the larger bag of a basket in which a man could sit. Man would then be pulled skyward, freed at last from his earth-bound existence.

Small wonder then that when Joseph Montgolfier conducted his initial experiment in 1782 with a paper bag filled with hot air, he was excited. The inflated bag rose quickly to the ceiling of the room, to the intense delight of Montgolfier, and the utter astonishment of his hostess.

Montgolfier did not need a Sir Isaac Newton to tell him why the inflated bag rose into the air. Suffice it for the moment that the bag did rise. He did not care whether it was smoke, hot air, or some other substance that gave the bag its lift. The important point was that man had apparently discovered how to defy the law of gravity.

Now man could fly. Excitedly, Montgolfier penned a note to his brother, Etienne, at Annonay, and asked him to assemble an adequate supply of strong paper and thin silk, together with some cordage, and to meet him prepared to see the most astonishing thing in the world.

The scientific notions that prevailed at the time as to the qualities of the air and as a concomitant thereof, as to the problems that confronted man in his attempted conquest of the air, made the Montgolfier discovery revolutionary and far-reaching in scope and promise. Man believed that the earth was surrounded by a thin layer of heavy air and that this was the sole barrier to human flight. If man could lift himself up on top of this heavy air, he would enter into a realm of thin air on which he could sail just as he sailed on the navigable waters of the world. In other words, man believed that just as he could not then sail on the bottom of the ocean because of the heavy pressure of the deep water, so he could not navigate in the heavy air near the surface of the earth; but just as he *could* sail about on the surface of the oceans of the world by means of a hollowed-out tree trunk, if necessary, so he could also sail on the aerial oceans of the world once he had broken through the thin crust of heavy air near the surface. These are the notions or scientific hypotheses that prevailed at the time, and which, because they were currently accepted, made Montgolfier's elementary discovery seem the solution to the problem of human flight.

The first successful public experiment with flying was made on June 4, 1783, when the Montgolfier brothers inflated a linen bag over a fire of chopped straw and saw it rise rapidly to a great height, stay in the air about ten minutes and travel a distance of one and a half miles. Needless to say, the demonstration created great public excitement and stimulated widespread interest in flying.

A public drive was initiated immediately to raise money for the construction of another balloon. A new type was used, one that had been designed by a noted scientist, J.A.C. Charles, and one that utilized the newly discovered hydrogen gas instead of Montgolfier's hot air. The test was made from the Champ-de-Mars before a huge throng. It was a great success. The balloon rose to a great height, stayed in the air for about an hour, and traveled a distance of about sixteen miles. The only trouble was that when the balloon landed it so terrified the local peasantry that they attacked it with pitchforks and literally tore it to pieces.

Montgolfier made another flight on September 19, 1783, at Versailles, and it was witnessed by King Louis XVI, Queen Marie and the American envoy to France, Benjamin Franklin. Montgolfier used his fire-balloon again, by this time called a Montgolfier, as the hydrogen-type was called a Charlier.

The novel feature of the flight of September 19 was that it was the first occasion on which passengers were carried into the air. The passengers, a sheep, a cock and a duck, were suspended in a basket below the balloon. Whether these first passengers had names or not is not known, but 174 years later, On November 3, 1957, the Soviets launched Sputnik II carrying a live dog, Laika, as the first space passenger. The Soviet authorities announced the dog's death on November 10. No harm came to the passengers of the first air flight, however, except for the fact that when the balloon landed it was noticed that the cock had been injured by a kick from the sheep. However, it was determined that the accident had taken place before the party had become airborne, so the cock cannot be considered the first air casualty!

Now that it had been demonstrated that man could invade the air space successfully and that life could be sustained in the upper atmosphere, the next question was whether human life could also be preserved in the higher air. The same question arose after the first man-made satellite was launched by the Russians on October 4, 1957, Sputnik I. The death of Laika, announced on November 10, did not convince the Russians that human life could not be preserved in outer space, and on April 12, 1961, Major Yuri Gagarin became the first human orbital traveler. Ten months later Lieutenant Colonel John H. Glenn, Jr. became the first American in orbit. The first woman space traveler, Valentina V. Tereshkova, was launched into orbit on June 16, 1963.

In 1783, a noted Franch scientist, Jean Francois Pilatre de Rozier, set out to prove that it was possible for human life to be preserved in the upper atmosphere. He began his experiments with a balloon that was tied to the ground by ropes so that after it became airborne it would not drift about aimlessly. His purpose was to see whether the life-giving fire could be taken up into the air without setting the balloon on fire. This proved to be possible and the way was paved for human flight.

On one of the test runs with a captive balloon, de Rozier took a passenger with him, Andre-Giroud de Villette. De Villette was very much impressed with the experience, and after his limited ascent wrote a letter to a Paris journal (October 30, 1783) expressing his conviction that flying could be made a very useful adjunct to both armies and navies. He foresaw the advantages, particularly in intelligence, that armies and navies equipped with air corps or air arms would have over armed forces not so equipped. De Villette stressed the tactical advantages that flying would make possible. Intelligence obtained as a result of flying, he said, could be transmitted instantly to the commanders of armies or fleets by means of signalling. De Villette thus foresaw with remarkable clarity the first use to which military aviation would be put, as the "eyes" of armies and navies.

The date set for the first human flight in a free balloon was November 21, 1783. However, there was a great hue and cry over who the guinea pig would be in this first human flight. At first King Louis XVI refused to allow de Rozier to take the risk, and suggested that in his place there should be sent two French criminals condemned to die. However, the King was at last prevailed upon by one of his noblemen the Marquis d'Arlandes, to allow de Rozier to become the first human aerial voyager.

What the Marquis did not tell his King, however, was that he planned to go with Rozier on the first flight. De Rozier had no liking for this idea because the flight was dangerous, but did not see how he could block it since he was under debt to d'Arlandes for having interceded with the King. Thus when the flight took place, the Marquis went along as the first human passenger on an aerial flight.

The experiment was a huge success. The flight de Rozier had planned for so long came off without a hitch. The balloon rose very rapidly to an elevation of some three hundred feet, floated a distance of about five miles and passed directly over the Ecole Militaire before it came down without crashing. While in the air, however, the balloon caught on fire. But since de Rozier had been more or less expecting this, he had carried a sponge and water with him and was able to extinguish the blaze before it got out of control. Thus the first free flight on the part of a human being in modern times was saved from a disastrous ending by calm foresight plus a sponge and water.

Now that man had conquered the air, to what advantage and for what purposes would he put this new invention? A contemporary writer suggested one highly imaginative use of the airship. He seemed to be afraid that its primary effect would be to corrupt human morals. He visualized the highly romantic occasions for innumerable immoralities that airplanes would invite and make possible. Suitors, he was afraid, would chase their mistresses about through the air, and pounce on them like a hawk after a lark. Additionally, he foresaw that the roofs of all the public buildings, and even historic monuments, would be common meeting places between the sexes for immoral purposes, and that the bedrooms of even the most chaste might be casually invaded by ardent males entering through the windows at any time of the day or night.[2]

Another writer, in the same vein sixty-five years later, wrote a fictional account of a class of flying prostitutes. These were women who had involuntarily grown wings as a result of their moral indiscretions. All in all, it was thought by some of the earlier opponents of aviation that the airship should be suppressed in its infancy in order to protect public morals.[3]

Horace Walpole, in a letter to Sir Horace Mann on December 2, 1783, wrote of his apprehensions concerning flying. Not mentioning the moral aspects, and ignoring almost entirely the scientific and commercial factors, Walpole commented philosophically on the tendency of man to turn all his inventions to destructive purposes and prayed that this would not be so in the case of the airplane: that instead the new invention would be used only for peaceful and constructive purposes.

The poet, William Cowper, was not so philosophic in his reaction to the conquest of the air. The airplane made him yearn to be a dictator for the sole purpose of issuing decrees, world-wide in scope, on the subject of aviation. By his decrees, he would make flying a crime punishable by death. More than that. Cowper would make flying comparable to piracy in that every flyer, like every pirate, would be fair game for anybody. No man would be subject to punishment for killing an aviator.

His reason? Cowper was sensible that his decrees would retard progress. He knew that he would be condemned by philosophers, scholars and historians. But he believed that his decrees were justified. For, as he concluded, without aviation the world's affairs would go on more serenely, and if man's liberties were slightly curtailed, this infringement would be more than compensated for by the greater security that the world would enjoy.[4] Subsequent events have certainly borne out Cowper's contention that the world would be a safer place to live in if the airplane had never been invented.

The military use of airpower seemed to be the uppermost thought in people's minds concerning the usefulness of the new invention. The *Journal de France* printed a very lively cartoon in 1783 under the interesting title, "Wars in the Air." The scene is a French town being invaded, apparently successfully, by what appears to be an airborne army. This was not all. A British author, William Cooke, published a treatise in 1783, "The Air Balloon: Or a Treatise on the areostatic Globe, Lately Invented by The Celebrated Mons. Montgolfier of Paris." The author proposed that an air patrol be established around the British Isles, much as was actually done a century and a half later in World War II, to warn the country of impending attack. This warning, he said, could be communicated swiftly to the defending forces, and they would thus be enabled to mount the counter-attack with greater assurance of success.

Cooke went further. He visualized generals in battles with command posts in the air, where they could see the enemy, observe their own troops in action and issue the necessary commands. Cooke said that such generals could literally ride in the whirlwind and direct the storm.

William Cooke did not limit his observations in 1783 to the subject of Army Intelligence and command on the battlefield; he foresaw that flying could be used as an adjunct to naval fleets at sea, to become their eyes in time of war and to prevent fatal accidents at all times. He thus foresaw clearly the brilliant and successful system of air-sea rescue used so extensively in World War II. Cooke also touched indirectly on the passenger-cargo aspects of military flying, and like Montgolfier, specifically referred to the siege of Gibraltar in order to emphasize how it might have been broken more swiftly and with less loss of life by the utilization of the flying machine.

1783 was the year in which air ship designers all over Europe, particularly in France, let their imaginations run wild and invented all kinds of contraptions, at least in their minds and on paper, contraptions that man has still not perfected, though in some instances he is still trying. The prevailing notion was a flying boat--an airplane that could also navigate the seas on the world. One of the earliest and most ambitious projects sponsored in France was a "Voyage to Other Spheres," a sort of space shuttle, that would make the long journey with a crew of scientists for the purpose of studying the solar system.[5]

Another apparatus in Madrid, Called the "Nave Atmosferica," was completely equipped for air or sea navigation by means of a ship's hull, rudder, sails and oars. There were many others, including the "Globe Volant" in Paris, the "Ellipsodial Airship" of General Jean-Baptiste-Marie Meusnier and a very racy-looking French model that the advertisements said would ride the air waves and race through the clouds in addition to sailing the high seas. Altogether, speculation was rife throughout Europe, and flying was confidently heralded as the harbinger of a new age, though it was not then a year old, and man was totally unacquainted with any of the basic problems concerning aerial navigation and mechanical flight.

1783 was a banner year for aviation, to be sure, and the fairer sex was not left behind. Despite the absence of a vigorous women's liberation movement in France at this time, the first woman in the modern world to fly, not the witches on broomsticks that dominated the Middle Ages, was Mme. Thible. She made history by ascending with Count de Lawrence in a Montgolfier balloon at Lyons on June 4, 1784. The first woman balloon pilot and also the first of her sex to be a parachute jumper was Jeanne Genevieve Garnerin. Vive la France.

Benjamin Franklin had some thoughts on the new craze of flying in the year 1784. Franklin suggested that ballooning might be the key to world peace enforced by an international police force. He also suggested that this could be done at no great expense, certainly at far less cost than would be the case with naval forces. Writing to his friend in Vienna, Jan Ingehaus, Franklin said that five thousand balloons would be capable of transporting ten thousand troops nearly anywhere in the world, and this force would be sufficient to wreak incalculable damage before an opposing force could be assembled and thrown against the aerial invaders. He seriously doubted whether any country could keep enough men permanently under arms to repel an airborne invasion. As to the cost of maintaining such an international police force, Franklin gave it in such a way as to disparage the idea of maintaining costly naval fleets. He said that a rapid deployment force of five thousand balloons could not cost more than five ships of the line.[6]

The concern that had been felt in France in 1783 about the dangers of an airborne invasion were reflected in England in 1784. British newspapers began to carry cartoons on subjects such as "War in the Air," and popular songs began to find acceptance on the theme of an invasion of England by air. One such, "The Air Balloon," speculated as follows:

> Should war 'gain break out,
> As it is not a doubt
> With some that it may happen soon;
> The French will invade us,
> Their troops all parade us,
> Brought over in an Air Balloon
>
> Their ships will appear
> Not in water but air,
> And come in a twinkling down,
> From Calais to Dover,
> How quick they'll be over,
> Blown up with the Air Balloon.

However, the British in 1784 refused to accept at face value the danger to their islands from an aerial invasion for the popular song concluded as follows:

> 'Tis the tale of some humbugging loon;
> So I say, Botheration
> To the frog-eating nation,
> Likewise to the Air Balloon.[7]

A british naturalist, Thomas Martyn, published a tract on military flying in 1784, the same year in which the British cartoons and songs appeared ridiculing the idea of invasion by air, and in it he speculated on the use of balloons in conjunction with naval fleets. His speculations concerning the launching of balloons from the decks of ships is probably one of the earliest suggestions of modern naval task forces based on carrier aviation.

That Martyn's ideas on the military aspects of flying, including naval aviation, began to make an impression on the services, is attested to by the fact that in June 1785 a joint service test was made in England, apparently unofficially, to determine the practicability of flying in connection with military and naval operations. One officer from the Army was chosen, a certain Major General Money, who was destined to become the Billy Mitchell of his day; one from the Navy, a George Blake, Esq. and a civilian pilot, Jonathan Lockwood, who was also the owner of the balloon.

The flight began rapidly and dangerously. Apparently too much gas was released in the ascent, and the balloon descended quite unexpectedly, finally lodging itself in a tree. The pilot tried to get her up again but found that he had lost too much gas, and that, therefore, the balloon would not rise until one of the three passengers bailed out. A terrific dispute arose as to which passenger should take leave of the balloon. The Army man argued that it would be unbecoming an officer of the Army to desert in a situation fraught with peril. The officer of the Navy emphasized the tradition of his service that made it impossible for him to abandon the ship. And the civilian pilot said that, not being officially connected with the Army or Navy, made it mandatory that he prove his courage in such a dangerous situation by staying with the ship.

Finally, it was agreed to settle the matter by lot. The officer of the Navy lost. Before abandoning ship, however, he insisted on having lunch, pointing out that his celestial flight had in no way diminished his terrestrial appetite. Then, and only then, did George Blake, Esq. from the Navy climb out of the tree and permit the flyers to continue their voyage. Thus, in what was probably the first inter-service squabble on the subject of aviation, the Navy lost out.

The first crossing of the English Channel, harbinger of the thousands of sorties of the Luftwaffe one hundred and fifty years later, occurred in 1785. Two things are notable about this first crossing: one, that it was accomplished at all in view of the extreme measures that the intrepid flyers had to take to make their journey a successful one; and secondly, that the flight was accomplished by the combined efforts of a Frenchman and an American. The Frenchman was J.P. Blanchard, known as the first professional pilot, and the American was a Bostonian, John Jeffries, who besides being a scientist and an aerial enthusiast was also a patron, furnishing the not-inconsiderable sum of money required to make the cross-Channel voyage.

The voyage itself was an ordeal. Blanchard wanted to go alone, claiming that the balloon would not be able to carry more than one person, but Jeffries was adamant and literally forced his way into the aerial ship. No sooner were they out over the Channel than they began to descend, compelling them to throw overboard all their sand ballast, and when they came in sight of the coast of France, they realized that unless they could get rid of a lot more weight they would never make it. Consequently, they methodically began to throw overboard everything in the balloon that was detachable until they finally got down to their landing cords, ropes and anchors. These too went over the side. Still they continued to descend. Becoming desperate, they took off and discarded all their clothing, including their trousers, and yet failed to check their downward direction. At this point they put on their life jackets and braced themselves for the plunge into the sea.

Jeffires had one last idea. He proposed that they spew their food. The elements came to their rescue, however, before this last measure had been undertaken, and the balloon suddenly began to rise, passing over the coast of France at the highest altitude they had reached during the entire flight. They finally landed successfully by bringing the balloon down over a forest close enough so they could reach out and actually grab onto the tops of the trees. Thus ended the first successful crossing of the English Channel by air between the ever-tempting cliffs of Dover and the invasion port of Calais.

For the next twenty-five years, from the storming of the Bastille on July 14, 1789, to the settlements reached at the Congress of Vienna on June 8, 1815, Europe was in a turmoil, and in this turmoil, and to a certain extent because of it, military aviation got its first real test in the fiery crucible of war.

Very early in this period a leading French revolutionary, companion of Robespierre, predicted that aviation would make radical changes in the whole course of mankind. In 1793, Montgolfier proposed that the city of Toulon be bombed because of its defection from the Republican authorities. More than that. He urged the destruction of the entire city using a bomb weighing over fourteen tons.

As France girded for war against the whole of Europe to defend the gains of the Revolution and to export her ideas of liberty, equality and fraternity, a Committee of Public Safety was formed to exercise the supreme executive authority in the country. This was April 6, 1793. To this Committee a proposal was made by a noted French scientist that captive balloons be sent to the Republican armies in the field. As a result, a young French scientist, Jean-Marie-Joseph Coutelle was sent to determine whether General Jourdan, in command of the Army of the Moselle, could make effective use of an air force in his operations against the Allies.

Captain Coutelle's first duty assignment was at Maubeuge, where the French were besieged by the Austrians. The "Aerostiers," as he and his men were called, were received very coldly and suspiciously by the Army. Even the common soldiers murmured against this unwelcome interference with operations. In fact, it was not until Coutelle and his men had broken away from their work and joined in a ground sortie against the enemy, during the course of which three of the men lost their lives, that the soldiers of the Army were convinced of the loyalty of the air force.

The first ascents were made under enemy gunfire, and the intelligence gained was said to be of some value to the defense. General Jourdan, however, ordered Coutelle to bring his air force out of Maubeuge to assist him in an investment of Charleroi. This Captain Coutelle did, even though it meant that he and his men had to take the balloon and walk twenty miles with it, holding onto the floating balloon by stay ropes. They also had to scale the walls of a town. However, they arrived in time for the investment, and despite his fatigue, Captain Coutelle made an ascent immediately to scout out the enemy position. The following day he went up into the air again in the company of one of Jourdan's generals and stayed up for eight hours. The next day, Austrian reinforcements arrived, and the Battle of Fleurus was on, during the whole of which the air force was extremely active.

Historians differ widely as to the usefulness of the air force during the critical Battle of Fleurus on June 26, 1794. Airmen than and now are inclined to give full credit to the air force for the intelligence that made the victory possible. The generals, on the other hand, seem to have taken a keen pleasure in disparaging and ridiculing this interloper in military affairs. General Jourdan, in particular, referred to the balloons as inflated taffetas. He admitted that their presence undoubtedly astonished the Austrians but said that he doubted whether they had any other effect. In fact, some years later, General Jourdan recalled with grim satisfaction how the air force had signalled him that his right wing had advanced successfully when in fact it had been driven back in retreat across the Sambre.

The next important occasion in which the air force participated was the Battle of the Ourthe, near Liege, when the French under Jourdan again defeated the Austrians. Major General John Money was evidently present at the time, for he later recorded that he was told on the spot by an eye witness in the Austrian position that the air force intelligence the French possessed was the deciding factor in the Austrian defeat. "One would have supposed the French general's eyes were in our camp," Money quoted the Austrian as saying.[8]

Meanwhile, a new star was rising in France, the star of Napoleon Bonaparte, now commander-in-chief of the Army of Italy. He left for Italy in March 1796, and his brilliant victories there were in sharp contrast to the fate of the French generals in other parts of Europe. So, upon the conclusion of his first Italian campaign, which ended in the treaty of Campo Formio, Bonaparte emerged as the idol of France. As the Directors squirmed in their respective seats at the public acclaim accorded the ambitious general, Napoleon, to their intense satisfaction, proposed the Egyptian campaign to disrupt communications between Great Britain and India.

At the beginning, the Egyptian campaign was a huge success. The "little corporal" sailed in May 1798 and succeeded in evading the British fleet under Lord Nelson. He surprised and took Malta on the way to Egypt on July 1. Alexandria was captured on July 2, and on July 21, Napoleon fought and won the famous Battle of the Pyramids, after which he called upon the Pyramids to witness the valor of French soldiers.

All this read well in the newspapers of France, and Napoleon's stock continued to rise. But then, on August 1, 1798, Lord Nelson pulled the rug out from under the "little corporal" by locating the French fleet in a harbor near Alexandria. The French fleet was destroyed, and Napoleon was cut off from France. He left Egypt on August 24 and returned to France a sadder, but wiser, man.

At the beginning of the Egyptian campaign, Napoleon was persuaded to take with him, as part of his force of thirty-five thousand, a corps of scientists, including Captain Coutelle and his balloon company. However, what good they might have done will never be known because the ship carrying the balloon equipment went down in the Battle of the Nile. Subsequently, the aviators were disbanded and assigned to other duties and then the Company itself was abolished. This turned out to be a mistake. The blunder, Liddell Hart has said, cost Napoleon his final victroy and his empire at Waterloo.[9]

Prior to Waterloo, however, the subject of military aviation was revived in the public's mind. The peace that was signed on March 27, 1802 between France and England, the peace of Amiens, lasted less than a year, and by May 16, 1803, France and England were at war again. All the while, however, and going back to before the invasion of Egypt, Napoleon had kept a large force, *The Army of England*, poised at Boulogne, for the invasion of England. Now the subject of the cross-Channel assault was once more in the limelight, and this was a tonic to those who thought in terms of Air Power.

Just as "We're Sailing Against England" was the most popular song in Germany in 1940-41, so the public imagination in France in 1803-04 was fired with the invasion of England. The most fantastic of all the projects for the invasion of England was one entitled "*Grand* DESCENT *on* ENGLAND," wherein it was proposed that the invasion should be made by air through the use of thousands of balloons; by sea, in literally hundreds of small transport ships; and, best of all, under the sea, by means of a tunnel large enough for an entire army.[10]

Another project for the invasion of England, this one proposed in all seriousness, was for the construction of a huge balloon, one capable of carrying up to four thousand fully equipped troops, called the "Grand Republican Balloon Intended to Convey the Army of Invasion to England from the Gallic Shore."[11] This proposal is reminiscent of the larger balloon project of Galien in the previous century. However, his dream had been that of a balloon capable of transporting whole armies, not a mere fraction of a division, in voyages between continents.

But all these proposals came to naught and military aviation ceased to be a topic of lively interest in France as Napoleon, like Hitler after him, forgot about the invasion of England and turned his legions to their doom in Russia. One of the things that influenced Napoleon, of course, was again something that sprang out of the genius of Lord Nelson. Just as he had ruined the Egyptian campaign by his great victory at the Battle of the Nile, so Lord Nelson dashed all Napoleon's hopes again by the defeat of the Franco-Spanish fleet at Trafalgar on October 21, 1805. "At Trafalgar it was not Villeneuve that failed," Admiral Mahan said, "but Napoleon that was vanquished; not Nelson that won, but England that was saved."[12]

While the Napoleonic Wars were in progress, a devoted advocate of Air Power in England, Major General John Money, made an appeal to the British authorities to support the idea of military aviation. He reviewed the success of military aviation in the French Republican armies before the corps was disbanded and urged the benefits on the English authorities.

General Money quoted some verses on the subject, written as he said, by a noble Lord:

Great use, he thought, there might be made
Of these machines in his own trade;
Now o'er a fortress he might soar,
And its condition thence explore;
Or when by mountains, woods, or bog,
An enemy might lay incog,
Our friend would o'er their station hover,
Then change his course, and straight impart
Glad tidings to his chieftain's heart;
Such lights convey, such knowledge gain,
As might decide the whole campaign.[13]

In conclusion, General Money cautioned the British authorities not to consult the old generals on anything concerning military aviation. He warned that just as they had always been opposed to any slight innovation in war, so they would also fight against military aviation. He cited, to prove his point, their opposition to artillery, to riflemen, to telegraphs and so forth. The old generals would answer with respect to military aviation, Money said, what they had always answered to every new invention: "that as we have done hitherto very well without them, then we may still do without them."[14]

How right he was! For the next fifty years the stodgy military authorities in England took no official cognizance of the fact that the country was in the air age and muzzled the zealots who sought to effect changes in policy. When these could no longer be resisted and policy had to be changed, the recognition of military aviation was phrased in such a way as to be almost a contradiction in terms. The official report that recognized the advent of military air power said that it was doing so only because "the subject is no longer an experimental one, there is now no novelty in the idea, or in the application of it."[15] In other words, the British military authorities were not willing to accept the idea of military aviation until they were compelled to do so by necessity. Such was the paralysis of orthodoxy in the British military mind at this time.

Even the Russians were ahead of the staid old English authorities in the matter of military aviation. Under the technical guidance of the Germans, the Russians were busy at this time with the construction of a huge military balloon that was designed with "an armament which would make possible the sudden destruction of armies on the earth below."[16] In other words, the Russian balloon was their answer, or at least it was hoped it would be, to the Grand Army of Napoleon. However, they were doomed to be disappointed. As a British aeronaut later explained, the Russian "expectations rose higher than their balloon, which refused to move off the ground."[17] As a result the soldiers of Napoleon overran the huge Russian balloon, complete with thousands of pounds of gunpowder, in the Castle of Voronzoff. This ended the Russian dream of stopping Napoleon with air power but marked another important milestone in the steady progress of military aviation.

A recurring theme throughout the period was the tactical one: Is there no defense against warfare from the air? Lord Stanhope expressed his concern with this aspect of the subject in a letter to Sir George Cayley dated December 1809. Cayley was an English inventor and scientist who, in a series of articles in *Nicholson's Journal*, discussed for the first time the scientific principles of mechanical flight.

Stanhope was not as much interested in the scientific principles as he was in the idea of there being no defense against this kind of warfare, and in what that probably meant to the future of mankind. Will it tend to advance the happiness of mankind, he asked,

> or to produce a contrary effect, by multiplying the means of annoyance whether between nations or between individuals, while the means of self-defence would remain very imperfect, by creating general insecurity, by rendering still more difficult than at present the detection and apprehension of criminals, by destroying many established Institutions and by depriving of

> their support many industrious individuals. Upon this point in particular I shall be desirous of being favoured with your opinion.[18]

Sir George Cayley, in his reply, sought to assure his Lordship that his fears might not be well founded:

> Your Lordship seems to think that the means of annoyance will be increased whilst the means of defence remains the same; surely, the defendant may use the same instrument as the assailant--I conceive we are looking much too far in attempting to unravel all the workings of a power like this -- Society may improve--Society may improve commensurate with the improvement of this power and we cannot estimate the moral influence of such improvement.[19]

What Sir George was touching upon was what Major General J.F.C. Fuller later called the Constant Tactical Factor. This was the idea that there had never been nor would there ever be an absolute weapon. "From the offensive to the defensive and then back again to the offensive," General Fuller said, "is the normal rhythm in tactical development, and it is this rhythm which has always had the profoundest influence in shaping history."[20] In other words, flying as it was conceived of then was, to Cayley, an offensive weapon, obviously, but that did not mean that it would remain that way, that the defense would never catch up with the offense. That is why Sir George told his Lordship not to worry and not to attempt to unravel all the workings of a power like this. It is also why Cayley was optimistic about the future and told his Lordship that the power of the air weapon might be such as to bring about an improvement in society itself, an improvement commensurate with the power of the air weapon.

Cayley and Stanhope were not the only ones to speculate on the tactical aspects of military aviation during this period of history. In 1818, a Frenchman who was evidently living in England at the time, publisted a tract on military aviation entitled, *A Word for my King and Country, Treatise on the Utility of a Rocket Armament, assisted by Balloons, where Ships of War cannot be accessible, both defensive and offensive, to the annoying of the enemy's harbour. That are stubborn and delight in War.*[21]

Charles Rogier's idea was the use of free balloons not manned by men, but loaded with combustible spiked rockets that could be released to float over enemy strongholds, particularly harbors, where rockets could be discharged automatically and even the balloon itself destroyed by means of a slow match. Rogier believed that this Rocket Armament assisted by Balloons would be an absolute weapon because even if one were shot down it would be over enemy territory and would wreck incalculable damage, particularly in a harbor. To the argument that his Rocket Armament assisted by Balloons would be at the mercy of the wind, which might change its course and carry the missile back to the ship that launched it, Rogier replied:

> Should the current of air change and return
> the balloon back
> Our tars can slip cable and get out of their
> track,
> But in a harbour they are confined as in a
> basin of water,
> And to flee will be in vain--they must submit to the slaughter.[22]

Despite his claim that his was an absolute weapon and that there was no defense against it, Rogier's pamphlet included elaborate drawings showing how enemy attacks could be forestalled. He said that rockets could be mounted and launched from ships far out at sea for use against the piloted balloons or guided missiles of the enemy so that the rockets would destroy them before they had traveled far enough to do any damage to their intended targets.

The problem that Rogier straddled, that of having an absolute weapon that was not absolute, is not a problem that can be straddled today. Rogier was dealing with a weapons-system that floated gently in the air like a cloud. His so-called solution to the problem of early detection and timely destruction was to have the counter-force at sea on patrol. This would not do any good today, and the patrol fleet would have had to have been awfully lucky to have done any good in Rogier's day.

Today's intercontinental ballistic missiles armed with nuclear warheads can reach any point on earth fifteen minutes from launching, and there is no defense today once the missiles are in flight. They are not like birds that can be shot down, or even balloons floating in the sky. There was not any defense against the relatively harmless German V-2 rockets until their launching sites were overrun. This is why SALT-1 was probably a spectacular achievement in the cause of peace. The agreement in 1972 limiting the United States and Russia to no more than two antiballistic missile sites each made mutually assured destruction the inevitable consequence of an all-out nuclear war. In other words, by this agreement both Russia and the United States, if not the whole of mankind, became the hostages of the peace.

Beyond Rogier and his absolute weapon, the Rocket Armament assisted by Balloons, speculation took another turn on the subject of military aviation in the 19th Century. In 1824, a philosophic doctor in Ireland, Joseph MacSweeney, wrote an "Essay on Aerial Navigation" in which he suggested that man's conquest of the air, coming as it did as the last of a long line of technological inventions that had been bent to serve the ends of war, was so terrible in its implications that its ultimate effect would have to be beneficient, as it would banish wars forever from this earth as being too horrible to contemplate. Therefore, he predicted that "Aerostation will aid in diffusing Christianity over the earth. The Gospel will be floated over mists and clouds, and will descend on distant lands to spread light and peace."[23]

While man was in this happy frame of mind about military aviation, his thoughts naturally turned to flying for pleasure--and profit, to the tourist opportunities opened up by the new mode of travel. One such entrepreneur was William Samuel Henson, an Englishman, and another was his countryman, John Stringfellow. In their research and experimentation both men were moving in the direction of the modern airplane, a craft that was heavier than air, which could be steam-driven by propellers. Henson called his machine an aerial steam carriage. Together, Henson and Stringfellow organized what is probably the first commercial airline, and the name they gave their undertaking was the AERIAL TRANSIT COMPANY. The advertisement was that they offered an "aerial steam carriage to Convey Passengers, Troops and Government Despatches to China and India in a Few Days -- to Travel from England to China in 24 hours--and to Distant Parts of the Earth"[24]

The public was not impressed, and the inventors found themselves unable to raise the funds necessary to carry on their business. In fact, the public was so hostile that they eventually took to carrying out their experiments only by night and in secrecy. This is said to have been extremely disadvantageous. For on one occasion, at least, their machine might have flown had it not been tried at night when the excessive moisture in the air shrunk and warped the silk covered wings. Eventually, Henson became discouraged and quit aviation altogether with the derisive words of a popular verse ringing in his ears:

> It matters not, I understand, whichever way the wind is,
> They'll waft you in a day or so, right bang into the Indies!
> Or you may dine in London now, and then, if you're romantic,
> Just call a ship and take a trip right over the Atlantic.[25]

The year 1847 was notable in the history of military aviation because of the emergence of another very modern idea, this time in Germany, and it was an idea that was to take nearly a full century to come to fruition. When it did, it was too late to save Germany from defeat in World War II. However, if the idea had been developed to its maximum, it might have saved the German people and their cities the battering and suffering they were subjected to in World War II. The idea was of a rocket plane, a peculiarly designed ship with warped wings that could shoot almost straight upward with great speed. The propellant was supposed to be the explosive force of gunpowder, and the designer was a German inventor, Werner von Siemens, later famous for his work in connection with the electric dynamo. There is no record, however, that any practical experiemnts were made on the basis of his idea for a rocket plane.

John Stringfellow continued with his experiments in England, but carried on alone after Henson's departure. Success finally crowned his efforts. In June 1848, a model airplane that he had made actually flew. He took his invention to London, where he again demonstrated successfully that a steam-driven, heavier-than-air craft could fly. However, both Stringfellow and the audience before which he made his demonstration seemed to miss the significance of his invention. For Stringfellow's son, who witnessed the exhibition, records that his father, "having now demonstrated the practicability of making a steam engine fly and finding nothing but a pecuniary loss and little honor,"[26] gave up his experimental work and left for an extended pleasure trip to the United States. The modern airplane, however, had been born, even though man was destined to flounder around for another half-century before he realized the vast potentialities in Stringfellow's quiet exhibition.

While Stringfellow was inventing, in principle, the modern airplane, Henry Tracey Coxwell made a continental tour demonstrating how the balloon could be used for bombardment purposes. He discussed the subject very extensively with the Prussian military authorities and put on several exhibitions. Returning to England, Coxwell tried to get the War Office to activate an air corps and to employ it in the investment of Sebastopol, but was unsuccessful. That year Coxwell published a pamphlet wherein he unburdened himself on the subject of the military mind. Coxwell entitled his pamphlet, "Balloons for Warfare: A Dialogue between an Äeronaut and a General." The General, needless to say, came off second-best in the Dialogue.

In 1849, while Coxwell was making his futile efforts to influence the British military on the subject of military aviation, the Austrians in their siege of Venice organized and actually employed a corps of bombardment aviation. The Austrian artillery was not within range of Venice, so several battalions of balloons were formed, and each was equipped with a supply of small bombs.

The technique and results of the subsequent bombardment are summed up in an Official Report to the British Institute of Royal Engineers as follows:

> When an attack was to be made the Austrians first chose a station to the windward of that part of Venice which they wished to bomb. They then released a small balloon to determine the speed and direction of the wind at various heights. With this knowledge the course of the bombing balloon was then laid out trigonometrically on a map.

> The full-sized balloon and its bomb were then allowed to ascend after first timing the fuse. Generally speaking, the use of these balloons would seem to have justified itself, for a number of the bombs fell within the precincts of the city, but the moral effect was said to be much greater than the damage inflicted by the explosives.[27]

Ten years later, in 1859, the French military authorities hired a civilian aviator to assist in the Italian campaign. Eugene Godard was the pilot's name, and he seems to have been very active, making ascents from Milan, Gorgonzola, Castenedolo and the Castiglione Hills. How much good he did, though, is an open question. Just as the stories are sharply conflicting as to how much help Coutelle gave in the French Revolutionary Wars, so the reports differ as to the military value of the intelligence that Godard obtained in the Italian campaign.

The newspapers of the time accredited Godard with obtaining vital intelligence that was of direct assistance, and this view is supported by some of later studies made by professional historians. But the French Government does not seem to have agreed, at least not the legal eagles. For they refused to pay Godard's bill for his services, and the case wound up in the law courts. However, Napoleon III was on Godard's side. He named him "The Aeronaut of the Emperor."

Godard also made an ascent during the Battle of Solferino on June 23, 1859. During this ascent it appears that some photographs were taken of the battlefield and terrain. These are believed to be the first aerial photographs taken from a balloon on active service. Later, when the British were timidly experimenting with the use of balloons in field operations, Captain Frederick Edward Beaumont, after an ascent, wrote that he believed

> some useful results might be obtained by photography from a balloon. A series of panoramic views might be taken by moving the machine along, which would be sufficiently intelligible to enable a draughtsman to make a sketch from,

> and which would have been taken far more rapidly than any survey on the ground could have been executed.[28]

The Captain seems to have been thinking not so much of photographic intelligence for military purposes as of photographic mapping. In any case, three-quarters of a century later, in World War II, aerial photographs, both for the purpose of military intelligence and for mapping, were among the most useful services airplanes performed.

Even today, world events can turn on aerial photography. Not to mention what the satellites are reporting on the missile sites of Russia today, it was the activities of a U-2 reconnaissance plane over Russia in 1960 that broke up the planned summit conference scheduled for that year, and again in 1962 it was the aerial photography that revealed the Soviet missile build-up in Cuba which almost led to a World War III.

But going back to the 19th Century, the United States, while not in the forefront of the race to develop military air power, nevertheless had its own pioneers. The noted French aeronaut Jean Pierre Blanchard, gave an exhibition of ballooning in the United States in 1793 that was witnessed by the First President and America's first aeronaut, Charles Ferson Durant, made an ascent on September 9, 1830. Later, some of his exploits were witnessed by President Andrew Jackson. Ballooning caught on with the public very quickly in America, and one of the first to become a professional pilot by today's standards was John Wise of Philadelphia. He made over four hundred ascents during his life, and when the war with Mexico broke out in 1846, he immediately tried to interest the War Department in military aviation. He offered to organize a Balloon Corps and "to bomb the Castle of San Juan de Ulloa, the key to the defenses of Vera Cruz."[29]

Aeronaut Wise was wasting his time. The American military authorities were just as hidebound and conservative as their opposite numbers in England and refused to have anything to do with something as novel as flying. The offer was refused.

The problem of military aviation raised its ugly head again upon the outbreak of the War Between The States. The culprit was a Yankee with the outlandish name of Thaddeus Sobieski Coulincourt Lowe. Not content with being a scientist and an astronomer, Professor Lowe was also a professional aeronaut and was not in the least shy. Immediately after the firing on Fort Sumter, Professor Lowe telegraphed President Lincoln and Secretary of War Stanton offering his services. These were accepted, and after the Union defeat at Manassas, Lowe was up over the Confederate lines to report the position of the victorious Southerners. General McClellan accepted Lowe and his balloons without much argument and discomfort and used him to good purpose in several important military engagements. However, he was never comfortable with the arrangement, and on the occasion of one incident, practically took the oath on military aviation.

General McClellan recorded the incident in his journal by saying,

> I am just recovering from a terrible scare. Early this morning I was awakened by a despatch from Fitz-John's headquarters, stating that Fitz had made an ascension in the balloon this morning and that it had broken away and come to the ground some three miles southwest, which would be within the enemy's lines. You can imagine how I felt. I at once sent off to the various pickets to find out what they knew and tried to do something to save him, but the order had no sooner gone, then in walks Fitz, just as cool as usual. He had luckily come down near my own camp after actually passing over that of the enemy. You may rest assured of one thing, you won't catch me in the confounded balloon, nor will I allow any other General in it.[30]

Lowe's activities were very extensive, and he participated in many important engagements, making ascents and descents sometimes under the heaviest artillery fire. On one occasion he reported an important change of position on the part of the Confederate general, and later the same month his balloon was used for directing artillery fire. On another occasion, Professor Lowe was in the air and "telegraphing, reporting and mapping the position of the Confederate

Army." As one reporter described it, Lowe was able to deliver "the information on which General McClellan saved his Federal Army."[31]

Henry Coxwell in England said of Lowe's observation work and communications system, "earthworks could be distinguished at a distance of five miles, while the piquets and supports of the enemy were distinctly seen. A telegraphic wire was sometimes attached to the balloon so that the aeronaut could at once communicate with the general, or even, as was done one time, to the Government at Washington. Some photographs were also taken of the enemy position."[32]

Coxwell's account continued by saying that Lowe and the general each had maps divided into small squares, which were numbered in order to simplify the communication. He also said that the *Times* correspondent said of the Battle of Chickahominy, "during the whole of the engagement, Professor Lowe's balloon hovered over the Federal lines at an altitude of 2,000 feet and maintained successful telegraphic communication with General McClellan's headquarters."[33] Coxwell's account concluded by saying that "in an attack on Mississippi Island, No. 10, Engineer Aeronaut Allan ascended and directed the artillery fire, communicating the effect of each shot."[34]

This was not all. Professor Lowe also had a captive balloon fastened onto a railway locomotive so that it could be moved about quickly from place to place. Finally, Lowe also obtained the services of one of the naval vessels of the union, the U.S.S. *Parke Custis*, and used her as an aircraft carrier in the fall of 1861. However, three months before this another Federal aeronaut, John La Mountain, had performed the same feat, using the U.S.S. *Fanny* for the purpose. In other words, U.S. naval aviation was also born in the fiery crucible of the War Between The States.

Professor Lowe as a professional aeronaut in the service of the Federal Army was one thing, quite another was whether the Army or the Navy could be persuaded to adopt military aviation as part of its regular Establishment. The question was raised squarely by a Dr. Solomon Andrews of Perth Amboy, New Jersey, in August 1862. He urged the Secretary of War to authorize the construction of an airship designed for use in war. The Secretary declined to consider it on the grounds that, even if one could be built, it would probably have very little "practical utility."[35] Andrews was not to be put off so lightly, however, and informed the Secretary, "I intend to build one immediately on my own account and if successful I shall present it to the United States Government in the hope that it may shorten the war."[36] Andrews then proceeded to construct a model of a fighting airship. He had it placed in the basement of the Capitol and then made a personal call on President Lincoln. The President received him in spite of the opposition of Stanton and most of the generals of the Army. Following this, Andrews addressed a formal petition to Congress informing the legislators of his intention and calling their attention to the presence of the model in the basement of the Capitol.

The one-man "pressure group" got action. Under pressure from the Congressmen, Stanton was forced to take some notice of the use of military aviation. He did the least he could do, however, which was to appoint a Commission to determine whether there was any merit in the idea. On July 22, 1864, the Commission reported in favor of Andrews' claims, but the idea got nowhere because of the subsequent "red tape and opposition."[37] In fact, the war ended before Dr. Andrews was authorized to proceed with his construction of the world's first war airship.

Before the greatest of America's tragedies, The War Between The States, had run its full course, the European powers, particularly Prussia, were busily engaged in war preparations of their own. In fact, the year Fort Sumter was fired upon, 1861, a very detailed and scholarly report was presented to the Austrian Imperial Ministry of War on the subject of military aviation by Lieutenant-Colonel Baron Ebner of the Austrian Corps of Engineers. After learnedly discussing the differences between, and the relative merits of, the various types of balloons in existence, Baron Ebner said that he felt that all the different types had some use in war. However, he was quick to add that when and if they were used with an Army, great care should be exercised to see that the people who were hired to use them were "respectable" characters.

In general, Baron Ebner said he felt that the present lighter-than-air craft were not the final answer to military aviation because "the model of the balloon is a cloud, and like it, it is always at the mercy of the wind, for no mechanism, however powerful, will be sufficient to overcome the enormous opposition which a movement in the air not in the same direction as that of the balloon and acting on its immense surface would present."[38] Baron Ebner concluded his report with some very wise observations concerning the course that mechanical flight of the future might take, and with the added warning that "the history of inventions abounds with instances, when the seeds of golden harvests have lain prominently in the road and yet have remained unobserved for ages through the want of opportunities to bend down and pick them up."[39]

While Baron Ebner was making these prescient observations about human flight and the course of military aviation, Henry Coxwell in England was still at war with the military authorities in that country. This time, however, he had a new idea-- chemical and biological warfare. Coxwell's idea was that "chemical compounds of a highly poisonous character" could be carried on balloons in shells and dropped on an enemy, or that alternatively, "aerial shells" could be used that carried "agents calculated to produce stupefaction" on enemy troups. "If by this method," he said, "our warriors could secure prisoners instead of increasing carnage, humanity would rejoice at so desirable a consummation by such ingenious means."[40]

On the other side of the world, in 1867, during a war between Brazil and Paraguay, military aviation was being utilized very effectively. Brazil appreciated the value of balloons in war and used them for reconnaissance as well as for military intelligence, going so far as to obtain, on one occasion, sufficient details concerning the Paraguayan positions for the drafting of a Paraguayan order of battle map of remarkable accuracy.

The Brazilian experience was cited in England, though to no avail, in an endeavor to persuade the War Office to recognize the importance of military aviation. The only thing the War Office would concede was that military aviation might be of some limited value during a long siege, but it was felt that aviation had absolutely no place in a war of movement. On the whole in the War Office, it was agreed that the Army should not waste any of its valuable time and money on military aviation.

Meanwhile, Prussia, under the guiding hand of Otto von Bismarck, was about to embark upon the wars with Austria and France that would result in the unification of Germany. In this atmosphere, it was only natural for Coxwell to be recalled to Germany to assist in the organization of the German Balloon Corps. He came back with alacrity to the scene of his triumph in 1848 and assisted in the organization of the German Balloon Corps. Although military aviation played no part in the sensational German victories at Sadowa and Sedan, Coxwell was nevertheless pleased with his work, and some years later, remarked that "his instructions, however, were not lost as may be gathered from the present efficiency of the Prussian Military Aeronauts."[41] What might Coxwell have said if he had lived to see the German Luftwaffe operate in the victorious years of 1939 and 1940!

If Coxwell was on the German side in making the preparations for the Franco-Prussian War, he was on the French side when the siege of Paris began. No sooner had the German armies closed in on Paris than Coxwell appeared on the scene with an offer to raise the siege by means of military aviation. Apparently, a lot of other people had the same idea at the same time, for very soon the French resorted to ballooning for the purpose of observing the enemy and his military preparations against the capital and also for the purpose of communicating with the outside world.

An interesting footnote to the use of aviation during the siege of Paris was the proposal, evidently by Coxwell, "to use a search electric light,"[42] to observe the Prussians from the air. Three-quarters of a century later, the crews of Allied bombers over Germany found their targets obscured and themselves nearly blinded by the great searchlight beacons the Germans threw up to protect themselves against Allied bombardment. The electric light turned out to be a two-edged sword.

Another footnote to the period is the fact that, during the siege of Paris, the first aerial battle of modern times took place. A well-known French pilot, Felix Nadar, was flying into Paris from Tours when, probably by accident, he was intercepted by a German balloon at an altitude of about ten thousand feet. The German pilot opened fire with a rifle, and Nadar returned the fire. Apparently all the shots went wild, however, for Nadar later landed safely in Paris, and there is no record that he was successful in shooting down the German. Coxwell must have smiled as he contemplated his part in this first aerial battle of modern times.

The *Ballon Poste*, as the air defense of Paris was called, accomplished two important objectives. One was to open up communications with the outside world, and the other was to evacuate Leon Gambetta on October 6, 1870. Gambetta was the outstanding member of the provisional government of national defense that had been set up after the capitulation of Napoleon III at Sedan. He was flown from Paris to Epineuse in three hours and made his way to Tours, where he set up the nineteenth century prototype of World War II governments-in-exile. In World War II the French leaders were severely criticized for submitting to the harsh peace terms imposed on them by the Germans instead of following Gambetta's example.

The communications work done by the *Ballon Poste* was equally as spectacular as the transfer by air of the seat of the French Government during the siege of Paris. Altogether, during the year, sixty-five balloons were dispatched from the besieged capital and, except for seven, they all accomplished their missions of flying people and mail out of Paris plus the evacuation of carrier pigeons that could be used later to bring messages back into Paris.

The pigeon messages were reduced photographically and were produced in collodion film form, making it possible for a bird to carry eighteen films as a normal load. In round numbers, it has been estimated that the balloons themselves carried out of Paris some nine tons of dispatches, comprising about three million letters and that the fifty-odd returning pigeons brought back with them some one hundred thousand messages. It was not exactly a Berlin Airlift, but the *Ballon Poste* was certainly a spectacular success in its own small way.

The Franco-Prussian War of 1870-71 was the last occasion on which military aviation played any important part in the affairs of man in the 19th Century, though the subject was revived in the imperialistic scramble for Africa and in the Boer wars. An English imperialist, Frederick Gower, specifically urged that his aerial batteries be sent to the aid of General Charles George Gordon when he was being besieged at Khartoum in 1884. Nothing came of this, however, and General Gordon and the entire garrison were the victims of a massacre when the ground relief force failed to reach the city in time.

The failure of the War Office to pay any attention to Gower did not cause him to give up his fight to be recognized. He simply turned his propaganda guns on the Navy and urged that he be backed in another favorite project, the construction of a fast aircraft carrier. Gower's argument was that "a vessel may be constructed especially for carrying aerial batteries and, thus dispensing with weight of armament, may be given a rate of speed such as to make her capture practically impossible."[43] To clinch the argument, Gower said that such batteries can be constructed singly for little more than the cost of one shot from a large caliber gun.

The Admiralty was, if anything, more cool to Gower's ideas than the War Office had been. However, he set about the construction of two aerial batteries at his own expense, and when these were completed, invited the Admiralty to a demonstration. While the naval authorities debated whether it would be worth any of their valuable time to witness the exhibition, Gower died, probably to the great relief of the Admiralty.

Thus, by the close of the nineteenth century, while man had not yet learned anything about controlled flight in the sense of knowing how to navigate a ship in the air with any degree of accuracy, he was well acquainted with practically all the advantages of military aviation. The end of the century was also marked by the appearance of two rather remarkable books on the subject of military aviation. The first was by Henry Tracey Coxwell on *Military Ballooning*, and the second by General W.H. Hutchinson, entitled *The Navigable Balloon in War and Peace*.

Coxwell's treatise has a chapter on "Air Torpedoes and Bombshells." In this chapter the noted aviator addresses himself to the question of the extent to which flying has or will revolutionize the strategy of war. He repeats the story that was current at the time, just as it was at the time of the Napoleonic invasion, that the Russians were finally in possession of a flying torpedo that could be directed in battle with great accuracy, and which would be used to paralyze armies and destroy forts without itself being exposed to any danger because it would emerge from the clouds unseen or strike from above the smoke of battle.

What is the element of truth in this? Coxwell inquires. The answer he gave is none. Coxwell said that navigable balloons had not as yet been invented, that man had not, and probably would not, become the master of the art of steering and propelling balloons. However, even if he did, Coxwell maintained, it would not be the end of the world. Being specific, Coxwell said that if the Russians had a navigable balloon that could launch such torpedoes, then other nations will also develop them. The result will be aerial fighting on a scale the world has never seen. "One may swoop down like a hawk," Coxwell said, "but another may rise up like a rocket and bring its adversary down like the stick."[44]

In any case, Coxwell did not believe that the sky was going to fall in, not even if the Russians invented their flying torpedo. Coxwell compared the idea of the Russian development to the fanciful notions to invade England by digging a tunnel under the English Channel and finally said, "we know that John Bull and all true Britons would rather face an adversary from above than if he were to crawl and pounce upon us from below. But at present, we need not quake as to the high or subway route."[45]

General Hutchinson, however, with his eye on the future rather than the past, speculated not only on the military, but on the broadest social implications inherent in the development of navigable balloons. In this connection, he said that to

> regard the subject, however, not from the strategist's point of view, but from the *seemingly* opposed standpoint of the humanitarian, the benefits to be derived from servicable Navigables are incalculable, for at a time when most enlightened nations are so perpetually equipped for slaughter than it would seem as though the mutual dread of the

> deadly machines of our nineteenth Christian century were the sole restraining influence which, in their "armed unrest," stayed them on the brink of hostilities -- at such a moment efficiency in the war-engine becomes but another influence in the cause of peace.[46]

General Hutchinson also said:

> It surely stands in no need of demonstration that this condition would be well fulfilled by the navigable war balloon, which, in darkness, itself invisible, scatters, broadcast, chemicals of indescribable power, setting on fire, or by explosives effecting the rapid destruction of well-guarded coal stations, the best sheltered dockyards and arsenals and the most formidable fortresses.[47]

Finally, General Hutchinson said, "happily--yes, happily--this needs no demonstration, for the more the inevitably increasing horrors attendant upon war are made manifest, the greater the hope that the organization of large masses of armed men to slaughter and be slaughtered will be soon abandoned in favor of the settlement of disputes by friendly arbitration."[48]

From this rapid survey of the eighteenth and nineteenth centuries, it seems fairly clear that even in the realm of human ideas, there seems to be little reason to question the familiar French saying, "plus ca change, plus c'est la meme chose." The hopes. fears, and aspirations of mankind do not change over the centuries. Nothing is entirely new, not even the idea of Air Power.

FOOTNOTES--CHAPTER II

[1]Francis Trevelyn Miller, *The World in the Air* (New York: G.P. Putnam's Sons, 1930), 127.

[2]J.E. Hodgson, *The History of Aeronautics in Great Britain* (London: Oxford University Press, 1924), 61.

[3]C.F. Gamble, *The Air Weapon* (London: Oxford University Press, 1931), 26.

[4]C.F. Snowden Gamble, *The Air Weapon* (London: Oxford University Press, 1931), 27.

[5]Francis Trevelyan Miller, *The World in the Air* (London: G.P. Putnam's Sons, 1930), 127.

[6]*Ibid.*, 140.

[7]C.F. Snowden Gamble, *The Air Weapon* (London: Oxford University Press, 1931), 29.

[8]*Ibid.*, 19.

[9]Liddell Hart, *The Ghost of Napoleon* (London: Faber & Faber, Limited, 1933), 36-38.

[10]Francis Trevelyan Miller, *The World in the Air* (London: G.P. Putnam's Sons, 1930), 232.

[11]*Ibid.*, 235.

[12]Alfred Thayer Mahan, *The Influence of Sea Power Upon History 1660-1783* (Boston: Little Brown and company, 1895), 11.

[13]C.F. Snowden Gamble, *The Air Weapon* (London: Oxford University Press, 1931), 32.

[14]*Ibid.*, 33.

[15]*Ibid.*, 48.

[16]Francis Trevelyan Miller, *The World in the Air* (London: G.P. Putnam's Sons, 1930), 235.

[17]Henry Coxwell, *My Life and Balloon Experience, with a Supplementary Chapter on Millitary Ballooning* (London: W.H. Allen Co., 1887), 176.

[18]C.F. Snowden Gamble, *The Air Weapon* (London: Oxford University Press, 1931), 36.

[19]*Ibid.*

[20]Major General J.F.C. Fuller, *Armament and History* (New York: Charles Scribner's Sons, 1945), 21.

[21]C.F. Snowden Gamble, *The Air Weapon* (London: Oxford University Press, 1931), 36.

[22]*Ibid.*, 37.

[23]*Ibid.*, 38.

[24]Francis Trevelyan Miller, *The World in the Air* (London: G.P. Putnam's Sons, 1930), 297.

[25]J.E. Hodgson, *The History of Aeronautics in Great Britain* (London: Oxford University Press, 1925), 357.

[26]Francis Trevelyan Miller, *The World in the Air* (London: G.P. Putnam's Sons, 1930), 303.

[27]C.F. Snowden Gamble, *The Air Weapon* (London: Oxford University Press, 1931), 46-47.

[28]*Ibid.*, 53-54.

[29]Francis Trevelyan Miller, *The World in the Air* (London: G.P. Putnam's Sons, 1930), 279.

[30]*Ibid.*, v. II, p. 22.

[31]Henry Coxwell, *My Life and Balloon Experience with a Supplementary Chapter on Military Ballooning* (London: W.H. Allen Co., 1887), 192.

[32]*Ibid.*, 178.

[33]Henry Coxwell, *My Life and Balloon Experiences with a Supplementary Chapter on Military Ballooning* (London: W.H. Allen Co., 1887), 178.

[34] *Ibid.*

[35] Francis Trevelyan Miller, *The World in the Air* (London: G.P. Putnam's Sons, 1930), v. II, p. 23.

[36] *Ibid.*

[37] *Ibid.*

[38] C.F. Snowden Gamble, *The Air Weapon* (London: Oxford University Press, 1931), 50.

[39] *Ibid.*, 51.

[40] *Ibid.*, 45.

[41] J.E. Hodgson, *The History of Aeronautics in Great Britain* (London: Oxford University Press, 1924), 267.

[42] Henry Coxwell, *My Life and Balloon Experiences with a Supplementary Chapter on Military Ballooning* (London: W.H. Allen Co., 1887), 179.

[43] C.F. Snowden Gamble, *The Air Weapon* (London: Oxford University Press, 1931), 68.

[44] Henry Cozwell, *My Life and Balloon Experiences with a Supplementary Chapter on Military Ballooning* (London: W.H. Allen Co., 1887), 186.

[45] *Ibid.*, 189.

[46] C.F. Snowden Gamble, *The Air Weapon* (London: Oxford University Press, 1931), 69.

[47] *Ibid.*

[48] *Ibid.*

CHAPTER III

"THERE SHALL BE WINGS"

Samuel Eliot Morison said that there were "three stages in the public attitude toward a great discovery: first men doubt its existence, next they deny its importance, and finally they give the credit to someone else."[1] The invention of the airplane was certainly not an exception to this rule. The airplane was invented, as it were, on December 17, 1903, when the Wright brothers made their historic first flight at Kitty Hawk, North Carolina. At first there was skepticism and doubt that man had really conquered the air; then there was the denial that the invention or discovery was of any importance; and finally the credit was given to someone else.

The "someone else" in this case was Samuel Pierpont Langley, who constructed a power-driven model aircraft in 1896, which had proved that mechanical flight was possible, and as a result had been commissioned by the War Department to construct a practical flying machine. To support him in this purpose, he was given a subsidy of $50,000 drawn from a secret fund in the War Department.

Professor Langley did his part, only to end in failure. The first test of the machine was on October 7, 1903, and the second on December 8. Both were unsuccessful, and Langley attributed the failure to a defect in the launching apparatus, not to the machine itself. For years after, Dr. Langley's supporters continued to insist that his machine had been capable of controlled flight, and that therefore he should be remembered as the inventor of the airplane, not the Wright brothers. One of Dr. Langley's strongest supporters was General William ("Billy") Mitchell, then the youngest captain in the Army.

Both trials of the Langley machine took place at sites on the Potomac River near Washington, and when both failed, there was a loud outcry from Congress and the press about the folly of wasting the people's money on the crazy idea of a flying machine. That this idea was crazy was fully attested to at the time by the cautious dean of science Simon Newcomb, who published an article on October 22, two weeks after the first failure, saying,

> There are many problems which have fascinated mankind ever since civilization began which we have made little or no advance in solving. The only satisfaction we can feel in our treatment of the great geometrical problems of antiquity is that we have shown their solution to be impossible. The mathematician of today admits that he can neither square the circle, duplicate the cube, or trisect the angle. May not our mechanicians, in like manner, be ultimately forced to admit that aerial flight is one of that great class of problems with which man can never cope, and give up all attempts to grapple with it."[2]

In this atmosphere of denial, doubt, and failure, the Wright brothers pushed ahead with their experiments, and their efforts were crowned with success. The first successful flight was made on December 14, 1903, almost a week to the day after Langley's second and last failure. However, since the Wright Flyer stayed up only three and a half seconds, December 17 is the day that went down in history as the one on which man conquered the air.[3] On December 17, Orville and Wilbur took turns at being at the controls, and on the first flight Orville flew the heavier-than-air powered biplane a distance of 120 feet in 12 seconds. The wingspread of a modern large airplane is more than 120 feet, but in 1903 the distance of 120 feet was considered a spectacular achievement.

As a date, December 17 is interesting because it was only fifty-six days after Simon Newcomb had predicted that man would never be able to solve the problem of human flight and, accordingly, had advised him to "give up all attempts to grapple with it." However, nothing succeeds like success, and December 17 is the great success date in man's long attempt to free himself from his earth-bound existence.

The Wright brothers' success on December 17 did not mean that their troubles were over. Far from it. Inventing the airplane was child's play compared to the job of getting their government interested in it, and this was important to the Wright brothers for two reasons. One was they were convinced of the military value of their invention, and the other was because they wished to see the United States get ahead of other countries in the matter of military aviation.

The invention of the airplane came at a very bad time for the War Department. The generals were still smarting from all the bad publicity connected with the Langley affair, and the last thing they wanted to do was to invite another scandal by giving support to another crazy idea connected with a flying machine, Langley's, Wright's, or anybody else's. Accordingly, the Wright brothers ran into a stone wall of resistance when they sought to interest their own government in military aviation. It did not seem to matter to the government that the airplane had been invented and that steady progress had been made since that time in all aspects of flying. In 1904, more than a hundred flights were made at Dayton, Ohio, and not only had a complete circle been made for the first time on September 20, but flights of two miles each had been made on November 1 and December 1. Still the United States did not believe that the airplane had been invented.

Finally, on January 13, 1905, the Wright brothers penned a letter to their Congressman from the Dayton district and asked him to intervene with the War Department, saying that the airplane had been invented, that their accomplishments included "numerous flights in straight lines, in circles, in calms and in winds," and asked whether the government would be interested. The letter concluded with the Wright brothers making a specific offer, saying, "We shall be pleased to take the matter up either on a basis of providing machines of agreed specification, at a contract price, or of furnishing all the scientific and practical information we have accumulated in these years of experimenting, together with a license to use our patents, thus putting the government in a position to operate on its own account."[4]

The Wright brothers' letter was routinely passed to the Board of Ordinance and Fortification in the War Department. This was the agency that had been scarred by the bad publicity in the wake of the Langley failure. It is not surprising, therefore, that the Board was wholly negative in its reply to the Wright brothers. The Wright brothers had every reason to be discouraged by the nature of the reply from the War Department. It was obvious that the policy was one of once bitten twice shy, and that the Board did not want to repeat the mistake it had made with the distinguished scientist, Dr. Samuel Pierpont Langley. In fact, the Board was so cautious that it not only used a form letter to communicate with the Wright brothers, but one in which it was stated that it was not the policy of the War Department to grant financial assistance to would-be inventors. In other words, the Board wanted to go on record with the fact that the airplane had not been invented.

On October 9, 1905, the Wright brothers went directly to the Secretary of War with a letter about their airplane. This was also to no avail. This time the reply came to them over the signature of an officer of the General Staff, Major General J.C. Bates, and the substance was essentially the same. The War Department repeated its disinterest, and again said it had no money to subsidize would-be inventors. By this time, the Wrights had made over one hundred successful flights and had put their machine to every possible test, but still they could not arouse any interest on the part of the War Department. What did they have to do to convince the War Department that the airplane had been invented?

On October 19, 1905, the Wright brothers made their third attempt to penetrate the War Department, only this time they wrote directly to the Board of Ordinance and Fortification, and they also tried another approach to the subject. This time, instead of asking the Department whether it was interested in their flying machine, the Wrights asked the Department to state its requirements for a flying machine, and made it very clear that they were not asking for financial assistance. The Board sent its reply to the Wrights under date of October 27, 1905, and it was a shocker. "The Board," the Wrights were informed, "does not care to formulate any requirements for the performance of a flying machine or take any further action until a machine is produced which by actual operation is shown to be able to produce horizontal flight and to carry an operator."[5]

Meanwhile, the Wrights, tiring of the excessive obduracy of the War Department, began to put out feelers to foreign countries on the subject of their invention. France was the first government to make the

Wright brothers an offer, but the terms did not appeal to the Americans, and the matter was dropped. The French were not particularly concerned about their failure to come to terms with the Americans because flying machines were beginning to crop up all over France, and the French public went wild over their own birdmen. They didn't need the Americans.

The French machines were probably copied from the crude illustrations of the Wright brothers' machine that had been published in Europe, and the French machines were also probably inferior to the Wright Flyer, but they were just the thing for the French, proud as they were of their leadership in the field of flying, going back to Montgolfier's fire-balloon in 1783. Which only goes to show, as was later demonstrated with the "secret" of the atom, that science and advanced technology are restless citizens of any particular nationality.

In any case, the Wrights were fighting a losing battle when they attempted to make a secret of the details of their flying machine. It was not much of a secret to start with, any more than the atom bomb was, and once it was out, it was out, just as in the case of the bomb.

Like the French, the Germans were interested in the flying machine that the Wrights had invented, but not necessarily as a weapon of war. Lieutenant-General Walter von Eberhardt said, "I consider the thing at present to be merely a clever circus stunt without military value. It is, however, a great sport and after all, the principle is known and a great step forward has been made."[6] The passion of the Germans at this time was reserved for their great dirigibles. An exception to this, however, was a relatively obscure member of the airship section of the German Army, Alfred Hildebrandt. He was keenly interested in the Wright brothers' invention and made a special trip to the United States to find out all he could about the Wright Flyer.

It was not much. Hildebrandt learned enough, however, to convince him that the Wright Flyer was a workable flying machine, but since the Wright brothers refused to demonstrate it in advance of a signed contract, the German returned home, where he was caustic about the American invention, saying, "why do the Wright brothers now decline to exhibit in public to possible purchasers before the conclusion of an agreement? The Flyer is so simple that they fear no purchaser after viewing it would pay such an enormous sum as one million dollars for it."[7]

The Wright brothers continued to work on their flying machine to improve its performance, and at the request of the Aero Club of America submitted a report to them on the progress thus far. The report stated that over the last three years, beginning at Kitty Hawk in 1903, they, the Wright brothers, had made a total of one hundred sixty flights with their motor-driven flyer and had covered a distance of exactly one hundred sixty miles, or an average of a mile to each flight. However, the Wrights pointed out that the more recent flights had covered longer distances than the earlier ones, and said that the flight of October 5, 1905, had covered more distance (twenty-four miles plus) than the one hundred five flights of the year 1904 altogether. The Wrights said that the normal speed in straight flight was more than forty miles an hour. Finally, the Wrights submitted a list of names of persons who had witnessed one or more of the flights, since they had all been made in private. The report was dated March 12, 1906.[8]

During the spring of 1907 some prominent members of the Aero Club got a promise from President Teddy Roosevelt that he would make arrangements for the Wright brothers to get a fair hearing from the War Department. This was not difficult to do because Roosevelt, as Assistant Secretary of the Navy in 1898, had teamed with the Chief of the Signal Corps in the Army to get the War Department to appropriate the fifty thousand dollars to support the Langley project. Since then, while Langley had failed, the Wrights had greatly succeeded, so much so that they were the talk of the town, not only in the United States after a hundred sixty successful flights, but in Europe as well, where everybody seemed to be getting in on the act, particularly in France.

After the President's intercession on the part of the Wright brothers, the War Department had no choice but to reopen the negotiations with the pesky inventors of the airplane. However, the generals were under no obligation to be gracious about the renewed contract discussions, and they were not. But neither were the Wright brothers, who still refused to make a public demonstration of their flying machine without having a signed contract ahead of time. It was not the publicity that the Wright brothers were after, not the plaudits of the generals, but a binding obligation on the part of the War Department for an important weapon of national defense. The Wright brothers did not want to be left with the German conclusion that the airplane was merely another clever circus stunt without military value.

After much bickering back and forth, an agreement was finally reached that satisfied both parties, the conservative and cautious Wright brothers and the skeptical and reluctant generals. The agreement was that the War Department would draw up a list of specifications for an airplane, and that these would be made public so that the contract could be issued to the lowest bidder. This covered the War Department against another scandal of the Langley type, and also satisfied the Wright brothers, who had proposed a year earlier that the Board do just that, namely tell them what requirements they had to satisfy in order to convince the War Department of the utility of their flying machine.

As soon as the "Advertisement and Specification for a Heavier-Than-Air Flying Machine" (No. 486) was made public, which was on December 23, 1907, the editor of the *American Magazine of Aeronautics* criticized the Army for stipulating performance requirements that could not possibly be met and said that the resulting failure would delay the development of mechanical flight in the United States for an extended period. And looking at some of the 1907 specifications from the vantage point of later-day progress in aviation, some of the requirements do look a bit much. For example, the army wagons and ploughed fields concept, the requirement that the flying machine should be so designed that it could be transported in army wagons, and that it should be able to take off and land safely from ploughed fields.

There were two other somewhat remarkable requirements in the 1907 specifications. One was that the flying machine be simple enough to permit an intelligent man to become proficient in its operation within a short period of time, and the other was that the flying machine be safe enough so that in the case of a power failure, the occupants could descent without personal injury. In addition to these somewhat far-out requirements, the specifications stipulated that the flying machine should be designed to carry two persons with a combined weight of three hundred fifty pounds; that it have a range of one hundred twenty-five miles; that it have a speed of at least forty miles per hour in still air; and that it be capable of remaining in the air for at least one hour. Also, it was provided that the designer should be entitled to a specified extra sum of money for speeds in excess of forty miles per hour, and that in any case the price quoted in the proposals should include the training of two men in the handling and operation of the flying machine. Finally, to weed out the crackpots, the specifications required the bidders to furnish with their proposals a certified check amounting to 10 percent of the price stated for the forty mile speed. The last requirement caused one disappointed inventor to say, "where there is money there are no brains and where there are brains there is no money."[9]

The Wright brothers delivered their flying machine to Fort Myer on August 20, 1908, in accordance with the terms of the contract they signed with the War Department on February 10. The airplane was delivered in knocked-down condition so it could meet the requirement for transportation in army wagons, but the military authorities had to give a little bit on the additional requirement that the machine be crated in such a way that it could be made operational in one hour. Actually, it took several days for the machine to be uncrated and assembled for flying. The tests began on September 9, and continued until September 17, when there was an accident that put the pilot, Orville Wright, in the hospital for several weeks and killed his passenger, Lieutenant Thomas E. Selfridge. The War Department then extended the contract delivery time until the summer of 1909.

Wilbur and Orville Wright arrived together in Washington on June 20, 1909 and began a series of test flights before the official trials in the presence of the Board took place. These began on July 27, and the first flights were designed to demonstrate the capacity of the flying machine to take off and land in open fields, to remain in the air for at least one hour, to carry two persons, and to be steered in all directions. The flying machine passed these tests and established a new world's record for a two-man flight, one hour, twelve minutes and forty seconds. Orville was at the controls.

Then came the speed test. This was important because the government had specified that it would pay additional sums of money for each mile-per-hour the plane achieved over forty. The starting point for the speed test was the Fort Myer drill field, and the test run was over a five-mile measured course to the site of the present Masonic Memorial in Alexandria, Virginia. This was called a cross-country flight.

The Wright Flyer passed this test also. It met the required speed of forty miles-per-hour and exceeded it. The final determination was that the average speed of the Wright Flyer over the measured course was more than forty-two miles-per-hour, but less than forty-three. This raised a big question: should the Wrights be paid for a forty-two mile-per-hour airplane, or a forty-three mile-per-hour, or should the difference be split? The opinion of the Board, according to the records of that day,

> was that, as the Specification called for a separate quotation by bidders for each mile of speed, it was not practicable legally to figure payment on the basis of fractional miles. To do so would have introduced the speed bonus idea which was prohibited by law with the certain prospect that payment of the voucher would not be approved by the auditor for the War Department. Therefore, the amount due had to be limited to the contractor's quotation for the 42-m.p.h. figure, which was $30,000, 20 percent above the base quotation of $25,000 for the 40-mile speed.[10]

During the progress of the cross country flight to determine the speed of the Wright Flyer, a journalist who was present asked Wilbur Wright why he continued using the catapult-monorail launching device instead of the wheeled chassis construction used by most of the European inventors? In reply, Wilbur referred to the requirement in Specification No. 486 for the Flyer to be able to operate from ploughed fields. Wilbur said that his airplane could similarly start and land on wheels, but "he believed that the starting rail and skid chassis were more suitable than wheels for compliance with that special requirement."[11]

The ploughed-fields and army-wagons concept of military aviation died hard in the War Department. In fact, it was not until 1912 that the War Department even got around to including a separate item in the budget for military aviation. Funds for this purpose had been taken from money allocated for the maintenance of military telephone and telegraph installations, or from any other readily available special funds. However, the act for the fiscal year 1912 contained a special provision for one hundred twenty-five thousand dollars "for the purcahse, maintenance, operation and repair of airplanes and other aerial machines."[12]

One of the amusing stories that circulated in Washington at the time of the Congressional debate on the appropriation bill for 1912 was that a member, when asked to support the item for military aviation, replied petulantly, "why all this fuss about airplanes for the Army. I thought we already had one."[13] Another story that made the rounds at the time (also unsubstantiated) was the one about the gondolas on the Chesapeake and Ohio Canal. The Bill was for the purchase of several of these to add some of the charm of old Venice to the historic waterway, and the counter-suggestion of the member was to limit the purchase to two--and breed them. The atmosphere in Washington at the time, as two of the old hands at the business said, was that economy was the watchword, "and vision was lacking in those who held the purse-strings."[14]

The passage of time and the outbreak of World War I did not improve the vision of the diehards, and even those who were more flexible in their thinking found it hard to adjust themselves to the new thinking. For example, one Chief of Staff in the Army said he saw no necessity of extending the range of aircraft beyond thirty or forty miles because that was as far as the Army could march in a single day. Also, Commander-in-Chief of the British Army, Field Marshal Douglas Haig, is reported to have said that the only trouble with the airplanes "was that they always frightened the horses."[15] He was a friend of the airmen.

When Orville Wright was putting the Wright Flyer through its paces at Fort Myer in 1908, Captain William ("Billy") Mitchell was just completing his Staff College studies with the customary inspection trip of the Virginia battlefields. Mitchell was in Washington in August and met Orville Wright on one of his visits to Fort Myer.

However, he was not one of the army officers who were officially assigned to attend the trials that led to the purchase of the Wright Flyer in 1909.

Captain Mitchell was dispatched to the Philippines for a two-year tour of duty in 1909 and returned to the United States in 1911, where he was immediately sent to the Mexican border as a result of troubles in that area. A year later, in March of 1912, Mitchell was notified of his selection to serve on the General Staff. At age thirty-two Mitchell was the youngest officer ever to be assigned to the General Staff, just as he had been at age twenty-four the youngest captain in the United States Army.

As Captain Mitchell was the only career Signal Corps officer on the General Staff, he automatically became the "authority" on military aviation (a Signal Corps responsibility). Then, after the outbreak of war in Europe in August of 1914 and the crisis with Germany over the submarine campaign of 1915, Mitchell suddenly found himself thrown into military aviation.

The occasion was the "Preparedness Controversy" in the United States over the country's defenses, a controversy that President Woodrow Wilson entered into with extreme reluctance. However, when it became a hot political issue he could not avoid it and ordered the services to prepare studies of the country's military needs. The job of preparing the study on military aviation naturally fell to the Signal Corps representative on the General Staff, Captain Billy Mitchell.

The report Mitchell prepared for the President was a good one, and it was also a turning point in his life. Military aviation subsequently became a great "cause" for the young Army captain, and it was one to which he devoted the rest of his life. It was a cause that endeared him to the American people and made him a national hero. Finally, however, it was a cause that broke him.

Mitchell thought of the future and of his own military career, when, shortly after his assignment to the General Staff, he wrote to his mother on the subject of his assignment. "If fortunate," Mitchell said, "I may be a general before many years have passed."[16] It was not that Billy Mitchell was insincere about his cause, military aviation, not that personal ambition was his sole motivation in becoming a crusader for Air Power. Not this at all. It was simply the fact that Fortune, as Machiavelli would say, put this very able young officer in the right place at the right time to grab on to what was becoming a very vital issue and to make the best of his opportunity. From that time on, Air Power and Billy Mitchell were one and the same thing. Things stayed this way until the great airman's death in 1936, the year that Hitler marched into the Rhineland.

From the War Department's General Staff, Mitchell's next assignment was to the Aviation Section of the Signal Corps, where he became the Acting Chief. This got him more deeply involved in military aviation. Meanwhile, Mitchell was promoted to Major, and with characteristic vigor threw himself into the job of building up Army aviation. As a corollary, Mitchell felt it was high time he learned to fly himself, and on his own free will and motion began his pilot training at the Curtiss Aviation School in Newport News, Virginia. At the age of thirty-six, it was a little bit late to start learning the art of flying, but the young Major had a career to think of, and it was better late than never. Mitchell's practice was to steal away from his desk on Saturday nights, take a boat down the Potomac River to Newport News and fly all day on Sundays. This enabled him to be back in his office on Mondays.

The instructors at the Aviation School did not allow the young Major to make a solo flight until after four full days of instruction, and when he did, the future "stormy petrel" of military aviation cracked up the plane in making a bad landing. However, he eventually won his wings, though the training cost him $1,470, because when Mitchell submitted his voucher to the War Department it was disallowed. The reasoning was that, while Congress had authorized officers like Mitchell to fly, it had not appropriated any money for their instruction, and thus if they were not naturally endowed with this ability, they had to pay for their own training. Subsequently, Billy Mitchell had a fixation on the subject of getting pilot training included in all bills pertaining to military aviation.

The next career opportunity that opened up for Billy Mitchell was one of his own making, and it probably had something to do with the fact that he had won his wings. He managed to get himself ordered to Europe in 1917 as an aeronautical observer. This was just on the eve of the American entry into the great war. He arrived in Paris on April 10, 1917, which was four days after the United States declared war on Germany. At this time the air armada the United States had to throw into the conflict was conspicuous by its absence. Not a single combat plane was available. In fact, thirteen weeks after the United States entered the war, and the first American division arrived in France, Mitchell made a note in his diary, Our air force consists of one Nieuport plane which I use myself, and that is all."[17] The order-of-battle in the air was fairly even when hostilities began on the Western Front. Both the Germans and the Allies had about one hundred eighty airplanes each, while the Germans had twelve Zeppelins to the thirteen dirigibles held by France and England.

World War I was not a major conflict from the point of view of military aviation. In fact, it was a very minor one. During the whole course of the war, for example, a total weight of only 275,000 pounds of bombs was dropped on the enemy whereas in a single "Big Week" in February 20-25, 1944, twenty million pounds of bombs were dropped on the enemy, to say nothing of the atomic bombs at the end of the war. However, World War I was important because it was the melting pot, the crucible, in which the ideas of air warfare were tested, the concepts defined, and the doctrine or principles established. For the thoughtful, therefore, World War I answered the questions: when does an idea become a concept, a concept a doctrine, and a doctrine a principle?

World War I, to Billy Mitchell, was a time for ideas, concepts, doctrines, and principles. It was like what Clausewitz said of the Napoleonic era, in which he had been a participant, a time when "war itself, as it were, had been lecturing."[18] Mitchell could not wait to hear the lectures of the battlefields and to prepare himself to understand them. For him, one of the most important occasions of his life was the opportunity he had to discuss the subject with a highly experienced British tactician and strategist, Major General and later Air Marshal Lord Trenchard.

The occasion presented itself as a result of an extensive tour of the battlefields on the Western Front that Mitchell took almost as soon as he arrived in France. During this tour Mitchell did everything he could to familiarize himself firsthand with actual combat conditions, during the course of which he became the first American to fly over enemy lines. He was impressed with the valor of the French flyers and the quality of their aircraft, which was far superior to anything available in the United States at the time, but the learning process about concepts, doctrines and principles did not really begin until Mitchell reached the British sector of the battlefield in May and met Hugh Trenchard.

At the meeting between Mitchell and Trenchard in May 1917, it soon became apparent that the two men had a lot in common, and that the conference would go well. The two men were about the same age, Trenchard being the senior by about six years, and both were from the same strata of society, with patrician family backgrounds. Both men were regular Army officers; both had chosen to make flying a career in the Army; and both had come to this decision somewhat late in life, Mitchell being thirty-six before he began his off-duty pilot training and Trenchard was thirty-nine before he took up flying. In other words, the Englishman and the American were like two peas in a pod, in a certain sense and complemented each other in background and outlook to such an extent that communication was natural and easy. Mitchell confided to his diary on taking leave of Trenchard, "His judgment inspired my immediate confidence and his whole personality my deep respect." Mitchell also said, "we became fast friends at once."[19]

The exchange of views between Mitchell and Trenchard in 1917 was as one-sided as it was pleasant. It could not be otherwise. The war had been going on for almost three years before the conference took place, and it was a case of an experienced British commander talking to an inexperienced American of a lower military grade. There is usually a wide gap, or a big difference, between theory and practice, and this was doubly true in this case because of the newness of air warfare. Therefore, Trenchard had a tremendous advantage over Mitchell in their first contact, and it says a lot about the tact of the Englishman that the conference went off so amicably and well.

At this time Trenchard's concept of tactical air power had been forcibly stated in a Memorandum to the Commander of the British Army in France, Field Marshal Douglas Haig, in September 1916. He repeated to Mitchell what he had told Haig, that the airplane is an offensive and not a defensive weapon, and that therefore it must be employed aggressively to force the enemy to take a defensive position. General Trenchard went on to tell Major Mitchell (promoted to Lieutenant-Colonel in the same month) that the Field Marshal not only approved of his concept, but had quickly overruled British ground commanders who had other ideas about the employment of aviation.

There was a complete meeting of the minds between Trenchard and Haig on the tactical employment of military aviation. In other words, by giving full support to Haig's idea of winning the war on the ground, Trenchard obtained Haig's support of how the air war should be fought. It is not surprising, therefore, that the Lieutenant-Colonel who first came to France as a Wing Commander, rapidly rose to become a Major General. Trenchard's Royal Flying Corps in France was a model for the world's airmen.

In addition to his indoctrination by Trenchard on the tactical employment of military aviation, Mitchell became aware of the strong feeling on the part of some British military authorities at the time that "a radical air strategy"[20] should be adopted to break the stalemate on the Western Front. In accordance with this concept, the new strategy would leave the battlefield to the soldiers, and the air strategy would be designed to win the war by striking directly at the industrial wellspring of German military power.

The line of reasoning for the radical new strategy raised the question that Clausewitz had labored so mightily--the purpose of war. Armies, it was argued, came into existence to protect the vital centers of the country, and in the pre-aviation days they had to be defeated before the centers could be reached. Not so in the days of aviation, however, the argument continued, because airplanes could go straight to the vital centers and immobilize the armies by destroying their strength at its source. Therefore, the argument concluded, the enemy army should not be the primary object, as it was in Clausewitz's theory of war, but the vital centers themselves.

The advocates of this new and very "radical air strategy" got the spotlight of attention in England when London came under air attack on June 13 and July 7 of 1917. After the daring daylight attacks, and largely because of them, a decision was made in England to give the radical air strategy a trial, and along with other changes in the structure of national defense a British Independent Bombing Force was created. General Hugh Trenchard was designated to be the Commander of this new force, and its mission was to take "direct action against the heart of the German industrial system."[21]

The Independent Bombing Force was not a success from the point of view of accomplishing its mission of bringing about the progressive destruction and dislocation of the industrial system to the point where the will and capacity of the Germans for continued armed resistance would be fatally weakened. However, General Trenchard could not be blamed for the failure. He was promised a strategic bombing force of two thousand aircraft and got one of less than a hundred. Of these, he was compelled to divert over half his effort to the bombing of enemy airfields for the protection of his own force. Thus only five hundred tons of bombs could be delivered to industrial targets, barely a pinprick.

Nevertheless, an official in the Historical Division of the United States Air Force, Dr. Thomas H. Greer, would later write, "Trenchard became widely recognized as the leading prophet and pioneer of strategic aviation, and he strongly influenced the thinking of later air leaders like Mitchell and the Italian Douhet."[22]

Mitchell learned a great deal from General Trenchard about Air Power. In fact, one of Mitchell's early biographers, Isaac Don Levine, said that "probably no one throughout his life had a greater influence on Mitchell's aviation views than General Trenchard."[23]

From Trenchard, Mitchell gained the view that Air Power was a new kind of warfare with its own philosophy, its own strategy, its own tactics and its own concepts, doctrines and principles. It was also from Trenchard that Mitchell got his ideas about the unity of command in the employment of military aviation and the necessity of the Air Force's having a co-equal status with the Army and Navy at the highest levels of the government. Finally, it was from Trenchard that Mitchell got the other two ideas that played such an important role in his later life: one, that Admiral Mahan's celebrated idea of Sea Power was fighting a losing battle with Air Power, and two, that Air Power would be the dominant and decisive factor in any war of the future.

At the time Mitchell was absorbing his lessons about Air Power from General Trenchard and his British colleagues, the British were in the process of making some fundamental changes in their defense structure in response to the exigencies of war. The exigency of the moment, of course, was the German bombings of England, first with the Zeppelins and then with the four-engine Gotha airplanes. The Gotha attacks in 1917 produced the greater panic, the heaviest casualties and the strongest criticism of the government.

Less than a week after the second daylight attack on London, on July 7, the British Government, on July 11, directed General Jan Christiaan Smuts of South Africa, one of the Empire's leading statesmen, a friend of Winston Churchill, and a member of the Imperial War Cabinet, to examine and report on British military policy with respect to aerial operations and home defense. General Smuts' first report made history in two respects. In the first place it was completed in eight days, and in the second place his recommendations were immediately implemented. The report led to the end of an era in history when Air Power was subordinated to the Army and Navy in England, and to the beginning of an era in which Air Power was elevated to a unified and independent system co-equal with the Army and Navy.

The British set up a separate Air Ministry in December 1917, and in April 1918, the Royal Air Force was formed out of the Royal Flying Corps and the Royal Naval Air Service. The establishment of the Independent Bombing Force under General Trenchard at Nancy was another result of the Smuts Memorandum. This was the model that existed when Billy Mitchell returned to make his fight for the same thing in the United States: "Trenchard once again became Chief of Staff of the RAF. All of Britain's military aviation, including even the aircraft of the Royal Navy, now were under his control. His RAF in turn was a part of the Air Ministry, the governmental agency also charged with supervising civil aviation."[24] In other words, as a British airman expressed it, by 1917 British "ideas were maturing into theories and visions into forecasts. Air Power was given the chance to become a living, self-developing organism endowed with a voice, a brain and a limb."[25]

General Smuts' second report was submitted to the government on August 17, 1917 and has been called the "Magna Carta of British Airpower"[26] because it made the case for an Independent Air Force. General Smuts said that an Air Force was quite different from the Artillery Arm, for example, because the Air Force

> can be used as an independent means of war operations. Nobody that witnessed the attack of London on 11th July could have any doubt on that point. Unlike artillery an air fleet can conduct extensive operations far from, and independently of, both Army and Navy. As far as can at present be foreseen there is absolutely no limit to the scale of its future independent war use. And the day may not be far off when aerial operations with their devastation of enemy lands and destruction of industrial and populous centres on a vast scale may become the principal operations of war, to which the older forms of military and naval operations may become secondary and subordinate."[27]

Smuts differed from Winston Churchill, his friend and colleague, on the subject of the defense of London. At the beginning of the war the Royal Naval Air Service was charged with the responsibility for the aerial defense of Great Britain, and this included London, of course. At that time, Churchill took a dim view of giving a higher priority to London. "Far more important than London," he said, "are the vulnerable points in the Medway and at Dover and Portsmouth. Oil tanks, powerhouses, lock gates, magazines, airship sheds, all require to have their aerial guns increased in number. Portsmouth in particular requires attention now that enemy's territory has come so near. Aerial searchlights must be provided in connection with every group of guns."[28]

On the other hand, General Smuts believed that London should have the highest priority from the point of view of defense. He said that "London occupies a peculiar position in the Empire of which it is the nerve centre, and we consider, in the circumstances, that its defence demands exceptional measures. It is probable that the air raids on London will increase to such an extent in the next twelve months that London might, through aerial warfare, become part of the Battlefront."[29]

In conclusion, General Smuts said,

> It is important for the winning of the war that we should not only secure air predominance, but secure it on a very large scale; and having secured it in this war we should make every effort and sacrifice to maintain it for the future. Air supremacy may in the long run become as important a factor in the defense of the Empire as sea supremacy."[30]

The idea that air supremacy might be as important in the long run as sea supremacy was then might have been an origianl idea with General Smuts in 1917, but there was certainly no originality in the idea itself. It was the same with the idea of sea power. When Alfred Thayer Mahan put the term sea power into the vocabulary of the entire world through his writings in the 1890's, he specifically disclaimed any "rights of discovery" for the idea itself, and mentioned a long line of "predecessors," including Francis Bacon and Sir Walter Raleigh. However, Mahan also said that their ideas were unknown to him at the time. "For me, as for them," Mahan explained, "the light first dawned on my inner consciousness; I owed it to no other man."[31]

Perhaps it was the same with General Smuts in 1917. Even at that time the importance of air supremacy was something that had been recognized from practically the beginning of the human experience, and it had been given formal expression by Major J.D. Fullerton of the Royal Engineers in a paper presented to a meeting of military engineers in Chicago in 1893. In this paper, Fullerton expressed the opinion that the impact of aeronautics foreshadowed "as great a revolution in the art of war as the discovery of gunpowder." He also predicted that victory would go to the side that first succeeded in winning "command of the air."[32]

This was not all. In 1917 Lieutenant Commander Lord Tiverton presented a paper to the Air Board in England that expounded on a theory of target analysis which was essential to winning victory in the air. The emphasis here was on the concept of strategic bombardment, the individual factors that would have to be taken into consideration in any plan designed to achieve command of the air. This may be the first analysis of strategic vulnerability from the point of view of victory through Air Power. It also illustrates what the Italian General Giulio Douhet meant when he later said that target analysis was "the most difficult and delicate task in aerial warfare, constituting what may be defined as aerial strategy."[33]

Meanwhile, under the watchful eye of Winston Churchill, First Lord of the Admiralty from 1911 to 1915, the Royal Navy Air Service had been bombing targets in the zone of the interior calculated to weaken the German war effort. They had also been doing work on navigational instruments and bombsights and making plans for extending their strategic activities, including the development of more powerful bombers, as well as long range fighter escort aircraft and modified fighter-bombers. All this, of course, as one British historian put it, "depended on the support of Churchill and the enthusiasm of relatively junior officers; between them stood many admirals whose interests were far more traditional."[34]

In this atmosphere of questioning the old ways that had led to the frightful slaughter and stalemate of the trench warfare of World War I, and of trying to find some way out of the impasse, Winston Churchill wrote one of his greatest state papers, comparable to his "Iron Curtain" telegram to President Harry Truman on May 12, 1945, about which Churchill said, "of all the public documents I have written on this issue I would rather be judged by this."[35]

The subject of Churchill's state paper in 1917 was war in the air, and it was dated October 21. The paper took the form of a Memorandum to the War Cabinet on the Air Offensive that was being planned. In it the future Prime Minister dealt with the broader aspects of war, and with particular reference to the proposed Independent Bombing Force. Churchill began by saying that "war proceeds by slaughter and manoeuvre. Manoeuvre consists either in operations of Surprise or in operations against the flanks and communications of the enemy."

Applying this philosophy to the war of attrition then in painful progress, Churchill said,

> Owing to the lines now stretching continuously from the Alps to the sea, there are no flanks. But the Germans,

> striking under the sea at our vital communications, have threatened us with a decisive peril, which we are warding off only by an immense diversion of our resources. If we take on the one hand the amount of national life-energy which the Germans have put into their submarine attack, and compare it with the amount of national life-energy we are compelled to devote the meeting and overcoming that attack, it will be apparent what a fearfully profitable operation this attack on our communications has been to the enemy. Would it be an exaggeration to say that for one war-power unit Germany has applied to the submarine attack we have been forced to assign fifteen or twenty?

Coming back to the war in the air and the proposed British Independent Bombing Force, with particular reference to the "life-energy" and "war-power" Units the Germans would have to put into the defense of their industrial targets, Churchill said,

> even better than an operation against communications is an operation against bases. Air predominance affords the possibility of striking at both. It can either paralyse the enemy's military action, or compel him to devote to the defence of his bases and communications a share of his straightened resources far greater than what we need in the attack. All attacks on communications or bases should have their relation to the main battle.

From this, it is obvious that Churchill believed that the decision in the war would come through the advance of the ground armies rather than through Air Power, and that he did not subscribe to the "radical air strategy" which thought that "terror" bombing could or would destroy the will of the German people to resist. Churchill said it was not reasonable:

> to speak of an air offensive as if it were going to finish the war by itself.

> It is improbable that any terrorisation of the civil population which could be achieved by air attack would compel the Government of a great nation to surrender. Familiarity with bombardments, a good system of dug-outs or shelters, a strong control by police and military autorities, should be sufficient to preserve the national fighting power unimpaired.

In support of his opinion about terror bombing, Churchill went on to say,

> In our own case we have seen the combative spirit of the people roused, and not quelled, by the German air raids. Nothing that we have learned of the capacity of the German population to endure suffering justifies us in assuming that they could be cowed into submission by such methods, or indeed that they would not be rendered more desperately resolved by them. Therefore, our air offensive should consistently be directed at striking at the bases and communications upon whose structure the fighting power of the enemy's armies and his fleets of the sea and the air depends. Any injury which comes to the civil population from this process of attack must be regarded as incidental and inevitable.

In other words, what Churchill was saying was that wars could not be won by terror bombing and that therefore the target system should be devised so as to strike at the capability of the Germans to survive, not at their will alone. Churchill almost ruled out the "bonus" effect of bombing on the will to resist. "The supreme and direct object of an air offensive," Churchill said, "is to deprive the German armies of the Western Front of their capacity for resistance. It must therefore be applied and reach its maximum development in proper relation to the main battles both of Exhaustion and Surprise during thc culminating period of our general offensive."

Being more specific about the Air Offensive, Churchill said,

> German armies whose communications were continually impeded and interrupted and whose bases were unceasingly harried might still, in spite of all that could be done from the air, be able to maintain themselves in the field and keep the front. But if at the same time that this great difficulty and menace to their services in rear had reached its maximum, they were also subjected to the intense strain of a great offensive on the ground proceeding by battles both of Exhaustion and Surprise, the complete defeat and breaking up of their armies in the West as a whole might not perhaps be beyond the bounds of possibility.

Going on with the same thought, and again being very specific, Churchill said,

> There is an immense difference between merely keeping an army fed and supplied on a comparatively quiescent front in spite of air attacks, and resisting the kind of offensive which the British are delivering at the present time. It is imperative that the defending army should be able to move hundreds of thousands of tons of stores and ammunition within very limited times to the battlefront, and to maintain a most rapid circulation of hundreds of thousands of troops; and the double strain of doing this under a really overwhelming air attack might well prove fatal. More especially might this be hoped for if the form of our offensive were not confined simply to the main battlefront, but if it were so varied in locality and direction as to require from the enemy *an exceptional degree of material mobility*. For our air offensive to attain its full effect, it is necessary that our ground offensive should be of a character to throw the greatest possible strain upon the enemy's communications.

Churchill then got down to the nitty-gritty of the Memorandum. This was that the full potential of the air arm, which might be decisive, could not be realized as long as it remained in a subordinate position to the Army and the Navy. This was the point that General Smuts had made. "We have greatly suffered," Churchill said, "and are still suffering in the progress of our means of air warfare from the absence of a proper General Staff studying the possibilities of air warfare, not merely as an ancillary service to the special operations of the Army or the Navy, but also as an independent arm cooperating in the general plan. Material developments must necessarily be misguided so long as they do not relate to a definite War Plan for the Air, which again is combined with the general War Plan."[36]

From this, it is clear that Winston Churchill and Jan Smuts saw eye-do-eye on the subject of military aviation in 1917, and that they were both way ahead of their generation in appreciating the potential inherent in this new method of warfare. The emergence of Air Power as the dominant and decisive element in war owes much to the prescience of the great pioneers such as Churchill, Smuts, and Trenchard. Thus, as an English historian said,

> By 1914, five fundamentsl ideas of air power had been formulated within the U.K. First, air power could contribute enormously to land and naval operations; second, command of the air was as essential to Britain as was command of the sea; third, to achieve command of the air an independent service needed to be established; fourth, air power could reach out far beyond the lines of battle and strike at targets in the enemy's homeland; and fifth, by such offensive action we could force the enemy to divert essential resources to his own air defence.[37]

The American contribution to Air Power at this time was not very great. Much of it was in the form of bombast about aircraft production and the huge clouds of American planes that would darken the skies of Europe. The idea was that American industry could win the war all by itself by producing the airplanes that would break the stalemate of trench warfare, bomb Berlin and make the world safe for democracy.

This was the approach to war that Americans liked, the businesslike approach, the no-nonsense approach. Though the United States did not have a single combat plane fit for duty on the Western Front at the time, the plan was to build more airplanes in twelve months than France had produced in three years of warfare. It was thought that the only thing needed to accomplish this miracle was money, and the necessary appropriation bill was pushed through Congress on July 24, 1917, without even a roll call vote. The people were full of enthusiasm for this happy solution to the problem of war.

The ballyhoo about American aircraft production at the beginning of the war was just that--ballyhoo. Instead of the tens of thousands of combat planes that the world was led to expect from American assembly lines, the output was hardly a drop in the bucket and very late at that. The first American-made combat planes to arrive in Europe did not do so until May of 1918, and then they were unready for operations because many had been damaged in transit. Moreover, they were only versions or copies of a British model, and the American version quickly earned the reputation of being a "flaming coffin" because of the tendency of the airplane to explode under combat conditions.

Altogether, American industry's record in producing combat airplanes during World War I was less than impressive. A spokesman for the newly created Aircraft Production Board said for publication on June 13, 1917, "we believe we have worked out a program which will make it possible for the United States to secure to the Allies next year the permanent supremacy of the air, and with that we hope to become an immediate, decisive factor in ending the war."[38] Instead, by November 11, 1918, when the war ended, American industry had produced only 1,213 planes for the American Air Service in France. Of the 6,287 planes delivered to the American Expeditionary Force, 4,791 had come from the French, 261 from the British, and 19 from the Italians.[39] In other words, far from making it possible for the Allies to win the war with American aircraft, the production of American airplanes was so small that American flyers were almost totally dependent on the Allies for their aircraft. Even Italian production had to be used to beef up the meager American effort.

This was not all. During the war, American Flyers had to pay with their lives for these idle boasts about aircraft production in the United States. This came out in the testimony by Captain Eddie Rickenbacker before a Congressional committee. The captain was the outstanding American fighter pilot of World War I, with twenty-six kills. He said, "I think the fact that we had advertised that we were going to have 100,000 planes and 20,000 aviators on the front by April 1918, was the worst thing in the world we could possibly have done." Rickenbacker continued,

> The Germans got busy immediately and enlarged their industry and manufacturing facilities, and when April came around we did not have a plane on the front and they had twice as many as they would have had had we not given them advance warning.[40]

Sardonically, the Germans called their intensified production effort of 1917, the *Amerikaprogramm*.

In the fighting on the ground, General John J. Pershing led the first American Army in an attack against the Saint-Mihiel salient. In connection with this, he ordered Colonel Billy Mitchell to assemble what was then a truly massive aerial armada to support the operation. It consisted of some fifteen hundred aircraft, of which about a third were American with the rest being French and British, including a sprinkling of Italian and Portuguese. This was on September 12. By September 13 the Saint-Mihiel salient had been pinched out.

The tactics Colonel Mitchell used to support the ground action at Saint-Mihiel were identical with those that the Germans had employed in their win-the-war offensive in the preceding March. This was when the Germans had effectively demonstrated what massed air power could accomplish when they concentrated some three hundred aircraft for direct support of the ground advance. At Saint-Mihiel Colonel Mitchell said, "I so timed the attacks that several hundred would hit from the right side of the salient and then several hundred would hit from the left side. As soon as the combat became general, the main attack was launched by aircraft which had flown clear around both forces and attacked the enemy in the rear."[41]

In other words, during the battle of Saint-Mihiel, Colonel Billy Mitchell, taking his cue from the tactics the Germans had employed in the great Ludendorff offensive, demonstrated the great advantage of the concentration-of-force principle. General Pershing was delighted with the results, and not only recommended Mitchell's promotion, but said in a formal letter to Mitchell,

> Please accept my sincere congratulations on the successful and very important part taken by the Air Forces under your command in the first offensive of the First American Army. The organization and control of the tremendous concentration of Air Forces, including American, French, British and Italian units, which has enabled the Air Service of the First Army to carry out successfully its dangerous and important mission, is as fine a tribute to you personally as is the courage and nerve shown by your officers a signal proof of the high morale which permeates the service under your command.[42]

This was not all. When General Trenchard added his congratulations it was, to Mitchell, praise from the master himself. The British general reported to Pershing "that of all the air operations in his experience this was the first in which no hitch had occurred, no order had been misunderstood, and no mission had failed."[43]

The next ground action that involved American airpower on a large scale was the Meuse-Argonne offensive. The first action, at Saint-Mihiel, was designed to eliminate a German salient so that a subsequent drive might be launched into the Meuse-Argonne. At the Meuse-Argonne, of course, the Americans were in the position of attacking from, rather than against, a salient. This was quite a different thing, not only from the point of the basic principle involved, but also because at Saint-Mihiel the Germans were withdrawing when the American attack began, but in the Argonne they had to hold on because it was the pivot of the Hindenburg Line.

General Billy Mitchell was in command of the supporting aviation in the Meuse-Argonne offensive. The force was somewhat smaller than the Saint-Mihiel operation, never exceeding twelve hundred aircraft, but again it was international in character, including British, French and Italian units. General Mitchell again covered himself with glory, not only as a planner, but as a commander. The campaign itself lasted from September 26 to the conclusion of the Armistice on November 11, 1918.

In the campaign itself, as one of Mitchell's biographers relates,

> General Mitchell had his hands full supporting the ground forces. He had drawn a hint of the future on October 9 when he sent two hundred French and American bombers against a German troop concentration behind the lines at Damvillers. The aircraft dropped 39 tons of bombs, followed during the same 24-hour period by British units dropping some 40 more tons on other targets. The 79-ton total was impressive when balanced against the RAF total of 138 tons dropped during the entire war.[44]

By this time the heroic period in the history of military aviation, of course, was a thing of the past. The aces of Hollywood fame, the plumed knights of the air, glamorous in helmet, goggles and silk scarf, had to give way to mass. However, mass was not a principle that ground commanders found it easy to accept, and it was not one that could be applied without a unified air command. One of Mitchell's greatest achievements was his success in convincing Marshal Ferdinand Foch and General Pershing that this was a correct principle. It was also one of the secrets of General Trenchard's success with Field Marshal Douglas Haig.

The old principle of tactical aviation was just the opposite of the one that gained general acceptance in 1918. According to the old theory, all air elements overseas belonged to divisions, corps, armies and the GHQ Reserve. Even the GHQ Reserve, however, was not thought of as an independent force, but as a pool from which necessary replacements could be drawn when necessary. The new theory was that a minimal number of airplanes should be assigned to the ground units, and the great bulk of the total would constitute a central mass that could be used as circumstances required. To put this principle into effect meant, of course, that the ground commanders would lose control of the great bulk of available aviation. It was a revolutionary concept.

General Henry H. Arnold later (1949) said that the new theory of the command and employment of tactical aviation was one of the important lessons of World War I. He admitted that at the beginning of the war it was thought that "our first projected task was to provide every two ground divisions with one squadron of aerial reconnaissance and one balloon company." In addition to learning the lesson of command, and as part of it, General Arnold also said, "such things as formation flying, a new German development appearing on the Western Front that spring (1918), was unknown to us." Finally, General Arnold said, "for the moment, a complete lack of combat experience had left American aviation behind."[45]

The strides that Billy Mitchell made in learning the lessons of World War I, not only from the Germans, but also from the British and the French as well, plus his great ability to command troops in battle, are the qualities that account for the fame he achieved in World War I. They were also the qualities that were required in the one promotion that eluded him, This was to be the one supreme commander of all the Allied air forces in Europe.

There was nothing illogical about this. With the French in command of all the Allied ground forces, and with the British in charge of all the Allied naval forces, why should not an American be put in command of all the Allied air forces? "And it was proposed," Isaac Don Levine said, "that all the Allied air forces in France be put under the single leadership of General Mitchell."[46] However, the proposal came to naught because something else intervened-- the end of the war. In a way, this was a pity, at least from Billy Mitchell's point of view. "I was sure that if the war lasted," Mitchell said, "air power would decide it."[47]

There was good reason for Billy Mitchell to believe that Air Power would have decided the war had it lasted a few months longer. He was already convinced by his experience at Saint-Mihiel and in the Meuse-Argonne that Air Power was the dominant and decisive element even in a ground war, and the availability of

Allied air forces was increasing by leaps and bounds. By October 1918, the goal of one hundred thousand planes a year was being rapidly achieved. American industry was producing Liberty engines at the rate of 46,000 a year, as well as 23,000 planes; British aircraft production was up to 41,000 a year; and that of the French was up to 31,000. No longer was there any doubt about the fact that the skies of Germany would be blackened with wings of steel.

As for the impact of the air battle on the ground fighting, Mitchell had the example of October 9 to influence his thinking. This was when seventy-nine tons of bombs had been dropped on a German troop concentration behind the lines of the Argonne, and the planned German offensive was broken up. Mitchell said, "It was indeed the dawn of the day when great air forces will be capable of definitely effecting a ground decision on the field of battle."[48] It is significant that Mitchell used the word effecting, not affecting.

And Mitchell had other ideas for the use of military aviation in connection with the ground battle. He presented one of these to General Pershing during the course of a long conference on October 17. The idea was to transfer the First Division, the Army's best, to the Air Service for retraining as a Parachute Division. "Then," Mitchell said, "we could attack the Germans from our army on the front and support the whole maneuver with our great air force. This is a perfectly feasible proposition."[49]

By the time Mitchell had made the paratrooper proposal to General Pershing on October 17, 1918, and Pershing had given his approval to this highly imaginative use of Air Power, the war was over for all practical purposes. In fact, the Germans on October 4 had already appealed to President Wilson for an armistice on the basis of the Fourteen Points. Even earlier than that, on August 8, General Erich Ludendorff had more or less conceded defeat. This was when he had christened the great tank victory at Amiens as "the black day of the German Army."[50]

The paratrooper army was not the only idea Mitchell had for ending the war with the collapsing foe. One of his other ideas was "to blow up Germany,"[51] to use General Arnold's expression. However, these ideas pale by comparison with some being advanced by the British at this time. "Had the war continued but a few weeks longer," Professor George H. Quester said,

> the British Independent Bombing Force would have executed its first raids on Berlin. Raids from Britain and from bases near Prague in the newly independent Czechoslovakia, were to be executed with poison gas programmed as the major means of attack for the spring of 1919.[52]

For their part, the diehards in Germany were urging a plan on Ludendorff to destroy London and Paris by use of incendiary bombs and to throw a scare into the American people by having a transatlantic Zeppelin drop bombs on New York. However, by this time the peacemakers had taken over control in Germany, and the militarists on both sides had to lay away their well-made plans. The result was the Treaty of Versailles instead of poison gas attacks on Berlin, plus the incendiary raids on London, Paris and New York.

General Ludendorff said in his *War Memories* that he made the decision not to allow Paris, London and New York to be attacked with the new incendiary bombs because of the fear of reprisals and also because the attacks would have had no military value, the outcome of the war having been decided. The German position, he said:

> was now so serious that G.H.Q. could not hope that air raids on London and Paris would force the enemy to make peace. Permission was therefore refused for the use of a particularly effective incendiary bomb (expressly designed for attacks on the two capitals), sufficient supplies of which were ready in August. The large amount of damage that they were expected to do would no longer have affected the course of the war. Destruction for its own sake had never been permitted.[53]

World War I was an important milestone in the evolution of Air Power not so much because of what was accomplished, but because of what was revealed about the future of Air Power. It was not until the end of the war that the ground generals began to take the airmen seriously, and even then there were important reservations. However, the airmen were satisfied that they had made their point in World War I, and that the future belonged to them. And in one country at least, Germany, the airmen were successful, or partially so, in getting their ideas accepted. However, it was not until after the outbreak of World War II that the airmen in the democracies had any success in getting their main ideas accepted, and here again there was only a partial acceptance. Nothing like a full acceptance came until Hiroshima and Nagasaki upset all the older ideas about warfare. The period between the end of World War I and the beginning of World War II was one in which the philosophers of Air Power battled with the military, naval and political bureaucracies over the meaning and control of Air Power.

FOOTNOTES--CHAPTER III

[1]Samuel Eliot Morison, *Admiral of the Ocean Sea* (New York: Time Incorporated, 1962), 58.

[2]William Fielding Ogburn, *The Social Effects of Aviation* (New York: Houghton Mifflin Company, 1946), 53-54.

[3]John P.V. Heinmuller, *Man's Fight to Fly* (New York: Aero Print Company, 1945), 3.

[4]Charles deForest Chandler and Frank P. Lahm, *How Our Army Grew Wings* (New York: The Ronald Press Company, 1943), 141.

[5]*Ibid.*,142

[6]John R. Cuneo, *Winged Wars* (Harrisburg: The Military Service Publishing Company, 1942), 87.

[7]*Ibid.*, 83.

[8]Charles deForest Chandler and Frank P. Lahm, *How Our Army Grew Wings* (New York: The Ronald Press Company, 1943), 136-38.

[9]*Ibid.*, 174.

[10]*Ibid.*, 159-60.

[11]*Ibid.*, 158.

[12]George E. Stratemeyer, "Administrative History of U.S. Army Air Forces," *Air Affairs*, vol. I, no. 4 (Summer 1947), 510.

[13]Charles deForest Chandler and Frank P. Lahm, *How Our Army Grew Wings* (New York: The Ronald Press Caompany, 1943), 183.

[14]*Ibid.*, 193.

[15]Charles H. Gibbs-Smith, "Commentary" in *Air Power and Warfare* (Washington: Government Printing Office, 1979), 50.

[16]Alfred F. Hurley, *Billy Mitchell* (Bloomington: Indiana University Press, 1975), 15.

[17]Isaac Don Levine, *Mitchell: Pioneer of Air Power* (New York: The World Publishing Company, 1944), 100.

[18]Hans Rothfels, "Clausewitz" in Edward Mead Earle's *Makers of Modern Strategy* (Princeton: Princeton University Press, 1971), 96

[19]Isaac Don Levine, *Mitchell: Pioneer of Air Power* (New York: The World Publishing Company, 1944), 96.

[20]Alfred F. Hurley, *Billy Mitchell* (Bloomington: Indiana University Press, 1975), 26.

[21]USAF Historical Studies, No. 89, *The Development of Air Doctrine in the Army Air Arm, 1917-1941* (Air University, September 1955), 9.

[22]*Ibid.*

[23]Isaac Don Levine, *Mitchell: Pioneer of Air Power* (New York: The World Publishing Company, 1944), 95.

[24]Alfred F. Hurley, *Billy Mitchell* (Bloomington: Indiana University Press, 1975), 38.

[25]R. A. Mason, "The British Dimension," in *Air Power and Warfare* (Washington: Government Printing Office, 1979), 28.

[26]*Ibid.*, 27.

[27]George H. Quester, *Deterrence before Hiroshima* (New York: John Wiley & Sons, Inc., 1966), 38.

[28]*Ibid.*, 16.

[29]*Ibid.*, 38.

[30]R. A. Mason, "The British Dimension," in *Air Power and Warfare* (Washington: Government Printing Office, 1979), 27.

[31]Alfred Thayer Mahan, *From Sail to Steam* (New York: Harper & Brothers, 1907), 276.

[32]Alfred F. Hurley, *Billy Mitchell* (Bloomington: Indiana University Press, 1975), 142.

[33]Giulio Douhet, *The Command of the Air* (New York: Coward-McGann, Inc., tr. Dino Ferrari, 1942), 50.

[34]R.A. Mason, "The British Dimension," in *Air Power and Warfare* (Washington: Government Printing Office, 1979), 25.

[35]Winston S. Churchill, *Triumph and Tragedy* (Boston: Houghton Mifflin Company, 1953), 572.

[36]Winston S. Churchill, *The World Crisis 1916-1918* (London: Thornton Bitterworth, 1923), vol. II, Appendix V, 309-13.

[37]R.A. Mason, "The British Dimension," in *Air Power and Warfare* (Washington: Government Printing Office, 1979), 24.

[38]Wesley Frank Craven and James Lea Cate, *The Army Air Forces in World War II* (Chicago: The University of Chicago Press, 1948), v. I, p. 5.

[39]Robert Frank Futrell, *Ideas, Concepts, Doctrines: A History of Basic Thinking in the United States Air Force 1907-1964* (Alabama: Maxwell Air Force Base, Air University, 1975), 15.

[40]Isaac Don Levine, *Mitchell: Pioneer of Air Power* (New York: The World Publishing Company, 1944), 102.

[41]*Ibid.*, 135.

[42]*Ibid.*, 136.

[43]*Ibid.*

[44]Alfred F. Hurley, *Billy Mitchell* (Bloomington: Indiana University Press, 1975), 36.

[45]Henry H. Arnold, *Global Mission* (New York: Harper & Brothers, 1949), 52.

[46]Isaac Don Levine, *Mitchell: Pioneer of Air Power* (New York: The World Publishing Company, 1944), 145.

[47]*Ibid.*, 148.

[48] Thomas H. Greer, *The Development of Air Doctrine in the Army Air Arm, 1917-1944*, USAF Historical Studies: No. 89, p. 6.

[49] Isaac Don Levine, *Mitchell: Pioneer of Air Power* (New York: The World Publishing Company, 1944), 147.

[50] J.F.C. Fuller, *Armament and History* (New York: Charles Scribner's Sons, 1945), 141

[51] Henry H. Arnold, *Global Mission* (New York: Harper & Brothers, 1949), 52.

[52] George H. Quester, *Deterrence before Hiroshima* (New York: John Wiley & Sons, Inc., 1966), 44-45.

[53] Erich Ludendorff, My War Memories, qt. by George H. Quester, in *Deterrence before Hiroshima* (New York: John Wiley & Sons, 1966), 45.

CHAPTER IV

THE PHILOSOPHERS AND THEIR INFLUENCES

A great deal of thinking was done in all the major countries of the world on the subject of national security during the relatively short interval between World War I and II. The thinking differed, of course, depending on whether it was done in a status-quo country or in one of the countries where the people felt they had been cheated or wronged in the peace settlement. As a result, some highly controversial ideas began to spring up about the role Air Power could play in defending the status quo, or alternatively, in remaking the map of Europe.

The Italian General Giulio Douhet, put forward one of the most radical ideas. This was not just another new idea, but a concept of war that differed radically from those of the accepted theoreticians. Douhet's idea was that the airplane had become the dominant and decisive factor in war, and that as a result all the old theories had become part of the dustbin of history. The idea was that the objectives in war had been changed from armies against armies and navies against navies to air forces against all. In this concept, armies and navies were relegated to supporting roles, and the rival air forces would play the dominant and decisive role.

On the other hand, there was nothing entirely new in Douhet's theory of war. It had all been said before. From earliest times, as witness the classical legends, and all the great religious writings of the world, the will to fly is the will to conquer, and he who controlled the air space dominated the ground and sea spaces below. De Lana in 1670 and Galien in 1775, were "moderns" in this sense. All the Italian did was put the doctrine together.

Douhet was not put out because he did not have the cushion of modern history to support him, as did Clausewitz and Mahan. He was not only undisturbed by this, but expressed a certain impatience about history. On one occasion he said, "in the name of charity, let us forget the last war," and by that he meant history. In explanation, he said, "it is always dangerous to keep looking backward when marching forward, and still more so now when the path is full of sharp detours."[1] Douhet was a prophet looking forward, not backward, unless the backward glance was to the earlier dreams of mankind, that the will to fly is the will to conquer, and that he who controlled the air space above dominated the ground and sea below.

Clausewitz and Mahan were two of the great luminaries of the past that Douhet challenged and challenged on their own home grounds, military objectives and grand strategy. To Mahan and Clausewitz, it was apparent that the military objective of war was and could be only one thing, the destruction of the organized forces of the enemy. Clausewitz said that this, "the bloody solution of the crisis, the effort to destroy the enemy's force, is the first-born son of war."[2] Mahan quoted Jomini's saying "that the organized forces of the enemy are ever the chief objective" in war, and said that this dictim "pierces like a two-edged sword to the joints and marrow of many specious propositions."[3]

On the subject of strategy and tactics, Clausewitz and Mahan were of the same mind. Clausewitz defined tactics as skill in "the use of armed forces in engagements," and strategy as "the use of engagements to attain the object of the war."[4] In this connection, Mahan said that tactics were ephemeral and changed with the progress in the instrumentalities of war, but that "the considerations and principles which enter into the field of strategy belong to the unchangeable, or unchanging, order of things, remaining the same, in cause and effect, from age to age. They belong, as it were, to the Order of Nature, of whose stability so much is heard in our day."[5]

To Douhet, all this was nonsense. He rejected the idea that strategy was unchangeable, and unchanging, or that the principles belonged to anything like an Order of Nature. Instead, he maintained that the principles were in a constant state of flux and in the process of being revised constantly, particularly in his day. The advent of the airplane, Douhet argued, had changed the nature and character of war, and this necessitated a drastic change in strategy. In other words, the Douhet contention was that what had been true in the times of Clausewitz and Mahan was no longer true in an absolute sense, and would be even less so in the future. In particular, Douhet challenged the concept that the organized forces of the enemy, the army and the navy, would always be the chief objectives of war, and that wars in the future would be won or lost by the outcome of great battles such as Austerlitz or Trafalgar. Armies and navies were false objectives in the Douhet line of thinking. He argued that their roles would be supportive in the future, and the decision would come as a result of decisive action in the air.

Douhet's general theory of war was well stated by General Billy Mitchell in his testimony before the House Committee on Military Affairs on February 3, 1926. "There has never been anything that has come," General Mitchell said,

> which has changed war the way the advent of air power has. The method of prosecuting a war in the old days always was to get at the vital centers of the country in order to paralyze the Resistance. This meant the centers of production, the centers of population, the agricultural districts, the animal industry, communications-- anything that tended to keep up war. Now, in order to keep the enemy out of that, armies were spread in front of those places and protected them by their flesh and blood. You had mass killings there, sometimes for years before these vital centers were reached. It led to the theory that the hostile army in the field was the main objective, which it was.[6]

Mitchell continued by saying that now it was all different, now it was possible to get at the "vital centers" by going over the heads of armies and navies "straight to the vital centers, the industrial centers, through the use of air force and hit them. That is the modern theory of making war."[7]

As Mitchell pointed out, this was not only a modern theory of war, but a revolutionary one, a theory that subordinated land and sea operations to war in the air, and one that emphasized the importance of fighting in the air. He said that now for the first time in history there was a doctrine of air power with a strategy and tactics of its own, completely different from the traditional strategy and tactics of surface fighting. But as early as 1909, Douhet had pointed out that in the future air supremacy would be more important than sea supremacy because the sea touches only a small part of the territory of some nations, whereas the air envelopes tham all.

As to how this air supremacy would be established, Douhet took exception to the ideas of Clausewitz, Jomini and Mahan, that the organized forces of the enemy would have to be defeated in a great battle to achieve supremacy. Instead, Douhet argued that it would be far "easier to destroy the potentiality of the enemy in the air by destroying the nests and eggs of the flyers rather than to seek the flyers in the air for the purpose of bringing them down."[8] As to the exploitation of air supremacy, Douhet pictured a city burning under attack "by a single bombing unit," and said, "what, I ask you would be the effects upon civilians of other cities, not yet stricken but equally subject to bombing attacks? What civil or military authority could keep order, public services functioning and production going under such a threat? And even if a semblance of order was maintained and some work done, would not the sight of a single enemy plane be enough to stampede the population into panic?"

Finally, Douhet said,

> In short, normal life would be impossible in this constant nightmare of imminent death and destruction. And if on the second day another ten, twenty, or fifty cities were bombed, who would keep all those lost, panic-stricken people from fleeing to the open countryside to escape this terror from the air? A complete breakdown of the social structure cannot but take place in a country subjected to this kind of merciless pounding from the air. The time would soon come when, to put an end to horror and suffering, the people themselves, driven by the instinct of self-preservation, would rise up and demand an end to the war -- this before their army and navy had time to mobilize at all![9]

From the experience of World War II, it is not difficult to see where Douhet went wrong. The technology of his time was not set up to the task he set for his doctrine, and the urban population of the time turned out to be much tougher in spirit and morale than Douhet imagined they were or could be. In other words, Douhet's doctrine was way ahead of its time. It was not a doctrine for the pre-atomic age.

Douhet was not alone, of course, in his grasp of the fundamentals of air power, and of how this force could change the face and nature of war. He has his counterparts in America, General William ("Billy") Mitchell, and in England, Air Force Lord Trenchard, Billy Mitchell's mentor in World War I. All three were contemporaries. Mitchell was the youngest, being junior to Douhet by ten years and six to Trenchard. The only team player among the prophets of air power was Lord Trenchard. Mitchell and Douhet were just the opposite. They were the loners who did not know the meaning of compromise. As a result, the careers of Douhet and Mitchell were stormy, and both had to face court-martial proceedings before they were through. Douhet had to serve a term of imprisonment for his indiscretions, and Mitchell was demoted to colonel for his public criticism of military leaders.

Douhet was spared the fate of most prophets, that is to be a man "not without honour, save in his own country and in his own house"[10] because right reason finally prevailed, and his court-martial was reversed. Not only that. Douhet was restored to active duty and promoted to the rank of general. In addition, after the march on Rome in 1922, Mussolini had Douhet appointed to a high position in the Italian government. However, after a short time in office Douhet resigned in order to continue his study and writing. These were the strong points of Douhet. He was primarily a scholar, not a political or military leader. In this respect, he was different from both Trenchard and Mitchell. The latter had ambitions that were both political and military, and both he and Trenchard were first-rate soldiers, that is combat leaders. All three men were fighters for what they considered right, but only one was really qualified as a philosopher, Douhet. In this respect, he stood alone in his field. In fact, it could almost be said that it was difficult to be original in this field after Douhet. His writings are classics.

One of the recurrent themes in Douhet's writings is the importance of target selection. While this may seem completely obvious to laymen, operational types (excluding Douhet, of course) have never seemed to grasp the idea that targets are at the base of all combat operations. They will greatly succeed if the right targets are destroyed, and miserably fail if they are not. Douhet's perception and insight on this score are remarkable. It is the heart of the problem of air power.

"The selection of objectives," Douhet said, "the grouping of zones and determining the order in which they are to be destroyed is the most difficult and delicate task in aerial warfare, constituting what may be defined as aerial strategy. Objectives vary considerably in war, and the choice of them depends chiefly upon the aim sought, whether the command of the air, paralyzing the enemy's army and navy, or shattering the morale of civilians behind the lines. This choice may therefore be guided by a great many considerations--military, political, social and psychological."[11]

Finally, Douhet said, "the choice of enemy targets, as I have already pointed out, is the most delicate operation of aerial warfare." He then went on to explain what lesser minds have failed to grasp ever since, that "no hard and fast rules can be laid down on this aspect of aerial warfare. It is impossible even to outline general standards because the choice of enemy targets will depend upon a number of circumstances, material, moral and psychological, the importance of which, though real, is not easily estimated. It is just here, in grasping these imponderables, in choosing enemy targets, that future commanders of Independent Air Forces will show their ability."[12]

Douhet's theories of absolute air war were put into effect when Mussolini came to power in 1922. A separate Air Ministry was created with the Duce as the Air Minister, and the Italian Air Force was reorganized to reflect the new doctrine. It consisted of an Independent Air Force for strategic bombing, army cooperation units, a naval air service and a colonial air force. Marshal Italo Balbo was the head of the Air Force, and Marshal Pietro Badoglio became Chief of Staff of the United Armed Forces.

The new Italian dictator had a distinguished American visitor in 1927. Billy Mitchell, now a private citizen after his resignation from the Army, visited Italy in the course of a European trip undertaken for the purpose of gathering information he could use in his campaign to get an Independent Air Force in the United States. In an interview with Mussolini the American airman was greatly impressed with what he saw and heard. Mitchell said that Mussolini "stands as one of the greatest constructive powers for good government that exists in the world today."[13] This was not the only thing about the new Fascist state that impressed Mitchell. "With a good military air force," Mitchell said, "Italy can very easily control communications by vessels or ships across the Mediterranean Sea because they have the central position and a narrow stretch of ocean to guard by which all the shipping has to pass."[14] In other words, Mitchell was taken in by Mussolini's boast, as a lot of other people were, that the Mediterranean was an Italian lake--*Mare Nostrum*.

However, this was only true to hear Mussolini tell it; otherwise it was a sham. Despite going in for some stunt aircraft involving a few amazingly long-range flights in the early 1930's, the Italians settled for an air force designed to support operations on the ground, not a long-range strategic bombing force. The only thing Douhetan about the Italian Air Force was Mussolini's rhetoric. In other words, Douhet was admired as a theoretician, but his words did not carry great weight with Balbo, and none at all with Badoglio. The only battle Douhet won in Italy was with the general public. The generals were not convinced, and the so-called revolutionary Fascist air force was neither revolutionary nor Fascist: it was an ordinary, conventional air force, and not a very good one at that. "The Douhetans," as one keen observer said, won the debate in Italy and elsewhere with the general public "but lost it was the professional military."[15]

Italy was not the only place where this happened. The public in nearly every country, with the possible exception of Japan, was totally disenchanted with the conduct of World War I, with its frightful cost in blood and the very meager results obtained. As a result, the prestige of the military sank to a new level nearly elsewhere. Douhet's theories promised a change. He held out the hope for a short war, one that might be settled in days, if not in hours, and one that was incredibly cheap in comparison with the enormous outlays required to support and maintain the conventional armies and navies. Therefore, it is not surprising that Douhet's theories were popular in nearly every country during the interval between the great wars of the 20th Century. It is also not surprising to find that they were not accepted and made the basis for planning in any major country. Such is the paralysis of orthodoxy in most professional military establishments. It is usually the dominating factor.

The new Italian Air Force was put to a test, at least in the headlines of the papers, in the Ethiopian campaign of 1935-36. Against no opposition, its success was assured. Pretty much the same thing happened in the Spanish Civil War, the only difference being that there were some Soviet aircraft in this conflict. However, the Germans and the Italians had everything pretty much their own way, and all the Russian aircraft were pulled out before the war ended.

During the course of the Spanish Civil War the Douhet theorists had an opportunity to try out some of their ideas. The result was the bombing of Barcelona in March of 1938. The Spanish generalissimo, Francisco Franco, was opposed to this, but he had no control over the Italian Air Force, and a number of attacks were carried out. The results were inconclusive, but the Duce was elated. He gave the order for the bombings himself and told his Foreign Minister, Count Galezzo Ciano, he was delighted that the Italians should "be horrifying the world by their aggressiveness for a change, instead of charming it by their skill at playing the guitar." Count Ciano also quoted Mussolini as saying, "this will send up our stock in Germany, too, where they love total and ruthless War."[16]

The real test of the Italian Air Force came after Italy entered World War II, just a few days before the collapse of France. At that time the Regia Aeronautica had a front-line strength of 2,600 aircraft. The test came in Italian East Africa, and in the first retreat in Libya. There the Italians lost roughly one-half of their total front-line strength. It was a blow from which they never really recovered. The Italian aircraft industry couldn't begin to keep up with combat losses; training was bad and morale sank to a new low. In fact, it was not until much later, when Mussolini's airmen came under the influence of the Germans, that the Italian Air Force began to show the first signs of efficiency. Mussolini's posturing and Douhet's theorizing could not begin to make up for all the things wrong with the Regia Aeronautica. It was essentially only a hollow shell.

Douhet's ideas had some circulation outside of Italy during his lifetime, particularly in the United States and England, but it was only after the author's death in 1930 that they became directly available to large foreign audiences. In some ways this was like the experience that Machiavelli had with his little volume, *The Prince*. Circulated widely in manuscript form during his lifetime, and often plagiarized, *The Prince* was not published until five years after Machiavelli's death, and then it went through twenty-five editions in twenty years.

In 1932, extracts from three of Douhet's works were published in a French magazine, *Les Ailes*. This article was translated, in turn, into English, and made available to officers in the United States Air Corps. A German translation of Douhet's works appeared in 1935, just as Hitler was repudiating the conscription clauses of the Treaty of Versailles. A Russian translation was published in the same year. In April of 1936 the Royal Air Force Quarterly published an abstract of Douhet's major works. This was followed by the publication in New York in 1941 of Louis A. Sigaud's *Douhet and Aerial Warfare*. Finally, a single volume of Douhet's major works was published in New York in 1942. This is the standard reference in English to his writings. It includes one of the best known and widely reprinted of Douhet's tracts, "La Guerra del' 19--," a fictional account of an imaginary war between Germany and France. This was first published as a book-length article for *Rivista Aeronautic* in March 1930. Thus, by the middle of the 1930's Douhet became almost as well known as Mahan had been in the late 1890's or Clausewitz after Sadowa and Sedan. The one-volume edition of Douhet's works was translated by Dino Ferrari and published by Coward McCann in New York.

France was one of the countries where a considerable interest in Douhet was shown in the 1930's. Not only were his writings published, as in the French magazine *Les Ailes*, but a number of critical studies made their appearance. One of these was by Colonel P. Vauthier, *La Doctrine de Guerre du General Douhet*

(Paris: Berger-Levrault, 1935). No less a person than Marshal Henri Petain expressed a friendly if skeptical interest in the views of the Italian airman when he said that "the study of Douhet is an inexhaustible source of reflection. The notable doctrine that he has established may have a decisive influence on coming events. Conventional in his initial assumptions and in his methods, he shows the conclusions to which they lead. Let us take care not to treat lightly, as a Utopian dreamer, a man who may later be regarded as a Prophet."[17]

Marshal Petain to the contrary notwithstanding, the French military authorities did not take kindly to the theories of Douhet. Nevertheless, in the period following 1919 the French Air Force was steadily increased in size until it finally reached the point in 1923 where it, together with France's Eastern European Allies, was more than twice the size of the remaining powers in Europe combined.

This was not all the French did in furtherance of the ideas of Douhet. In 1928 a separate Air Ministry was established, and in 1936 an Independent Air Force was created. But these moves in the right direction, as Douhet would have interpreted them, did not bring the desired results. The Air Ministry was limited in what it could accomplish by the frequent changes in the French government. With nine cabinet changes in ten years, there was no opportunity for the Air Ministry to establish anything in the way of policy. As for the Independent Air Force, at about the same time it was established the aircraft industry was nationalized. And all the while, and in keeping with French traditions, the air force remained closely attached to the army and the naval units to the navy. The reorganizations merely papered over the fact that France had ceased to exist as a first-class power in the air.

By the time Hitler came to power in Germany in 1933, the French air force was in a state of obsolescence, and nothing could hide it. From bad, it got worse, and by the time of Munich in 1938 the output of military airplanes in France was down to the negligible rate of ten a month, one-tenth of the rate of monthly production in Italy! The Luftwaffe, therefore, did not have its hands full in overcoming the French Air Force in 1940. Not only was there no Douhetan force in France capable of striking at the vitals of Germany, but there was not even an air force of sufficient size to give the French ground forces adequate air support. There were not enough bombers for effective offensive action against the Germans, and not enough fighters for defense against the Germans.

After the French surrender on June 22, 1940, the Luftwaffe was faced by only one challenger, the Royal Air Force. It was the only arm that stood between Hitler and his planned domination of the world. As the Battle of Britain reached its first peak, in the fighting between August 24 and September 6, Churchill said,

> We must regard the next week or so as a very important period in our history. It ranks with the days when the Spanish Armada was approaching the Channel, and Drake was finishing his game of bowls; or when Nelson stood between us and Napoleon's Grand Army at Boulogne. We have read all about this in the history books; but what is happening now is on a far greater scale and of far more consequence to the life and future of the world and its civilisation than those brave old days.[18]

The success of the Royal Air Force in the Battle of Britain was due in large measure to four things. The first of these was the failure of the Germans to understand the Douhetan principles of strategic vulnerability, of which the key element was target analysis. The second was the failure of the Germans to have at hand a bombing force capable of executing a strategic mission. The third was the advantage of radar, detection and early warning. And the fourth was the qualitative superiority of the aircraft in British Fighter Command, to say nothing of the courage and determination of the British airmen. Fighter Command was really the element of the Royal Air Force that Churchill had in mind when he said, "never in the field of human conflict was so much owed by so many to so few."[19]

What did the Battle of Britain prove? Nothing really, and certainly nothing to gainsay or disprove the basic principle of the Douhet theory, that no country can live long under a full-scale and free exploitation of air weapons (even pre-atomic) over the heart of its territory. All the Battle of Britain proved was that the blitz technique so successful in Poland, Norway, France and the Low Countries, was unsuccessful against a major opponent equipped with a qualitatively superior air force; that an air force built primarily to operate in short, sharp actions in conjunction with and in pursuance of the mission of the ground forces, was no force with which to wage strategic air war; and that as long as a country maintains mastery of or even superiority in the skies above, it can never be invaded from its sea approaches. The Battle of Britain was the Battle of Norway in reverse.

In a way, there had been a preview of the Battle of Britain in the skies over Dunkirk only three months before. In May of 1940 the British were forced back to Dunkirk, which was the last port available for their escape from Europe, and they could not have reached the port had it not been for the folly of the Führer, for the fact that Hitler, for some unexplained reason, held up the panzer divisions that were ready to sweep into Dunkirk ahead of the British. Hitler was convinced by Hermann Goering, his Air Chief, that the Luftwaffe could deny the port to the British and that it would be a good idea to hold the panzer divisions in reserve. So Hitler decided. The result was what the British call the "miracle of Dunkirk."

What made Dunkirk a miracle was the Royal Air Force, and within it Fighter Command. The Luftwaffe was chewed up in the skies over Dunkirk, losing over six hundred planes to a British loss of one hundred thirty, while a quarter of a million British, French and Belgian soldiers were evacuated to England. Churchill drew the right lesson from the air battle over the beaches of Dunkirk and Hitler the wrong. The right lesson was that the Luftwaffe was not invincible, and that as long as it was not, England could not be invaded from her sea approaches. "We must be very careful," Churchill said, "not to assign to this deliverance the attributes of a victory. Wars are not won by evacuations. But there was a victory inside this deliverance, which should be noted. It was gained by the Air Force."[20]

Dunkirk and the Battle of Britain were surprise victories for England and the Royal Air Force, and they were also defensive victories in every sense of the word. Neither would have been possible, of course, without a strong Fighter Command. This raises the question of Fighter Command itself, and it is a question because at the end of World War I British policy was tied to a Douhetan concept, which put the emphasis on bombers, not fighters. It was also tied tightly to the Douhetan concept.

It will be remembered that when the war ended the British had already gone from theory to practice with the Douhetan concept. That is to say, they not only had their separate Air Ministry in the government, but the Royal Air Force was also in existence as a third and coordinate branch of the Armed Services. Not only that. It will also be remembered that had the war lasted a few weeks longer the British Independent Bombing Force would have taken the war in its own hands by attacking Berlin and other German cities, not only with their explosives and incendiaries, but also with poison gas. This was pure Douhetan--absolute air war--and the concept carried over into the post-war years.

In 1932 Prime Minister Stanley Baldwin made a statement on British policy that is representative of the extent to which the Douhetan theory of absolute air war was incorporated in British thinking. He said,

> The speed of air attack, compared with the attack of an army, is as the speed of a motor car to that of a four-in-hand and in the next war you will find that any town which is within reach of an aerodrome can be bombed within the first five minutes of war from the air, to an extent which was inconceivable in the last war, and the question will be whose morale will be shattered quickest by this preliminary bombing? I think it is well also for the man in the street to realize that there is no power on earth that can protect him from being bombed. Whatever people may tell him, *the Bomber will always get through.*[21]

The Prime Minister went on to draw the obvious conclusion and to indicate that the basis of British policy was unchanged from the end of World War I. Baldwin said, "the only defence is in offence, which means that you have to kill more women and children more quickly than the enemy if you want to save yourselves."[22] This could have been written by Douhet himself or taken straight from the pages of *The Command of the Air*.

From 1932, over the next several years, British philosophy continued to be Douhetan, but as a result of budget slashes, one on top of another, the Air Force became less and less effective. The watchword in British thinking was not defense, but disarmament, and there was much wild talk of banning the bomber, or turning it over to the League of Nations as an international enforcement weapon. While the Air Force remained Douhetan in concept, it was not Douhetan in capabilities. Not even the advent of Hitler was enough to change the mood in the country. It was still hell-bent on self-destruction.

Despite the downward spiral of British armaments and the orgy of internationalism that followed in the wake of World War I, including pacts to abolish war altogether, the British continued, in theory at least, to hold fast to the basic precepts of Douhet. However, with the rapid rise of the Luftwaffe in Germany, the thinking in England began to undergo some changes. The stronger the Luftwaffe got, or seemed to get, the more the British seemed to back away from Douhet. This was not the case in the Air Ministry, of course, since Trenchard was still in the saddle, but it was the line of reasoning followed by many who sought above anything else to keep the country out of war.

A showdown of sorts came in February 1935 when Germany publicly proclaimed the existence of the Luftwaffe. This was followed by the visit of Anthony Eden, Lord Privy Seal and Sir John Simon, Foreign Secretary, to Berlin. They were received by Hitler in the Reichs Chancellery and then informed that Germany had now achieved parity in the air with England. This was a bald-faced lie, of course, but there was no way to prove it, and the British were shaken to their foundations.

Churchill had been the Cassandra of German Air Power in the Commons and of Britain's declining fortunes. As early as 1934, he had warned that parity between the two countries would be reached by 1935; that by 1936 Germany would be in the lead; that by 1937 Germany would be twice as strong as England; and that "once they have got that lead we may never be able to overtake them."[23] In a celebrated "confession" on May 22, 1935, the British Prime Minister, Stanley Baldwin, accepted the blame and responsibility for having let British defenses fall behind the Germans in aircraft production. The error was not as great as Baldwin believed, and nowhere near as great as Churchill and Hitler (albeit for different reasons) led him to believe. But it was an error nevertheless, and Baldwin promised a catch-up program to rectify it. However, the program was not one that satisfied either side. The liberals wanted more retrenchment in the armaments program, no catch-up at all, and the conservatives wanted an all-out production program.

Finally, in early 1936, just at the time of the Rhineland crisis, a new Ministry was formed, called the Coordination of Defence. To Churchill's disappointment, he was not named to head up the new Ministry, but the appointment instead was given to an unknown political figure, Sir Thomas Inskip. Inskip was more or less of a pledge on the part of the British not to take a hard line against Germany. It was a gesture of appeasement.

The Rhineland crisis reached its peak on March 7, 1936, when Hitler tested the will of the Allies by denouncing the Locarno pacts of 1925 and occupying the Rhineland. The test was a complete success from the Führer's point of view. The Allies squirmed and expostulated, but did not raise a finger to oppose the occupation. The shadow of the Luftwaffe had clouded the vision of the Allies.

The Rhineland occupation was the key to the next decade of history in Europe, and in the world for that matter. It gave Hitler a defensive line in the West that left him free to sally forth in the East and southeast. The conquests of Austria, Czechoslovakia, and Poland would not have been possible without the occupation of the Rhineland. Professor Adam B. Ulam did not exaggerate when he said, "March 7, 1936, was the turning point in the history of Europe between the two wars."[24] In fact, it could also be said that this date marked the end of an era in history, the post-war years, and the beginning of another, the pre-war years. March 7, 1936 was a watershed in the history of the human experience.

Hitler was overjoyed at the happy outcome of his adventure in the Rhineland. It was all a gigantic bluff, and Hitler admitted as much. Looking back on the event, Hitler said, "The forty-eight hours after the march into the Rhineland were the most nerve-racking in my life.

If the French had then marched into the Rhineland, we would have had to withdraw with our tails between our legs, for the military resources at our disposal would have been wholly inadequate for even a moderate resistance."[25]

The British were not satisfied with putting up no resistance to Hitler's occupation of the Rhineland. They also cautioned the French against taking any action, even a police action. The British Prime Minister told the French, according to the latter, that "England was not in a state to go to war" and that "if there is even one chance in a hundred that war would follow from your police operation, I have not the right to commit England." Churchill's conclusion was, "weakness is not treason, though it may be equally disastrous. Nothing, however, can relieve the French Government of their prime responsibility. Clemenceau or Poincare would have left Mr. Baldwin no option."[26]

England's primary concern at this time was not with the German occupation of the Rhineland, but with the possibility that war might break out between Germany and England-France. In other words, the clear and present danger, from England's point of view, was the Luftwaffe, the danger that England might find herself on the receiving end of a knock-out blow. This was putting a little reverse-English on the Douhetan theory that was so popular with Trenchard and the bomber advocates in the Air Ministry. However, the theory had its chills when the defenseless condition of London had to be taken into consideration, considering the long lead the Germans were believed to have in bomber production.

The opponents of the hard line policy against Germany were critical of the bomber production program in England, not only because it was based on the Douhetan theory of warfare, which could work two ways, but because it only served to bait the Germans, not calm them down. Many believed that the German bomber production program was justified in the light of the English program. They argued that the English program should be cut back, not pushed to new heights. They also argued that it was the height of folly to be spending huge sums of money on these Douhetan monsters, and that if the money had to be spent at all, it would make much more sense to see that it went into a more rational program, one that would emphasize active and passive measures of defense, such as fighters instead of bombers.

Of course, all this talk about defense being the cornerstone of a production program aimed at the Germans was highly unpopular in the Air Ministry. Douhet's ideas still reigned supreme in that august body where fighters were considered very second-rate. Any thought of a revolutionary change in air policy brought cries of outrage from the bomber theorists. Lord Trenchard made it clear that in his opinion "any radical shift in strategy would defeat England in any new conflict."[27]

Nevertheless, the new Minister for the Coordination of Defence, Sir Thomas Inskip, thought otherwise and got the Cabinet to agree with him. He stood Douhet on his head by declaring that "the role of our Air Force is not to deliver an early knock-out blow" to the Germans, "but to prevent the Germans from knocking us out." Being specific, Inskip made it clear that in his opinion "the new RAF fighters, the Hurricane and the Spitfire, coupled with the secret development of radar, could defend England from bomber attack."[28] This was a reversal of the Douhet-Trenchard dictum that there is no defense against bombers and of Stanley Baldwin's statement, "the bomber will always get through."[29]

Thus when the chips were down, over the skies of Dunkirk and in the Battle of Britain, England was not found wanting. The Hurricanes and Spitfires paid off handsomely in those two great battles. So did the decision to concentrate on the production of fighters a few years back. Actually, at the time of Munich two years after Inskip took office, England had only five squadrons of Hurricanes, but at the time of Dunkirk and the Battle of Britain she had firty-seven squadrons of modern fighters available, Hurricanes and Spitfires. The edge was thin between quality and quantity in those two great battles, but it was the edge of victory.

In the interim between World War I and II, when England was deciding to abandon Douhet in favor of fighters, on the other side of the world Japan was having to decide some basic questions concerning the role of aviation. For Japan, the decision was easy. Douhet was known to the Japanese, but for them he did not make much sense and for obvious reasons. The potential enemies of Japan were Russia, England and the United States. Douhet's theories did not help the Japanese in their planning against these potential enemies. Geography and common sense ruled out any idea of bombing these countries into submission from Japanese bases. Also, the Douhet approach to beating China and establishing the Co-prosperity Sphere in East Asia did not seem to be practical. In other words, Douhet was simply irrelevant to the Japanese. Therefore, the Japanese gave no serious thought to the establishment of an independent air force coequal with the Army and Navy in accordance with the doctrines of Douhet. Instead, the Japanese air forces were put directly under the Army and Navy respectively. The functions of the two forces were to support the army ground forces on the one hand and the naval forces on the other. There was no cooperation between the two forces. The separation was complete, and the lack of cooperation even extended into the production realm, where there was fierce competition for scarce production facilities and raw materials.

Douhet would have been flabbergasted by the inability of the Japanese to understand, much less to put into practice, any of his fundamental ideas about Air Power. No doubt, he would have understood their lack of interest in his knock-out doctrine, considering their geopolitical position, but he would not have approved of the wall of separation they put between their air forces; their lack of interest in any kind of training for joint forces; and their failure to understand the principle of unity of command. Douhet, on the other hand, and like everyone else, would have been forced to admire what the Japanese were able to accomplish, not only at Pearl Harbor, but in the sinking of the battleship *Prince of Wales* and the battle cruiser *Repulse*.

In Russia, the situation was different. Not that Douhet reigned supreme--he didn't anywhere, not even in Italy--but his influence was felt more than in Japan. This was also quite understandable considering the close proximity of Russia to her potential enemies.

The Russian translation of *The Command of the Air* appeared in the Soviet Union in 1935, and the introduction was written by the Deptuy Air Force Commander, Lieutenant General V.V. Khripin. He was the leading air force theoretician in the Soviet Union and became the first commander of the organ charged with long-range bombing and airborne troops. That is, this was his job until he got caught up in the purges of the 1930's and disappeared along with his organization, doctrine and concepts.[30]

Even before the days of Khripin, however, the Russians had shown an interest in the theories of Douhet. As early as 1926, another theoretician, Lapchinskii, stressed the advantages of strategic bombing and underlined the importance of independent air operations. Even the government was impressed and began building a fleet of heavy, long-range bombers. Fifty of these were shown in the fly-by on May Day in 1933, and two hundred fifty a year later. By 1936 the Soviets had a very respectable strategic air force and an organization to go along with it, called the Army of Special Designation (A.O.N.).[31]

But then came the purges, and the roof caved in on the A.O.N., as well as on any other Douhetans in the armed forces. Before it was over, 75 percent of the senior officers in the air force had been eliminated, and the bloodletting was even extended to include top officials in the aircraft industry. For a time, the designer of the Russian long-range bomber, the TB-3, was placed under arrest.[32]

The senseless purges not only of the air force, but of all the armed services in the U.S.S.R., deprived the Soviet Union of leaders they badly needed when war finally came. Important lessons had been learned in Spain, in China and against the Japanese before the great conflict started in 1941, and many of these were lost. In the air, the Russians entered World War II with the most tactically-oriented air force in Europe, and one of the worst. Otherwise, the Germans would not have been so successful against the Red Air Force, practically eliminating it in two days, mostly in one.

The Russian flirtation with Douhet did not set the pattern for the German experience but was almost exactly the same, at least up to a certain point. That is to say, just as there were airmen in the Soviet Union (Lapchinskii and Khripin) who were fascinated with the radical ideas of Douhet, so there were Germans who drew their inspiration from the writings of the Italian theorist, and to a lesser extent from those of the American, Billy Mitchell. One of these, Colonel Walther Wever, was, curiously enough, not even an airman when he became the first Chief of the Air Command Office under Hermann Goering.

Walther Wever came to the Air Command Office from his post-war assignment to the *Truppenamt*. The *Truppenamt*, or Troop Office, was precisely the same as the outlawed General Staff, and its first chief was General Hans von Seeckt, the brilliant strategist and tactician who would lay the foundation for the resurgence of Germany as a great military power, with particular emphasis on the Luftwaffe. "The whole future of warfare," Seeckt noted in 1921, "appears to me to be in the employment of mobile armies, relatively small but of high quality, and rendered distinctly more effective by the addition of aircraft."[33]

General Seeckt--or the Sphinx, as he was known by his contemporaries--was better qualified than any other German general to be Paul von Hindenburg's replacement, and it was Seeckt who pushed Wever forward. When he was posted to the Air General Staff in 1932, the then Chief of the *Truppenamt*, General Werner von Blomberg, complained, "you're taking my best man!"[34]

By this time the foundations of the new German Air Force had been well laid. While German aviation had been virtually abolished by the Treaty of Versailles, ways to circumvent it had been found, and under the guise of commercial aviation, factories producing military airplanes had been located in Russia, Sweden, Turkey, Denmark, Italy and Switzerland.[35] All this had been done under the aegis of the German Republic's Ministry of Transport, which had had an aviation department headed by a former officer of the German Air Force. The airplanes from these factories were tested under mock combat conditions, and pilots were trained at a German base at Lipetsk in Russia six hundred miles from Leningrad. General von Blomberg made a secret visit to the base in 1928 and was highly pleased with what he found there.[36] So by the time Hitler came to power in 1933, the German Air Force, officially non-existent, was already a force in being, with four fighter, three bomber and eight reconnaissance squadrons.

One of Hitler's first acts, taken three days after he came to power, was to transfer out of the Ministry of Transport all civil and military aviation and to put these under the control of Hermann Goering. In May 1933, the German Air Ministry was established, and Goering was posted to the office of Air Minister. Wever's position under Goering was that of Chief of Staff, which meant that he was in working command of the burgeoning Luftwaffe.

Walther Wever was thoroughly familiar with the writings and bombardment theories of Douhet and believed in them. *The Command of the Air* had been published in Germany in 1929 and again in 1935. Some of Mitchell's articles had also appeared in the German press in 1933. Wever had not a doubt in the world that Douhet and Mitchell were right, and that Germany's need for a strategic air force was very great indeed, particularly in the light of the "soil policy" that Hitler had outlined in *Mein Kampf*. What Wever wanted was the creation of a Douhet-type bomber force within the Luftwaffe, including a heavy, long-range bomber that could be used, if needs be, against the Russians, or against any other potential enemies of the Fatherland, for that matter. Wever called his Flying Fortress the Ural Bomber.

The fact that Wever was a card-carrying Douhetan in his thinking, did not mean that he was any less of an advocate of close-support aviation. Quite the contrary. At this time he did not even have his wings, and continued to think of himself as a non-flying infantryman, as he had always been, and one with twenty-eight years of unbroken service marked by brilliant performances, not only in the field, but on the highest staffs as well, including those of Erich Ludendorff and Paul von Hindenburg. It was only that he could not see why it was not more important to destroy the enemy's weapons at the source than on the battlefield, and why it was not more economical to knock out the enemy's will to resist in the zone of the interior than on the battlefield. This made so much sense to Wever that he dedicated himself not only to the building of the four thousand plane air force required by the army, but to the building of a strategic air force at the same time, one that could put Douhet in practice against the potential enemies of the Fatherland, including Russia, and one that might in itself be decisive.

Walther Wever was not a day-dreamer. There was nothing in his program that was not entirely feasible and practical. At the outbreak of the general war in 1939, the German aircraft industry was actually producing only about five hundred combat aircraft each month, and was capable of producing over a thousand. This means that it was running at only about 50 percent of actual capacity. Wever's program could have been cranked in from the beginning without putting any undue strain on the German aircraft industry.

But it was not, and the result was a disaster for Germany. When a strategic air force was needed in 1940 against England, and in 1942 against Russia, all the Germans had was a tactical air force, the Luftwaffe. The medium bombers that were used against England had neither the bomb-load capacity nor the necessary range to accomplish their mission, and they were so lightly armed that they were sitting ducks for the British fighters. And against Russia the Germans were in an even worse position. Despite the vulnerability of Russian industry, the Germans had no long-range, heavy bomber force capable of destroying it. They could have had such a force. As early as 1942 Messerschmitt had built a four-engine bomber that could have carried more than a two-ton bomb load on a transatlantic flight. However, only one of these strategic bombers were built. This was the ME-264. More were not built because there were not any military "requirements" for them.

Walther Wever was the Douhet of the German Air Force, just as Trenchard was the Douhet of the Royal Air Force, and Mitchell the Douhet of the American Air Force. Like Trenchard and Mitchell, Wever was a non-flying officer up until a very late period in life. In fact, he was forty-eight years old and a major general, before he applied for flight training. When he undertook it, however, the Luftwaffe picked up the bill and did not, as in the case of Billy Mitchell, force the officer to pay for his own training. Nevertheless, in the case of Wever, it must have been a strange spectacle indeed for a major general to be taking orders from an enlisted flight instructor just like everybody else.

It was certainly not Walther Wever's fault that the Germans did not have a strategic air arm when one was needed against England in 1940 and Russia in 1942. No sooner was the aircraft production program for the army in full swing before Wever turned his attention to the design and production of the strategic bombing force, including the four-engine model that he called the Ural Bomber. All the while, however, he kept his superiors fully informed of what was going on. For example, in 1935 Goering and General von Blomberg were shown mock-ups of the new bombers at the Dornier and Junkers plants, and they even expressed a mild interest in them, particularly after being assured that work on the "second-generation force" would not interfere with the production schedules for tactical aircraft, especially dive bombers.

The Luftwaffe was barely a year old, officially when it was called on to play a major role in Hitler's most audacious and dangerous gamble-- the occupation of the Rhineland. It was alerted to the move as soon as Hitler decided on the gambit, and General Wever immediately began shifting his units around to provide air support for the three small ground battalions that were to do the occupying. It is hard to say which was the more ludicrous, the three thousand infangrymen armed only with rifles, carbines, and machine guns, or the hollow shell of a Luftwaffe. It is no wonder General Wever's airmen were nervous. Their airplanes lacked guns or ammunition, or both, and if any enemy airplanes appeared, all they could do was ram them. What a way to go--even for the Fatherland.

General Wever's "support" of the ground forces in the Rhineland occupation (since the enemy failed to show), was a success and a model of perfection. And that was characteristic of the man. His was a father image among the pilots and crewmen, and they would go to great lengths to earn his praise. In this respect, Wever was like Trenchard when he was serving under Haig, and Mitchell when he was serving under Pershing. All three were obviously leaders of men and brought out the best in them.

Wever was like Trenchard and Mitchell in another respect. While he gave his heart and soul to the job at hand, support of the ground forces and did it well, his mind was on other things, on the future. Just as Trenchard and Mitchell came out of the combat in World War I believing in strategic air power, so Wever all during the Rhineland occupation continued to think about his Douhetan bombers and of their role in laying waste to the industrial heartlands, whether they were in France, England, Russia, or even the United States, operating from appropriate bases. In other words, once the nightmare of the Rhineland occupation was over, General Wever picked up where he had left off and concentrated on getting the airplanes for his "second-generation force."

The prototypes of the first basic four-engine bombers of the planned strategic air force were just beginning to make their appearance in Germany when, on June 3, 1936, death struck down the German Douhet, Generalleutnant Walther Wever. Goering is said to have wept like a child on getting the news that Wever had been killed in an air crash. Whether he did or not, the death of Wever meant that the driving force behind the Luftwaffe as a strategic arm was gone, and that Germany would not have the wherewithal to fight the Battle of Britain in 1940, nor the airplanes with which to exploit the strategic vulnerability of Russia in 1942. For Germany, Wever's death in 1936 was a tragic loss. Goering would have cried harder if he had known how tragic it was.

As things turned out, on April 29, 1937, Goering ordered the cancellation of the Ural Bomber project. This was only ten months after General Wever's death. It was not because Goering found fault with Wever's strategic thinking, but simply because Goering wanted to remain in the good graces of Hitler. Goering said the Führer was not interested in what kind of bombers he had, but only in how many. On this basis, it made sense to scratch the Ural Bomber. All the work that had gone into its development was poured down the drain, and the two promising prototype bombers were broken up for scrap.

Later, there was a revival of interest in a heavier bomber, one that could carry four thousand pounds of bombs over a distance of one thousand miles, in order to blanket the British Isles and reach Moscow. Accordingly, specifications were let out to the German aircraft industry, but then someone in the Air Ministry had a brainstorm, and another four-engine bomber was dead. The brainstorm was that the thirty-ton four-engine bomber should be designed in such a way that it could also be used as a dive bomber![37]

This is reminiscent of Hitler's famous decision in 1944 to convert a jet fighter into a bomber.[38] The jet fighter might have been able to stand the Allied Air Offensive down if it had been produced in quantity. By January 1, 1939, the jet principle was well known in Germany, but the Luftwaffe High Command took no interest in the project. The jet fighter was allowed to go the way of the atomic bomb in Hitler's Germany, forgotten.[39] The theory of the dive bomber dominated everything else.[40]

Meanwhile, the Spanish Civil War began on July 18, 1936, and before the end of the month the Luftwaffe was in Spain transporting Francisco Franco's Army of Africa to the mainland of Spain. To Georing, it was a good opportunity to "blood" the Luftwaffe, and he had no trouble convincing Hitler that it was a good idea to support General Franco. Above everything else, the Spanish generalissimo needed support in the air. Outnumbered four to one in airplanes and two to one in pilots, the big requirement was for air support, and Hitler, for his own selfish reasons, was more than generous. The key to a quick victory in the war was Madrid, and this was held by the Republicans. Therefore, Madrid was the first target as soon as the Luftwaffe had gathered enough strength to make itself felt in the Spanish Civil War.

The German air units in Spain were called the *Legion Kondor*, or Condor Legion, and their commander was the German General Hugo von Sperrle. By the middle of November 1936, Sperrle felt strong enough to test the mettle of Madrid. The object was to find out whether terror bombing would work. The question was quite fundamental to the Douhet thesis: was the will to resist so fragile that it could be easily broken by air bombardment? The capital was bombed for three days on a quarter by quarter basis using a mix of high explosive and incendiary bombs. The results were disappointing. The Madrilenos refused to be cowed by the bombing, and after the test was concluded the password in the city continued to be, "no *pas-ar-an*."

Despite the stubbornness of Madrid, by the beginning of 1937 the Nationalists controlled 65 percent of Spain. However, it was still anybody's guess as to who would win the war. As the first move in the new year, and while there was a lull in the fighting around Madrid, the Nationalists decided on a major offensive against the Basque provinces in the north. In line with this objective, General Franco called on the Germans to provide the air support. The result was a frightening demonstration of what unopposed air power could do against ground defenses, and this was one of the main things the Germans wanted to find out, second only to the answer to the question about terror bombing. Therefore, it is not surprising that General Sperrle pulled out all the stops in responding to General Franco's request for air support.

One of the towns in the path of the Nationalist drive in the north was Guernica. With a population of under ten thousand, the town was important nonetheless because it was a transportation center and the headquarters of two battalions of Basque infantry. It was also important geographically as it shielded the strategic stronghold of Bilbao, the port city the Republicans were using to bring in supplies and reinforcements from France, Russia and other countries sympathetic to their cause. Therefore, it was against Guernica that the Germans launched their heavy attack on April 26 to pave the way for the ground forces of the Nationalists.

Guernica was completely destroyed. The German bombers came over at twenty-minute intervals. Using the quarter by quarter techniques they had employed against Madrid, the bombers sought to destroy every building in town with their high explosives and then to set the rubble on fire with their incendiaries. The bombing lasted three hours and fifteen minutes, and when it was over there was nothing left of Guernica. The last wave of bombers had to come down to six hundred feet to distinguish the town from the countryside. Three days later the Nationalist troops swept through the area that had been Guernica. They were on their way to Bilbao, and this fell to the Nationalists on June 19. The Condor Legion could very well boast that the Nationalist victory in the north was "made in Germany."

The destruction of Guernica did more than facilitate the capture of Bilboa by the Nationalists. It sent seismic shockwaves around the world and confirmed the worst fears some people had about the role the Luftwaffe would play in the next war. They could see that the world itself, at least to some extent, was in hock to Hitler's Luftwaffe. The fact that the German government tried to deny the responsibility for the horror of Guernica did not assuage the fears of these people, but made them all the more certain that Douhet was in the saddle in Germany, and that Hitler's basic plan was to "Guernica" the rest of the world in the event of war, beginning with Paris and London.

Ever since February 1935, when Hitler proclaimed the existence of the Luftwaffe and later solemnly (and falsely) assured the British that it had achieved parity with England, the Führer lost no opportunity to impress the world with the fact that Germany was supreme in the air. Charles A. Lindbergh, unwittingly of course, helped to perpetuate this myth. After an inspection of the Luftwaffe and its facilities in Germany, by invitation of Hermann Goering, Lindbergh sent in a series of "expert" reports on the Luftwaffe. These were grossly exaggerated and helped create the impression that it was more than twice its actual size. For example, when the Germans invaded Poland, the Luftwaffe was equipped with approximately four thousand first-line aircraft,[41] and the estimate carried in the United States was something in excess of ten thousand.[42] The actual German production figures at that time were five hundred aircraft a month, and the American estimate was that the Germans were producing more than a thousand planes a month. This is exactly what Hitler wanted the world to believe.

Another unwitting ally of Hitler in perpetuating the myth of the Luftwaffe was the Spanish painter and scluptor, Pablo Picasso. His famous painting, Guernica, not only conveyed an allegorical condemnation of fascism and war, but made Guernica a household word throughout the world. Picasso's portrayal of the death of the Spanish town was exactly the thing to put the fear of God into the hearts of Hitler's enemies. Nothing could do the job better. If fear of the Luftwaffe was the objective, Picasso's Guernica was worth more than a thousand bombers and more even than the horror movies that were being churned out to depict the power of the Luftwaffe. Picasso was a godsend to Hitler.

Hitler made skillful use of the trump cards he was handed, not only by Picasso and Lindbergh, but by all the other dupes who contributed to the legend of the Luftwaffe. Before long, and of course with the help of clever propaganda, Hitler had the leading statesmen in all the key countries convinced that the world was more or less in hostage to the German Luftwaffe, that if they did not toe the German line Guernica would be the fate of their cities. Hitler used this universal fear of the Luftwaffe against friends and foes alike. He used it against his friends to keep them in line and against his foes to keep them at bay. For all practical purposes, the Luftwaffe was his sword of Damocles.

This was never more evident than at Munich when Neville Chamberlain begged the Führer to promise that if he had to use force against Czechoslovakia, he would not bomb Prague. Hitler readily agreed to this and added the thought that if the Czechs compelled him to "take forcible action, he would always try to spare the civilian population and to confine himself to military objectives. He hated the thought of little babies being killed by gas bombs."[43] Hitler's denial of intent to use the invincible air weapon against the people of

Czechoslovakia convinced most people that if war came the first thing Hitler would do was to Guernica Prague.

While Neville Chamberlain was trading off Czechoslovakia for "peace in our time," or at least peace in his part of the world, the clouds of war were gathering in the United States. At first, it was only a tiny speck of a cloud, but one of those who could see it clearly was Franklin D. Roosevelt. Hearing Hitler's famous Nuremburg speech in his private railroad car in Minnesota on September 12, 1938, the President concluded that the German dictator was bent on war and that it would be one which would involve the United States. Accordingly, Roosevelt directed Harry Hopkins to make a personal investigation of the aircraft industry in the United States, and as he did so Hopkins made a note, "the President was sure we were going to get into war and he believed that air power would win it."[44]

This was a little bit late for the United States, and for Roosevelt in particular, to be waking up to the fact that Air Power could be decisive in war and that the author of *Mein Kampf* in 1925 meant it when he said that Germany "must some day become lord of the earth."[45] However, in fairness to Roosevelt, it must be noticed that his was almost a lone voice in 1938. By this time the great pioneer of Air Power in the United States, Billy Mitchell, was on the scene no longer. He contracted influenza, complicated by heart trouble, in January 1936, and went to a hospital in New York, where he died on February 17. Nevertheless, the great Mitchell tradition lived on, and there were those in the United States who were still fighting the good fight. However, one of these, and the greatest by later events, was Henry H. Arnold, who recorded in his *Memoirs*, strangely enough, that it was not the Mitchell doctrine that guided Air Force thinking, but the "Douhet theories." He said that the Tactical School "had been teaching the Douhet theories as an abstract science for several years."[46]

Be that as it may, the thoughts that prevailed in air circles in the United States at this time was that of strategic air power. The idea was that the peak expression of Air Power was in the heavy, long-range bomber, which was the only weapon capable of cutting off the enemy's supplies in the rear and of destroying the real sources of military power, if not of compelling a decision by breaking his will to resist. The

British expression of this idea is found in one of their Air Ministry Manuals: "the bomb is the primary weapon of air power; the bomber is the chief means of conveying it to its target; an air striking force composed of bombers is the chief means by which a nation wields its air power."[47]

All this, of course, is pure Douhet. It is also pure Mitchell. And pure Trenchard as well. But it is not pure German, pure Russian, or pure Japanese. The doctrine, except for the founding father, is British-American, no matter what was being taught at the Tactical School. The doctrine remained a doctrine and did not become a practice until atomic bombs were put in the quiver of Air Power.

President Roosevelt did not let any grass grow under his feet once he had declared that a world war was coming, that the United States would be drawn into it, and that Air Power would decide it. Therefore, shortly after the forced capitulation at Munich, he summoned his military chiefs and principal Cabinet members to a meeting at the White House. This was on November 14, 1938, at 2 P.M. The President was not in the mood for a dialogue on this historic occasion. Doing most of the talking himself, he informed the group that he had decided to ask Congress for twenty thousand planes, with a back-up capacity of twenty-four thousand more. He followed up by saying that he did not want to hear anything about the problems of the Army, and was not interested in hearing anything about the requirements of the Navy. Explaining, the President went on to say that the problem in the world was to stop agression and that planes were the only things that would do the trick. He amplified this by pointing out that Hitler had no fear of armies and would not be impressed by sea forces, but that superior air forces would give him pause very quickly.

Naturally, this was music to General Arnold's ears, and he later called the decisions arrived at, or rather announced at the White House meeting the Magna Carta of American Air Power. These decisions were certainly no less than a charter of freedom for the American Air Forces, and it is interesting to note that the idea for the freedom charter came from the Great Satan himself. If it had not been for the fact that when Hitler threw the Luftwaffe on the table at Munich there was no gainsaying his slightest wish, there would not have been a

Magna Carta for American aviation.

At the end of the meeting in the White House, and almost as a courtesy, President Roosevelt turned to General Marshall and asked him what he thought about the program. The reply came as something of a shock. General Marshall said, "I'm sorry, Mr. President, but I don't agree with that at all."[48] What Marshall meant, of course, was that he did not agree with the idea that a huge augmentation of the Air Force would be enough to defeat Germany. In fact, the Army at this time was in such a deplorable state of preparedness that it could muster no more than five poorly-equipped divisions whereas the Germans at the time of Munich had ninety first-rate divisions with ample reserves.

Six months after the surrender at Munich, the Germans moved to complete the conquest of Czechoslovakia. This was on March 15, 1939. The invasion and occupation was unopposed. Goering drove the final nail into the coffin of Czechoslovakia when he summoned the President of the hapless state to Berlin on the eve of the invasion. Dr. Emil Hacha was told that the choice was his between peace and war. This meant that he could sign the surrender document or see Prague subjected to the fate of Guernica. After fainting at the prospect, Dr. Hacha signed the fatal document. The only consolation the Czechs had, if it could be called that, was that it would soon be the turn of England and France.[49]

At roughly the same time that Hitler was entering Prague in triumph a tiny economic analysis unit was established in the intelligence section of the Office of the Chief of the Air Corps in Washington. This was the brainchild of Major General Haywood S. Hansell. Its function was to study the vulnerability to air attack of the potential enemies of the United States, notably Germany.[50] This was a small step in the right direction, and it was a step pursuant to Douhet's first commandment, that a skillful selection of enemy targets was the first and most important part of the air force task, "constituting what may be defined as aerial strategy."[51] This was what Goering found out in the Battle of Britain-- that no amount of air power is enough to win a war if it is used against the wrong targets.

On the basis of this small start in the right direction and the President's determination to have an air force second to none, events moved rapidly. On

May 16, 1940, while the Western Allies were crumbling under the German steamroller, the President sent a message to Congress on the subject of air power. This time he demanded not twenty thousand but fifty thousand planes, and this time he was backed up by his new Secretary of War, Henry L. Stimson, who declared that "air power has decided the fate of nations; Germany, with her powerful air armadas, has vanquished one people after another. On the ground, large armies had been mobilized to resist her, but each time it was additional power in the air that decided the fate of each individual nation."[52]

Following the President's leadership on the subject of Air Power, the Army Air Forces were created on June 20, 1941, and one of its first projects was to produce a plan outlining what would be required to bring about the defeat of Germany. The plan was known as Air War Plans Division 1 (AWPD-1), and has earned the reputation of being one of the most remarkable documents in the history of American Air Power. One of the four authors of the plan, Major General Haywood S. Hansell, Jr., later wrote a book about the plan that is a classic in the field.[53]

The remarkable thing about AWPD-1 was not the philosophy which was a curious jumble of Douhet tempered by the combat lessons of the current conflict, but the actual projection of the forces required to bring about the defeat of Germany. Success in the air war depended, according to the plan, on the creation of an Air Force of nearly two and a half million men, some two hundred fifty combat groups and nearly eighty thousand aircraft. This force level was reached by April 1944, but the irony is that at almost the exact minute the Air Force reached the level where it could accomplish its mission, the mission itself faded into the background, as everything was subordinated to the ground invasion on June 6, 1944.

One of the interesting footnotes to AWPD-1 and to history, is the fact that the Chicago *Daily Tribune* published verbatim extracts of the highly secret war plan on December 4, 1941, just three days before the Japanese attack on Pearl Harbor.[54] Three years before this the New York *Herald Tribune* had done an even greater disservice by publishing a very accurate and comprehensive description of radar, the British system of early warning and fighter control that saved England during the Battle of Britain. The importance of the radar network guarding the approaches to England was such that Churchill said, "unless British science had proved superior to German, and unless its strange sinister resources had been effectively brought to bear on the struggle for survival, we might well have been defeated, and, being defeated, destroyed."[55]

The far-sighted planners in the Air Force did all they could when they predicted with accuracy the force requirements to defeat Germany. But this was not all that was needed. "We had the plans," General Arnold said, "but not the planes."[56] Getting the planes, training the crews and establishing the bases to bring into reality the awesome power needed to defeat Germany required over three years of the most tremendous work imaginable. After all, it took a dedicated Germany six years to perform the same feat--mobilization for war. The miracle of beating plowshares into swords in the United States would have taken far longer and might not have been accomplished at all, had it not been for the fact that the country had not only an air-minded President and Commander-in-Chief but a few great leaders among its airmen who were willing to risk their lives and careers on a conviction that they were right about air power.

To men of vision it was perfectly clear by the end of World War I that a new era of military thinking and planning had been born. It was the air age, or the age of military Air Power. The best expression of it was in the heavy, long-range bomber. The concept of the visionaries was that the airplane was not merely a new "weapon" of war, but something altogether different, a new *method* of waging war, and one that required a new way of thinking on the part of man.

Strategic air power was a concept that the public could grasp easily in every country. Not so with the professional military authorities, however. They were too hidebound in their thinking to be able to compromise with the laws of their profession, and hence were inclined to disparage the radical minorities within their midst who thought in terms of a vision. The visionaries in the United States were ridiculed and ignored to a great extent, and in England they lost power as the country moved to the brink of war. On the other hand, the fate of the visionaries was much harsher in the totalitarian countries. There they were persecuted, banished and even executed. Of course, it is one of the virtues of a democracy that freedom of thought and expression are tolerated even under the stress of war.

This paid off handsomely in World War II, at least for England and the United States. In the interregnum between World Wars I and II, while the idea of strategic air power was in disfavor with the military high commands, the idea itself was not uprooted in the democracies, and its advocates were not banished or executed. True, neither country had any Air Power in the strategic sense when the war began, but the potential was there. Consequently, and unlike the case in Germany, Japan and the Soviet Union, when World War II began there were men in the United States and England who were able to transform their ideas into tremendous and overwhelming realities. Victory in the air had much to do with the outcome of World War II, and may have dictated the course of civilization-- at least until the outbreak of a World War III.

FOOTNOTES--CHAPTER IV

[1] Bernard Brodie, *Strategy in the Missile Age* Princeton: Princeton University Press, 1965), 79.

[2] Karl von Clausewitz, *On War* (New York: The Modern Library, 1943, tr. O.J. Matthijs Jolles), 30 (Bk. I, ch. 2).

[3] Alfred Thayer Mahan, *From Sail to Steam* (New York: Harper & Brothers, 1907), 283.
Br

[4] Karl von Clausewitz, *On War* (New York: The Modern Library, 1943, tr. O.J. Mattihijs Jolles, 62 (Bk. II, Ch. 1).

[5] Alfred Thayer Mahan, *The Influence of Sea Power Upon History 1660-1783* (Boston: Little Brown and Company, 1895), 88.

[6] Robert Frank Futrell, *Ideas, Concepts, Doctrines: A History of Basic Thinking in the United States Air Force 1907-1964* (Alabama: Air University, Maxwell Air Force Base, 1974), 28.

[7] *Ibid.*

[8] *Ibid.*, 22.

[9] Giulio Douhet, *The Command of the Air* (New York: Coward-McCann, Inc., 1942, tr. Dino Ferrari), 50.

[10] St. Matthew, xiii, 57.

[11] Giulio Douhet, *The Comand of the Air* (New York: Coward-McCann, Inc., 1942, tr. Dino Ferrari), 50.

[12] *Ibid.*, 59.

[13] Alfred F. Hurley, *Billy Mitchell* (Bloomington: Indiana University Press, 1975), 115.

[14] *Ibid.*, 114-15.

[15]Edward Homz, "The Continental Experience," in *Air Power and Warfare* (Washington: Government Printing Office, 1979), 45.

[16]George H. Quester, *Deterrence before Hiroshima* (New York: John Wiley & Sons, 1966), 93.

[17]Edward Warner, "Douhet, Mitchell, Seversky: Theories of Air Warfare" in Edward Mead Earle's *Makers of Modern Strategy* (Princeton: Princeton University Press, 1971), 496.

[18]Winston S. Churchill, *Their Finest Hour* (Boston: Houghton Mifflin Company, 1949), 330.

[19]*Ibid.*, 340.

[20]*Ibid.*, 115.

[21]George H. Quester, *Deterrence before Hiroshima* (New York: John Wiley & Sons, Inc., 1966), 66-67.

[22]*Ibid.*, 67.

[23]Winston S. Churchill, *The Gathering Storm* (Boston: Houghton Mifflin Company, 1948), 117.

[24]Adam B. Ulam, *Expansion and Coexistence* (New York: Praeger Publishers, 1971), 235.

[25]William L. Shirer, *The Rise and Fall of the Third Reich* (New York: Simon and Schuster, 1966), 293.

[26]Winston S. Churchill, *The Gathering Storm* (Boston: Houghton Mifflin Company, 1948), 197-98.

[27]DeWitt S. Copp, *A Few Great Captains* (New York: Doubleday & Company, Inc., 1980), 416.

[28]*Ibid.*, 415-16.

[29]George H. Quester, *Deterrence before Hiroshima* (New York: John Wiley & Sons, Inc., 1966), 67.

[30]Bernard Brodie, *Strategy in the Missile Age* (Princeton: Princeton University Press, 1965), 76.

[31]Edward Homz, "The Continental Experience," in *Air Power and Warfare* (Washington: Government Printing Office, 1979), 44.

[32]Kenneth R. Whiting, "Soviet Air Power In World War II, in *Air Power and Warfare* (Washington: Government Printing Office, 1979), 105.

[33]Herbert Molloy Mason, Jr., *The Rise of the Liftwaffe* (New York: The Dail Press, 1973), 90.

[34]*Ibid.*, 184.

[35]Wesley Frank Craven & James Le Cate, *The Army Air Forces in World War II* (Chicago: University of Chicago Press, 1948), I, 85.

[36]Herbert Molloy Mason, Jr., *The Rise of the Luftwaffe* (New York: The Dial Press, 1973), 159.

[37]*Ibid.*, 257

[38]The United States Strategic Bombing Survey--*Summary Report: European War* (Washington: Government Printing Office, 1945), 8.

[39]Albert Speer, *Inside the Third Reich* (New York: The Macmillan Company, 1970), 272.

[40]Herbert Molloy Mason, Jr., *The Rise of the Luftwaffe* (New York: The Dial Press, 1973), 253.

[41]Wesley Frank Craven and James Lea Cate, *The Army Air Fordes in World War II* (Chicago: University of Chicago Press, 1948), 1, 89.

[42]DeWitt S. Copp, *A Few Great Captains* (New York: Doubleday & Company, Inc., 1980), 404 and 450.

[43]George H. Quester, *Deterrence before Hiroshima* (New York: John Wiley & Sons, Inc., 1966), 98.

[44]DeWitt S. Copp, *A Few Great Captains* (New York: Doubleday & Company, Inc., 1980), 445.

[45]Adolf Hitler, *Mein Kampf* (London: Hutchinson & Co., 1977), 629.

[46]*The Development of Air Doctrine in the Army Air Arm, 1917-1941*, USAF Historical Studies: No. 89 (Air University, 1955), 51.

[47]Wesley Frank Craven & James Lea Cate, *The Army Air Forces in World War II* (Chicago: Chicago University Press, 1948), 99.

[48]DeWitt S. Copp, *A Few Great Captains* (New York: Doubleday & Company, Inc., 1980), 456.

[49]Herbert M. Mason, *The Rise of the Luftwaffe* (New York: The Dial Press, 1973), 244.

[50]Wesley Frank Craven & James Lea Cate, *The Army Air Forces in World War II* (Chicago: University of Chicago Press, 1948), 353.

[51]Giulio Douhet, *The Command of the Air* (New York: Coward-McCann, Inc., 1942, tr. Dino Ferrari), 50.

[52]Robert Frank Futrell, *Ideas, Concepts, Doctrine: A History of Basic Thinking in the United States Air Force 1907-1964* (Air University, Maxwell Air Force Base, Alabama, 1974), 55.

[53]Haywood S. Hansell, Jr., *The Air Plan That Defeated Hitler* (Atlanta: Higgins-McArthur/Longino & Porter, Inc., 1972), xi.

[54]*Ibid.*, 96.

[55]Winston S. Churchill, *Their Finest Hour* (Boston: Houghton Mifflin Company, 1949), 381-82.

[56]Wesley Frank Craven & James Lea Cate, *The Army Air Forces in World War II* (Chicago: University of Chicago Press, 1948), 150.

CHAPTER V

AREA BOMBING IN EUROPE

What accounts for the fact that the Blitzkrieg was such an astounding success in the early days of World War II and the Maginot Line was such an abject failure? Is it simply that the Germans succeeded where the French failed to learn the right lessons from World War I, at least from the point of view of ground warfare? Every war is a laboratory or a proving ground of sorts, and the victor is usually the power that correctly interprets the last war from the point of view of the future. In this context, what did World War II prove about the future? Did it prove that air power was the dominant and decisive factor in war, or did it prove just the opposite? In a nutshell, was World War II a vindication of the Italian patron saint of air power, General Guilio Douhet, or did that war merely prove that the British philosopher General J.F.C. Fuller was right about Douhet when he called him "a wonderful salesman," but also "a prophet of the ridiculous"?[1]

Long before the last bomb was dropped in World War II, the airmen in the United States attempted to get a head start on the historians by setting in motion a series of events that, hopefully would provide some answers to the questions that would arise concerning the role of air power in World War II. The result was the establishment of the United States Strategic Bombing Survey pursuant to an order from the White House dated September 9, 1944, and an implementing directive from the Secretary of War on November 3, 1944. This was when the Germans were in full retreat to the Siegfried Line, and long after the liberation of Paris. The Third Reich was obviously in a state of collapse, and hopes were running high that the boys would be home for Christmas.

The United States Strategic Bombing Survey was a very high-powered and high-level organization. Its presidential directive and independent organization gave it the clout it needed to perform its important mission and to deal with all the armed services, Army, Navy and Air. The chain of command was from the President of the United States, through the Secretary of War, to the Chairman of the Survey.

The mission of the Survey was spelled out in the presidential directive, and it was to make an on-the-spot analysis of the role of Air Power in bringing about the defeat of Germany. This would not only set the historical record straight from the outset, but it was also important because the war in the Pacific was still going on, and it was hoped that what was learned about Air Power in the European War would be of great help in bringing about the speedy conclusion of the Pacific War. Finally, and perhaps of greatest importance, it was hoped that the findings of the Survey would point the way to the future and settle the question once and for all as to the proper place of Air Power in the national defense structure for the future.

When the Air Force took the lead in establishing the United States Strategic Bombing Survey it did so at considerable risk to its own future. While the airmen stood to gain a lot if the Survey came up with the conclusion that Air Power was the dominant and decisive factor in war, they also stood to lose a lot if the survey came up with the opposite conclusion, and found that Air Power was important not as an independent force, but only as an auxiliary of the Army and Navy. Since the stakes were so high, and because the Air Force would have no control of the project once it was launched, the search for a competent Chairman took on an added importance. The story of the search, a painstaking and diligent one, is well told and in detail by Professor Davie MacIsaac in his informative book on *Strategic Bombing in World War Two*.

At one point in the search for a Chairman of the United States Strategic Bombing Survey, or USSBS, as it came to be called, the idea was to find an institutional sponsor. In this connection, Harvard, M.I.T. and the University of California, were all considered ideal sponsors. If any one of them could be prevailed upon to accept the project, it would have several advantages, particularly from the point of view of public acceptance. Of course, in addition to this, having such sponsorship would guarantee complete objectivity in fact-finding and analysis, and it would also solve the problem of finding the right people to do the work, since the great universities could call on their specialized schools for qualified personnel. The White House generally favored this solution to the problem of the Strategic Bombing Survey.

For many practical reasons, not the least of which was the disinterest on the part of the great universities, this approach to the Bombing Survey had to be dropped. So the search went on for a suitable Chairman. Finally, the names narrowed down to one, Franklin D'Olier, a man who was sixty-seven years old and president of the Prudential Insurance Company of America. Upon his acceptance and with his approval, Henry C. Alexander, another leading citizen and vice-president of J.P. Morgan and Company, was prevailed upon to accept the position of Vice-Chairman. The selection of the Chairman and Vice-Chairman cleared the way for the United States Strategic Bombing Survey to begin its work.

Perhaps a word should be added at this point about how the movement for a Bombing Survey got underway in the Air Force and who was responsible for the drive behind the idea. The record is clear on this point, and there is no doubt that the brains behind the Survey were those of Muir S. Fairchild. "Santy," as he was called by his large following in the Air Force, was a Major General and a philosopher. In this respect he was like Clausewitz and Douhet, inclined to think beyond the moment, and to perceive things in a larger context. In the Joint Chief of Staff organization, General Fairchild was a member of the Joint Strategic Survey Committee, a think-tank group that reported directly to the Joint Chiefs and concerned itself only with the broadest problems of national policy. Within the Air Staff, General Fairchild was at the right hand of General Arnold and a member of a small group of officers called the Advisory Council. The job of these officers, and of Fairchild in particular, was as General Arnold said, to do the "blue-sky thinking."[2]

General Fairchild got the idea of a Bombing Survey as early as 1942, and in 1943 discussed the subject with key officers in the Air Staff. His thinking was that the Air Force had a responsibility not only to itself but to the country as a whole to have a full-scale evaluation of its performance as soon as it was possible to do so. Not only that. General Fairchild wanted it done by a competent civilian authority, not by the military. He often expressed the view that there was nothing mysterious about military strategy and that military officers should lean heavily on civilian brains, particularly when it came to such matters as air strategy. He was highly critical of the tendency on the part of the military to treat civilians "like firemen, calling on them only in an emergency and ignoring them the rest of the time."[3]

It is almost certain that without General Fairchild there would not have been a Strategic Bombing Survey. Not only did he provide the initiative for the project, but followed it through with determination and at the levels that counted. Thus it was that D'Olier found himself in the presence of Miur Fairchild within a week after he had accepted the job of organizing and conducting the Strategic Bombing Survey.

General Arnold gave D'Olier his important assignment on the morning of Saturday, October 21, 1944. Going back to the Chairman's World War I rank, General Arnold said, "Colonel D'Olier, I have a job for you, a tough one, as tough as any I've ever pinned on anybody. Bob (referring to Robert A. Lovett, Assistant Secretary of War for Air) and I have talked it over and we agree that you are the man for the job. Put simply, what we want is for you and a group of your own choosing to go over to Europe and conduct a full-scale evaluation of the effects of our strategic bombing against Germany."[4] To this, General Fairchild added a thought, saying "that the Survey's task was not to find excuses but to judge results, not to paper over difficulties but to call the shots as it saw them, to let the chips fall where they may."[5]

On October 28, 1944, a week after the meeting between Arnold and D'Olier, General Fairchild repeated the caution about a non-critical report. He told the Chairman that the job of the Survey should not be limited to evaluating the effect of the attacks "upon selected target systems, but also the effect of attacks upon secondary and other target systems. It should answer the question whether any target systems that were not attacked should have been attacked. One of the fundamental purposes of the Survey is to determine whether the results obtained were commensurate with the expenditure of men and material."[6]

The work of the Survey was done in a highly organized and very efficient manner. The first move was to establish a headquarters in London. As the work progressed, another general headquarters was activated in

Frankfurt, and of course there were several regional headquarters, or forward echelons, in various parts of Germany. The plan was to use field teams to make on-the-spot investigations of as many targets as possible, including all the important ones, and to feed the data back into a central file where it could be analyzed and appraised by carefully selected teams of highly qualified industrial and military specialists.

Considering the size of the task, the work-force of the Survey was not large. The authorized table of organization provided for a force of three hundred civilians, three hundred fifty officers and five hundred enlisted men. All the Directors of the Survey were civilians hand-picked by the civilian Chairman and Vice-Chairman. It was not an Air Force show, nor even a military one, but a wholly independent investigation and evaluation reporting to the White House through the Secretary of War.

In its Over-All Report on the European War, the Survey said, "the basic information was obtained from three main sources; first, close examination and inspection of several hundred German plants, cities and areas; second, masses of statistical and other material, ranging from records of individual companies and cities to top German government documents; third, interviews and interrogations of thousands of Germans including military and political leaders."[7] The field teams followed close on the heels of the advancing armies of Europe, so close in fact that Survey personnel suffered casualties, including four killed. "Germany," the report said, "was scoured for its war records. They were found sometimes, but rarely, in places where they ought to have been; sometimes in safe deposit vaults; often in private houses, in barns and caves; on one occasion in a hen house and, on two occasions, in coffins. Targets in Russian-held territory were not available to the Survey."[8]

The civilian members of the Survey, with particular reference to the field teams, wore uniforms and had assimilated military rank. The Chairman's assimilated rank was that of a Major General, and the Vice-Chairman had the rank of Brigadier General, but the field men had much lower ranks, of course. This was a severe handicap on occasions when rank was really needed to get at important records and personages.

Very junior officers do not cut much ice, as a rule, with battle-hardened colonels and generals. Even the jobs of interviewing, much less interrogating, important Germans was made more difficult by the junior rank of the field personnel. Even in defeat, the Germans had high respect for rank, and almost no respect for those on the lower end of the totem pole. The field teams had another disadvantage that impaired their effectiveness. This was their insignia--USSBS. USSBS was the shoulder patch on the military uniforms worn by members of the field teams, and USSBS was also painted on the Survey vehicles. The trouble was that USSBS was much too close to USSR for comfort, and many a German fled on sight when he caught a glimpse of the Survey insignia.

One of the blessings of the Survey was the appearance of J. Kenneth Galbraith to head up the important Economics Division. When he arrived on the scene in April 1945, he made his presence felt immediately. Galbraith insisted that the Survey was on the wrong track in its analytical work because it was putting too much emphasis on the masses of statistics. Rather than trying to prove everything by the use of numbers. Galbraith said the Survey should use its own good judgment and pay more attention to what the Germans themselves said about the causes of their defeat. He did not suggest that the search for German records should be called off, or even slackened, but said that the field teams should be encouraged to do more interviewing of high-ranking military and industrial officials.

An example of what Galbraith meant was the excellent work done in connection with the investigation of the damage done to the ball bearing plants at Sweinfurt. One of the team leaders entered Sweinfurt with the Infantry and forthwith captured the owner of the company that controlled forty percent of the German ball-bearing industry and was also head of the government's Ball Bearing Production Board.[9] Naturally, his testimony about the effects of the strategic attack on the ball bearing plants was valuable, and valuable far beyond what could be gleaned from the cold statistics.

Another imaginative coup was pulled off by another investigative team of the Bomb Survey at Flensburg. This time a member of the team found the secret hiding place of Hitler's Minister of Armaments, Albert Speer, and quickly rounding up other members of the Survey, submitted Armaments Minister Speer to an interrogation that lasted for ten days. Certainly from the German point of view, and this was what the Survey was all about, Albert Speer was probably better qualified to write the overall report than any other man alive. The whole air war, in a sense, was a battle between the Allied Air Force on the one hand and Albert Speer on the other.

The most audacious example of getting at the facts for the Strategic Bombing Survey was an incident that occurred at Essen. Two days after the fighting ended, a field team entered the city and made a beeline for the main Krupp Works. Despite all the damage, most of the important officers of the company were still on the scene, and the field captain promptly called a meeting of the board to answer questions about the effects of the strategic bombing. Among those who answered the questions were Friedrich and Alfred Krupp.[10]

The interrogations conducted by the field teams of the Bombing Survey were not "oral history" in the usual sense of that term. Oral history is usually of very dubious reliability because of the time factor, among other things. Such interviews are reminiscent, as one keen observer has said, "of the law of gravitation that it varies inversely with the square of the distance from the event."[11] On the other hand, the Survey interviews with the Krupps, with Speer, and with the head of the government's Ball Bearing Production Board, were contemporary interviews held under circumstances that did not invite variations from the cold truth. It might almost be said, as Kenneth Galbraith did, that these interviews were about as close to the truth as mortal man was likely to get.[12]

Based on these interrogations, plus hundreds of others, as well as on thousands of captured documents, plus reams of official and unofficial statistics, the Bombing Survey produced over two hundred individual reports. Some of these were on particular plants, others on cities, and still others on areas. In addition, reports were produced on whole segments of economic and national life, such as the effects of city bombing on the national will to resist. On top of the pile, and finally, the Survey produced an *Overall Report On the European War*, plus a *Summary Report*. It is probably safe to say that no war in history was better covered from the point of view of getting at the facts than the great strategic bombing offensive against Germany. The only disadvantage that the Survey operated under, if it can be called that, was the one stressed in the original directive, "that the value of the study would be enhanced by the speed of its completion."[13]

The civilian leaders of the Bomb Survey did not let the impatience of the military unduly influence them in arriving at conclusions. After all, they remembered their presidential directive and what General Arnold had said when he "pinned" the job on the Chairman, "this is your job, and when you've finished you report not to me, but to Secretary Stimson and the President."[14]

The more the Chairman and the Directors pondered the problem, and looked at the mountainous and forbidding nature of the data that had been gathered, the more it was apparent that theirs was a double-headed problem. One was to second-guess the airmen on the subject of target analysis; were the choices wise in the first place, considering the objectives sought and the available capabilities? Or was General Fairchild right when he suggested that there might have been target systems not attacked that should have been attacked, and put it up to the Survey not only to answer this question, but to speculate on whether the results obtained from the bombing were commensurate with the efforts.[15] The second question was more fundamental than the first, and this was as to the objectives of the airmen, were they at all reasonable? In other words, could the bombing airplane by itself force a decision in a great war?

On October 24, 1945, the Vice-Chairman of the United States Strategic Bombing Survey, Henry C. Alexander, gave a press conference at the Pentagon and released to the public copies of the two reports on the European War. One was called the Summary Report, and the other the Overall Report. The findings cleared the air on a number of important points, the most significant of which was the Survey's answer to the theoretical question posed by Douhet, whether a country could be bombed into submission. The answer was a clear-cut victory for the airmen. Answering the question yes, the Overall and Summary reports said, "the German experience suggests that even a first class military power -- rugged and resilient as Germany was -- cannot live long under full-scale and free exploitation of air weapons over the heart of its territory."[16]

Implicit in all the plans of the Allied air forces, British and American, was the philosophy of Douhet, that no country could survive in the face of a ceaseless pounding from the air, and that the bombing airplane was the dominant and decisive force in war.

The Royal Air Force was first in the field, of course, and its leader, Air Marshal Sir Arthur Harris, never wavered in his belief that area bombing at night would shatter the morale of the German people to the point where their capacity for armed resistence would be fatally weakened. The American air planners were in complete agreement with their British cousins in the fact that Germany could be defeated by Air Power alone, but did not agree that area bombing at night was the most promising method of proceeding. Instead, they put all their eggs in the single basket of precision bombing by day.

Right up to the end of the war, or at least almost to the end, the British and the Americans persisted in this disagreement on tactical procedures. It was not until 1945 in the attacks against Dresden, Berlin and the Ruhr, that the Americans seemed to abandon precision bombing in favor of area attacks. They did this not because they "saw the light" at the end of the tunnel, but because of what the official historians of the Air Force called the "numbers racket."[17] This referred to the pressure from Washington to step up the bombing sorties regardless of the weather. General Carl Spaatz was made to feel that he would be remiss in his duty if he failed to get more bombing missions out of the huge forces at his disposal. General George C. Marshall, late in 1944, had a discussion with the air leaders in Europe and suggested "that long-range objectives of strategic bombardment be abandoned for an all out effort to force an early victory."[18] This was the period when the ground offensive was stopped short of the Rhine by logistical shortcomings. The Air Force succumbed to the pressure, and the result was not precision bombing, or radar bombing, but blind bombing. The only question was whether blind bombing was better than no bombing at all?

Douhet would not have been of much help to the British and Americans in deciding the thorny question of bombing policy, the day versus the night, and precision versus area bombing. The only point Douhet would have insisted upon was the point of order in the bombing system. Even in the matter of area bombing, Douhet was a stickler for the systematic approach, one that would take into consideration all the pertinent material, moral and psychological factors. He said that "the grouping of zones and determining the order in which they are to be destroyed, is the most difficult and delicate task in aerial warfare, constituting what may be defined as aerial strategy."[19]

The very early air operations against Germany, of course, were on the part of the Royal Air Force, and in them Bomber Command experimented briefly with precision bombing by daylight against industrial targets. The experiment was not a success. Losses were heavy, and the bombing results were nil. Then came the switch to the night bombing of industrial targets. However, this was also unsuccessful. The targets were hard to find, and when this was done, they could not be bombed accurately with the equipment at hand. So out of operational necessity the British turned to area bombing at night. It was not until later that the British made a virtue out of this necessity.

When Winston Churchill replaced Neville Chamberlain as the British Prime Minister on May 10, 1940, the Royal Air Force found itself with a friend in court, and got a new lease on life. Using the accidental bombing of London by ten planes on August 24 as an excuse, Churchill on August 25 dispatched ninety-five bombers to Berlin for an urban area attack on the German capital. Though it had all the outward appearance of being an act of reprisal for the bombing of London, the large attack on Berlin was not for this purpose at all. Churchill's bold stroke was designed to accomplish something quite different. It was to bait the German Führer into an immediate shifting of their attack from the bases of Fighter Command to the city of London. Hitler made his decision on the London Blitz the morning after Berlin was bombed.[20]

This was during the decisive phase of the Battle of Britain. "In the fighting between August 24 and September 6," Churchill said, "the scales had tilted against Fighter Command."[21] If the Germans had persisted in their objective, Fighter Command would have been destroyed, and England would have been wide open to invasion. "There was always the possibility," Churchill said, "that victory over Britain in the air would bring about the end of the British resistance and that actual invasion, even if it became practicable, would also become unnecessary, except for the occupying of a defeated country."[22]

But this did not happen. Instead, Hitler took the bait, and on September 7 the Luftwaffe turned its attention to London as the primary bombing objective and let Fighter Command off the hook. This is the only thing that saved England, and by the same token is probably the one thing that cost Hitler the war he came so close to winning. It almost goes without saying that, without the English base from which to project power on the continent, not even the great strength of the United States would have been enough to defeat Hitler. And it must also be remembered that at this moment the United States was not in the war, or even close to being in it. Pearl Harbor was still four months away.

Of this sudden change in the German target plans, Churchill said, "Goering should certainly have persevered against the airfields on whose organisation and combination the whole fighting power of our air force at this moment depended. By departing from the classical principles of war, as well as from the hitherto accepted dictates of humanity, he made a foolish mistake."[23]

Emboldened by the victory of Fighter Command over the Luftwaffe in the Battle of Britain, Churchill's already great faith in the Royal Air Force was strengthened, and he turned his thoughts to Bomber Command as a possible war-winner against the German foe. It was the only way England had of coming to grips with Germany at this time, and the only question was how could it be used most effectively. The logic of operational necessity made it painfully clear that daylight bombing of precision targets was out of the question and also that night bombing of precision targets was not profitable with the techniques available. Thus Churchill did the only thing he could do, which was to become a convert to the idea of the night bombing of urban areas. Churchill rationalized the switch in his thinking by dwelling on the mistakes Goering had made in target selection, particularly the switch to London as a primary target before Fighter Command had been completely destroyed. In other words, the reasoning was that if Goering persisted against the airfields, "we might well have been defeated, and, being defeated, destroyed."[24]

Continuing in this vein, Churchill said it could be unwise to suppose "that if the attack had been ten or twenty times as severe -- or even perhaps two or three times as severe -- the healthy reactions I have described would have followed."[25] In fact, under those conditions, Churchill said, "we should have passed into conditions which might have pulverized all human organisation. However, everything happens in its turn and in its relation, and no one has a right to say that London, which was certainly unconquered, was not also unconquerable."[26]

While Churchill was making a rational analysis of why the Luftwaffe failed to conquer England with Douhet's theory of terror bombing, a report was submitted to the Prime Minister extolling the merits of area bombing and doing so on a scientific basis. The report was submitted by Churchill's principal scientific adviser and trusted personal friend, Lord Cherwell. The study was based on empirical data gleaned from the actual experience of the British under heavy German bombing, particularly in the attacks on Coventry, Hull, and Birmingham. It was found that the productive losses in these cities had been brought about not so much from the damage to their industrial facilities as by the general disruption in the lives of the working people, disruption brought about by the loss of housing, transportation difficulties, shortage of water and electric power failures. From this it was reasoned that the quickest way to defeat Germany was through a series of urban area attacks that eventually would result in the de-housing of twenty-two million people, enough to bring industry to a standstill and break the spirit of the German people. Lord Cherwell thought that this would be enough to force the German Government to sue for peace on any terms it could get. And it was all so scientific and convincing in its reasoning and logic.

At the time of the Cherwell report, the February-March period of 1942, Churchill's political fortunes were at a low ebb. On February 12, three German battleships, the *Scharnhorst*, the *Gneisenau* and the *Prinz Eugen*, supposedly bottled up at Brest, made their escape right under the nose of Bomber Command, up the

Channel toward Germany. Three days later, on February 15, the great fortress of Singapore was forced to surrender. Churchill called the surrender of Singapore "the worst disaster and largest capitulation of British history."[27]

Also at this time, Bomber Command's unimpressive record thus far not only in allowing the German warships to escape, but also in the matter of bombing Germany, was such as to prove an embarrassment to Churchill. In fact, so disenchanted were some of the planners in the British defense department that plans were actually afoot to disperse Bomber Command by transferring the bombers to Coastal Command, to the Middle East and to India. The thought was that winning the Battle of the Atlantic and holding the Middle East and India, were far more important objectives than the foolish notion of trying to bomb Germany into submission.

Churchill's reply to all this adversity was to come out fighting. Concentrating on Bomber Command, Churchill made the Cherwell report the basis of his directive even though this meant reversing himself on the stand he had taken in 1917 on terror bombing. Thus he ordered a full-scale bombing of the German cities on a systematic basis and put Arthur Harris on as the new head of Bomber Command.

At about the same time, a new generation of four-engined heavy bombers, Stirlings, Halifaxes and Lancasters was coming on line in England, and in Arthur Harris Churchill had a man who not only accepted the idea of area bombing, but one who would go Churchill one better in the organization and implementation of the idea. However, the idea itself was Churchill's, not Harris's. Arthur Harris made this clear in his *Memoirs*, saying when he took office in February 1942, "it had already been decided (it was a decision with which I had nothing to do) that all our main attacks should be against large industrial areas, which meant of course, large industrial cities as a whole."[28]

Churchill's urban area strategy was aimed at the German *will* to fight a war, as opposed to a stragegy designed to destroy the *capability* of the Germans to fight. Of course, there can be no capability without will, whereas there can be a will to fight without capability. The latter is the stuff of which guerrilla movements are made. However, there is a clear difference between the two, and the basis of Churchill's or Harris' plan was the will to resist, that is, terror bombing aimed at the civilian population with the idea of causing a collapse of the German will to resist. In advancing terror bombing as the basis for policy in 1942, Churchill was not only reversing a position he had taken in 1917, but aligning himself with some of the earlier philosophers, notably Giulio Douhet and Hugh M. Trenchard.

Regardless of any positions he may have taken in the past, Churchill was set on his new course, and when Bomber Harris came up with the idea of a series of "pulverization" attacks on the German cities, he found the Prime Minister in a mood to listen. The straw Churchill was grasping at was that Bomber Command could be the new "equalizer" on the Continent. He gave expression to this idea in a memorable speech to the Commons on August 20, 1940, when the Battle of Britain was raging. On that occasion, Churchill said, "there seems to be every reason to believe that this new kind of war is well suited to the genius and the resources of the British nation and the British Empire and that, once we get properly equipped and properly started, a war of this kind will be more favorable to us than the sombre mass slaughters of the Somme and Passchendaele."[29]

The first thousand-plane bombing attack in history took place on the night of May 30, 1942, and the target was Cologne. Almost as important as the number of planes was the fact that the attack was completed in a period of ninety minutes, and thus all the defenses of the city were completely saturated. During these ninety minutes the bombs fell at the rate of one every six seconds for a total of three thousand tons. From the British point of view, the bombing of Cologne was a complete success. Churchill and Harris hoped that the blow would resound throughout Germany and give the Russians heart as well by convincing them that the Allies were giving the Germans more than they could handle. The magnitude of the blow against Cologne was without precedent. The total tonnage of bombs dropped on Cologne in only ninety minutes was three times the weight of high explosives dropped on London during the worst month of the Battle of Britain.[30] This made it clear that the Luftwaffe was a swindle as a strategic air force. It may also help to explain Churchill's volte-face on area bombing. In other words, it was clear to the Prime Minister that what was not possible in 1917, or even in 1940, was clearly possible in 1942. Equally clear was the fact that Germany might not be able to stand up against this kind of punishment and could very well be bombed into submission, making another invasion of the continent unnecessary, except for the occupation of a defeated country.

Shortly after the thousand-plane attack on Cologne Air Chief Marshal Arthur Harris was knighted. In addition, the bombing of Cologne had another salubrious effect as far as the Royal Air Force was concerned. The surface minded generals and admirals who were playing with the idea of dispersing Bomber Command to the four winds suddenly found themselves with no one willing to listen to their ideas. The successful outcome of the thousand-plane mission against Cologne put a damper on all such defeatist thinking, and made Bomber Command the toast of the day. So, in a sense, Cologne saved Bomber Command.

Two nights after the devastating attack on Cologne, and before the Germans had a chance to regain their equilibrium, Essen was struck with almost equal force. And then it was Hamburg's turn. Hamburg was another operation that was typical of what Churchill had in mind when he fought the battle to save Bomber command and keep alive the idea that Germany could be bombed into submission. Over a period of nine days in late July and early August of 1943, the most devastating city attack of the war took place against Germany's second largest city and biggest port, Hamburg. This was one of the few occasions when there was a Combined Bomber Offensive in fact as well as in name. The Americans bombed their precision targets by day, and the British their area targets at night. The combined weight of the attack was so heavy that one of the results was a fire storm in which the force of the winds was like that of a hurricane or a tornado. No human resistance was possible. Nearly half of the buildings in the city were burned to the ground by incendiaries or were collapsed under the high explosires. The death toll was over fifty thousand, equal to the total number of people killed in England during all the German bombing attacks.

The United States Strategic Bombing Survey said, "no subsequent city raid shook Germany as did that on Hamburg; documents show that German officials were thoroughly alarmed and there is some indication from interrogation of high officials that Hitler himself thought that further attacks of similar weight might force Germany out of the war."[31] Albert Speer, who assumed office as Germany's Minister of Armaments in 1942, was willing to admit defeat after the bombing of Hamburg. "I reported for the first time orally to the Fuehrer," Speer told the Survey interrogators, "that if these aerial attacks continued, a rapid end of the war might be the consequence."[32] In his *Memoirs*, published after the war, Speer said, "fortunately for us, a series of Hamburg-type raids were not repeated on such a scale against other cities. Thus the enemy once again allowed us to adjust ourselves to his stragegy."[33]

It was certainly not Churchill's fault that the Germans were given the time to adjust themselves to the strategy of the Allies. All along, he had been in the forefront of those who believed in Bomber Command and was the modern father of the idea of area bombing. In fact, Churchill had been among the first, along with Jan Smuts, to recognize the link between traditional British policy and the newly emerging phenomenon of Air Power. Churchill was so enamored of his idea of victory through air power that he could not understand why the Americans could not see the light, that is the light of area bombing and why they stubbornly refused to pool their growing resources in one gigantic effort to destroy, if needs be, every city in Germany of a hundred thousand or more. This would be in accordance with Harris's plan and objective, which was to undermine the morale of the enemy to the point where his capacity for resistance was fatally weakened.

So strongly did Churchill feel about this matter, and so outspoken was he on the subject, that some of the British airmen tried to calm him down before he antagonized the Americans into departing the United Kingdom altogether, to use their Flying Fortresses in the war against Japan. The fever to do just this was still running high in naval circles in the United States, as well as in some Army circles, particularly those of Douglas MacArthur. Nevertheless, when Churchill went to Casablanca in January 1943 for a long series of meetings with President Roosevelt, the Prime Minister took a note with him to make a personal appeal to the President to order the American flyers to join the British in their night bombing of Germany and to do so at once.

Churchill's intentions were so well known that General Arnold brought General Ira C. Eaker from England to talk the Prime Minister out of the idea of making this appeal and to convince him that the Americans were on the right track in their daylight bombing of precision targets. It was a whale of an undertaking for Eaker, and in the interview with Churchill, he faced one of the principal challenges of his life.

By the end of 1942 the American bombers in England had flown only twenty-seven missions against enemy targets, and none of these had been against targets in Germany. There was good reason for this, of course, the slow build-up of bombers in England because of the heavy demand for American resources in the Pacific, and the many costly diversions of strength, for example to support the North African invasion, to say nothing of the futile attacks on the sub pens at Brest, Lorient and St. Nazaire. But Churchill was not interested in this. The only thing that interested him was the fact that the British night bombers were bombing targets in Germany and the American day bombers were not. Hence he was determined to put an end to this foolishness by getting the American bombers teamed up with the British in bombing Germany by night.

Churchill minced no words in his conversation with Eaker on the non-role of the American bombers in the air war against Germany. The Prime Minister said that he had always been "skeptical" of the "American scheme of daylight bombing of Germany by the armoured Flying Fortresses,"[34] and that the record thus far had confirmed his worst fears. He went on to remind Eaker (quite unfairly) of the glowing promises that had been made in Washington of the heavy bombardment of Germany by this time and said,

> it was certainly a terrible thing that in the whole of the last six months of 1942 nothing had come of this immense deployment and effort, absolutely nothing, not a single bomb had been dropped on Germany. There must have been twenty thousand men and five hundred machines all laid out in East Anglia--and nothing so far, as it seemed, to show for it at all.[35]

General Eaker explained that the American bombers had not been more active against targets in Germany because they had been diverted to satisfy operational requirements elsewhere, including North Africa. However, he stoutly defended the record of the 8th Air Force. Eaker also promised that the 8th would be bombing German targets within a month, and bombing them by day. He then surprised Churchill by speaking in glowing terms of the night bombing by the Royal Air Force. Eaker followed this up by telling the Prime Minister that the two teams working together were on the right track for defeating Germany, the British bombing by night and the Americans by day. As Eaker later recalled it, it was this last argument that seemed to impress the Prime Minister, the argument of the "around the clock" bombing, the idea that the Germans would be bombed by the British at night and by the Americans during the daytime. General Eaker said that the Prime Minister rolled the words off his tongue like choice morsels--around the clock. The words also apparently settled it. "I decided," Churchill said, "to back Eaker and his theme, and I turned round completely and withdrew all my opposition to the daylight bombing by the Fortresses."[36]

This was not the only meeting Eaker had with the Prime Minister during the war. A year later he had another occasion to meet with him, and at the second meeting Eaker said that Churchill very graciously recalled the Casablanca confrontation and admitted he had been wrong about opposing the daylight bombing. Referring to Eaker's prediction that it would all work out, Churchill said, "the prediction you made to me at Casablanca last February about our combined bomber missions, including around the clock bombing, are now being verified. I no longer have any doubt that they will prove completely valid."[37] General Eaker went on to say that this confession of error on the part of the great Prime Minister was one of the greatest thrills he had all during the course of World War II. In fact, he said that it was second only to the joy he felt when the Luftwaffe failed to show on D-day.[38]

Arthur Harris' area bombing of Germany's great cities by night stunned the Germans at first, and one of the immediate effects was to emphasize the importance of home defense, an area that the Germans had been prone to neglect. As a result, not only was the production of fighters doubled, but airplanes and seasoned pilots were withdrawn from the fighting fronts and pressed into service in defense of the homeland. As Harris pointed out, "in 1941, when the Germans invaded Russia, the German army had the support of well over 50 percent of the whole German Air Force. At the end of 1943, the German Army had the support of less than 20 percent of the German Air Force."[39]

The night-defense system of the Germans grew increasingly efficient as the war progressed, and at the same time Bomber Harris became more and more frustrated by his inability to repeat the success he had enjoyed at Hamburg. Finally, out of desperation, he staked his all on a single great offensive against Berlin. Harris said that he might lose as many as five hundred bombers in the assault, but that he was sure the end result would be worth the price. The end result that Harris had in mind was the complete and utter destruction of the German capital and nerve center. Harris predicted that this would knock Germany out of the war and make an invasion unnecessary, except for the occupation of a defeated country. Anticipating how historians in the future would deal with the event, he called his great offensive the Battle of Berlin. Needless to say, Harris had the full support and backing of Churchill in this great attempt to win the war by making another Hamburg out of Berlin.

The battle began on November 18, 1943, with a massive assault on the seat of Nazi power and lasted four and a half months, or until the end of March 1944. It included sixteen missions concentrated on Berlin itself, and nineteen against other German cities, just to keep the enemy guessing. It was Harris' final all-out effort to win the war without an invasion. It was also a failure. The last mission in the series was against Nuremberg, and it was the worst from the point of view of Air Marshal Harris. Bomber Command lost ninety-five bombers in the attack, and the Germans lost no more than ten fighters. The lesson was clear. The Luftwaffe had regained control of the night skies over Germany.

Air Chief Marshal Harris did not leave Bomber Command in a blaze of glory. His failure to accomplish what he set out to do demonstrated either one of two things. One was that the Royal Air Force, without the solid support of the Americans, was not strong enough to accomplish its objective. Or, alternatively, that a war simply cannot be won without a sound theoretical doctrine as a foundation. This would suggest that all the theoreticians of terror bombing had been wrong from the outset. In any case, no great honors were showered on Harris at the end of the war, and not long after its conclusion the Chief of Bomber Command retired to Rhodesia.

The United States Strategic Bombing Survey did not come to the express conclusion that the area bombing of German cities had been a complete failure. However, like everybody else, the members of the Survey were amazed at the resilience and extraordinary ability of the Germans to recover from the effects of even the most ruinous attacks. "The recuperative and defensive powers of Germany," the Survey concluded, "were immense; the speed and ingenuity with which they rebuilt and maintained essential war industries in operation clearly surpassed Allied expectations."[40]

On the subject of civilian morale, the real target of the area attacks, the Survey said,

> The mental reaction of the German people to air attack is significant. Under ruthless Nazi control they showed surprising resistance to the terror and hardships of repeated air attack, to the destruction of their homes and belongings and to conditions under which they were reduced to live. Their morale, their belief in ultimate victory or satisfactory compromise and their confidence in their leaders declined, but they continued to work efficiently as long as the physical means of production remained. The power of a police state over its people cannot be underestimated.[41]

The percentage of the total air effort of the Allies that went into the bombing of city targets in the European War was impressive. Altogether, some two million seven hundred thousand tons of bombs were dropped, and of this total approximately 24 percent, or roughly one-fourth, was launched in the attack on the great cities. This was almost twice the weight of attack against all manufacturing targets, including the tonnages against oil, chemicals and rubber targets, as well as the aircraft factories, plus all the other targets, ball bearings, tanks, ordnance, ammunition, motor vehicles and steel.

The pay-off from the attack on the area targets was large in at least one respect -- physical destruction. "Tremendous physical destruction," the Survey said, "was caused and great areas of Germany's principal cities were laid waste. In sheer destructiveness, these raids far outstripped all other forms of attack."[42] The destructiveness was widespread as well as heavy. No major city of Germany, except in the extreme eastern part of the Reich, escaped being bombed. Altogether, each of the sixty-one German cities brought under attack had a population of one hundred thousand or more, and as a result about a third of the Reich's total population, including the five million of the country's labor force, were part of the battlefield.

The attacks on the German cities damaged or destroyed about 20 percent of the total residential units in the country, and rendered seven and a half million people homeless. They killed about three hundred thousand people and injured another eight hundred thousand.[43] "These attacks," the Survey said, "left the German people with a solid lesson in the disadvantages of war. It was a terrible lesson; conceivably that lesson, both in Germany and abroad, could be the most lasting single effect of the air war."[44]

However, a distinction should be made between the efficiency of Bomber Command as an agent of destruction and its military effectiveness. Obviously, it was not a war-winner. The merciless pounding from the air did not shatter the will of the German people to the point where their capacity for armed resistance was broken. The only remaining question is as to what extent, if any, did the attack on the great cities of Germany impair war production. In other words, did the disruption of life in the industrial centers of Germany significantly contribute to a weakening of the German armies at the front?

Hamburg is probably the best answer to the question of war production losses and city bombing. Not only was Hamburg the enemy's biggest port and second-largest city, but it was also the target for the single most devastating city attack of the war. In addition, it was the most successful attack from the point of view of material damage. "Yet," the Bomb Survey pointed out, "despite the deaths of over 60,000 people, the total destruction of nearly one-third of all houses in the city and the disruption of normal processes of living, Hamburg as an economic unit was not destroyed. It never fully recovered from the bombing but in five months it had regained 80 percent of its former productivity, despite the fact that great areas of the city lay, and still lie, in dust and rubble."[45]

The economic life of Hamburg was so much alive even after nine days of almost continuous bombing-- around the clock, as it were -- that the net result of the nine-day ordeal was a production loss of only fifty-four days. Translated into a percentage of annual production loss for the Reich as a whole, the total Hamburg loss amounted to only about one percent, barely a pin prick.

The Survey drew two obvious lessons from the experience of Hamburg. One was that the German cities were like the British cities (only more so, if possible) in that they "had a surprising resiliency and extraordinary ability to recover from the effects of ruinous attacks." The other lesson was that "as in the case of industrial plants, where it was found much easier to destroy the buildings than the machines within, so also it was much easier to destroy the physical structures of a city than to wipe out its economic life."[46]

Nor was this all. The Survey studied ten other major German cities that had been subjected to heavy area attacks and found the same resiliency and extraordinary ability to recover from area attacks. In fact, the other cities all had a better recovery record than Hamburg. The big majority of them recovered to 100 percent of their pre-attack capacity, whereas in Hamburg it had been only 80 percent, and in most cases the time required for full recovery was less than five months.[47]

The conclusion appears to be inescapable that the area bombing did not substantially affect the course of the German war production, one way or the other, and also that it did not affect the outcome of the war, again one way or the other. In fact, considering the cost of the area attacks--England devoted 40 to 50 percent of her war production to her air force -- a case might be made for the fact that the area bombing took more out of the British than it did out of the Germans. This is reminiscent of a comment Churchill made with respect to the German submarines toward the end of World War One. Looking at what it had cost England to fight them and what it had cost the Germans to put them to sea, Churchill said it was obvious "that for one war-power unit Germany had applied to the submarine attack we have been forced to assign fifteen or twenty."[48]

Can it be that this was also the case for the area attacks, that it took far more out of England's resources to mount them than they were worth? In other words, was the 40 to 50 percent drain on England's resources matched by a forty to fifty percent reduction of Germany's industrial strength? Of course the answer is no, and the conclusion must be that the results obtained by Bomber Command were not commensurate with the huge expenditures required to mount and sustain the attack.

The failure of Bomber Command was the result of a basic error in strategic planning. It is almost an axiom that wars cannot be won unless there is a sound doctrinal and theoretical foundation for the strategic concept, and as Douhet said, "the selection of objectives, the grouping of zones and determining the order in which they are to be destroyed, is the most difficult and delicate task in aerial warfare, constituting what may be defined as aerial strategy."[49]

Lord Cherwell's idea turned out to be an illusion and a failure, particularly considering the limited capabilities at the disposal of Bomber Command at the time the idea became the basis of directive. Had the capabilities of Bomber Command been ten times greater, or perhaps only two or three times greater, and had they been available at the outset of hostilities, the results might have been quite different. They might have been more like the one Churchill referred to when he said that "if the bombs of 1943 had been applied to the London of 1940, we should have passed into conditions which might have pulverized all human organization."[50]

This is a point that is often overlooked in connection with Douhet's theory or concept of strategic bombing, the point of capabilities, or force-in-being at the outset of hostilities. However, one of Douhet's most basic premises was that a large bombing force would be in existence at the beginning of the war, not a small, inexperienced Training Command, such as was the case with the British bombing force in 1939 and the Americans in 1942.

Massive and sudden were the two key words in the ideas Douhet expressed about Air Power. For example, in his *imaginary* war between France and Germany, Douhet had the issue being decided in the first 36 hours of combat. He portrayed the Germans as having fifteen hundred bombers at the opening of hostilities and of losing one-third of the total force in the first 36 hours of fighting. The outcome was a victory for the Germans. Douhet pictured France as clamoring for peace within 36 hours, after four of her great cities had been "transformed into flaming masses."[51]

There are shades here certainly of Hamburg's massive firestorms, of Hitler's fears that "further attacks of similar weight might force Germany out of the war," of Speer's oral report to the Führer, "that if these aerial attacks continued a rapid end of the war might be the consequence,"[52] of the failure of the Allies to follow through on more Hamburg-type attacks and on Speer's post-war reflections that by failing to follow through on the initial Hamburg attacks the Allies "once again allowed us to adjust ourselves to his strategy."[53]

What a far cry from Douhet's concept of a force-in-being at the outset of hostilities was the situation of England and the United States at the beginning of the war can be gleaned from some statistics compiled by the United States Strategic Bombing Survey. "It should be borne in mind," the Survey noted,

> that the air offensive did not attain major significance until the spring of 1944. Of the total tonnage dropped in the European war by the RAF and AAF, 83 percent was dropped subsequent to January 1, 1944. Of the tonnage dropped on Germany proper, in the neighborhood of 85 percent was subsequent to January 1, 1944. Perhaps even more significant is the fact that of all the tonnage dropped

on Germany, 72 percent was after July 1, 1944.[54]

The importance of these statistics is obvious. They show that the capabilities of both air forces prior to January 1, 1944, were not enough to enable either one or both of them to carry out a meaningful strategy, at least not as Douhet would define it. In other words, the Royal Air Force bombed along on a wing and a prayer for five years, and the American Air Force did the same thing for the first two years of its combat history. It is no small wonder that neither was able to accomplish anything meaningful, particularly in view of the fact that both bit off far more than they could chew in the way of effective target systems.

Britich Bomber Command never had the capabilities required for a Douhet-like offensive against civilian morale and had the added disadvantage of operating under a strategy that practically guaranteed failure against its secondary objective, the factories that supported the German fighting forces. The reason for this is that the British strategy required that the bombs be dropped as close as possible to the built-up centers of the cities attacked, and as a result the bombs did little or no damage to the city's industrial installations because, as the Survey pointed out, "as a rule the industrial plants were located around the perimeter of German cities and thus characteristically these were relatively undamaged."[55]

FOOTNOTES--CHAPTER V

[1]Major General J.F.C. Fuller, *Armament and History* (New York: Charles Scribner's Sons, 1945), 146.

[2]David MacIsaac, *Strategic Bombing in World War Two: The Story of the United States Strategic Bombing Survey* (New York: Garland Publishing, Inc., 1976), 179.

[3]*Ibid.*, 28.

[4]*Ibid.*, 51.

[5]*Ibid.*, 58.

[6]*Ibid.*, 51.

[7]The United States Strategic Bombing Survey, *Over-all Report: European War* (Washington: Government Printing Office, 1945), ix.

[8]*Ibid.*

[9]David MacIsaac, *Strategic Bombing in World War Two* (New York: Garland Publishing, Inc., 1976), 90.

[10]*Ibid.*, 90-91.

[11]Raymond G. O'Connor, Commentary in *Versatile Guardain: Research in Naval History* (Washington: Howard University Press, 1979, ed. Richard Von Doenhoff), 77.

[12]David MacIsaac, *Strategic Bombing in World War Two* (New York: Garland Publishing, Inc., 1976), 95-96.

[13]The United States Strategic Bpmbing Survey, *Over-all Report: European War* (Washington: Government Printing Office, 1945), ix.

[14]David MacIsaac, *Strategic Bombing in World War Two* (New York: Garland Publishing, Inc., 1976), 52.

[15]*Ibid.*, 51.

[16]The United States Strategic Bombing Survey, *Overall Report: European War* (Washington: Government Printing Office, 1945), 107; *Summary Report: European War*, 16.

[17]Wesley Frank Craven & James Lea Cate, *The Army Air Forces in World War II* (Chicago: University of Chicago Press, 1948), III, xi.

[18]*Ibid.*, 649.

[19]Giulio Douhet, *The Command of the Air* (New York: Coward-McCann, 1942, tr. Dino Ferrari), 50.

[20]George H. Quester, *Deterrence before Hiroshima* (New York: John Wiley & Sons, Inc., 1966), 119.

[21]Winston S. Churchill, *Their Finest Hour* (Boston: Houghton Mifflin Company, 1949), 330.

[22]*Ibid.*, 321.

[23]*Ibid.*, 331.

[24]*Ibid.*, 382.

[25]*Ibid.*, 380.

[26]*Ibid.*, 366.

[27]Winston S. Churchill, *The Hinge of Fate* (Boston: Houghton Mifflin Company, 1950), 92.

[28]P.M.S. Blackett, *Fear, War and the Bomb* (New York: McGraw-Hill Book Company, 1949), 17.

[29]Wesley Frank Craven & James Lea Cate, *The Army Air Forces in World War II* (Chicago: University of Chicago Press, 1948), I, 592.

[30]Roger W. Shugg and H. A. DeWeerd, *World War II* (Washington: *The Infantry Journal*, 1946), 59.

[31]The United States Strategic Bombing Survey, *Summary Report: European War* (Washington: Government Printing Office, 1945), 3.

[32]*Ibid.*, 3.

[33]Albert Speer, *Inside the Third Reich* (New York: The Macmillian Company, 1970), 339.

[34]Winston S. Churchill, *The Hinge of Fate* (Boston: Houghton Mifflin Company, 1950), 678.

[35]*Ibid.*, 679-80.

[36]*Ibid.*, 679.

[37]Ira C. Eaker, "Some Observations on Air Power, in *Air Power and Warfare* (Washington: Government Printing Office, 1979), 355.

[38]*Ibid.*, 353-55.

[39]M.S. Blackett, *Fear, War and The Bomb* (New: York: McGraw-Hill Book Company, Inc., 1949), 30.

[40]The United States Strategic Bombing Survey, *Overall Report: European War* (Washington: Government Printing Office, 1945), 107.

[41]*Ibid.*, 108.

[42]*Ibid.*, 71.

[43]*Ibid.*, 72.

[44]The United States Strategic Bombing Survey, *Summary Report: European War* (Washington: Government Printing Office, 1945), 4.

[45]The United States Strategic Bombing Survey, *Overall Report: European War* (Washington: Government Printing Office, 1945), 73.

[46]*Ibid.*, 72-73.

[47]*Ibid.*, 73.

[48]Winston S. Churchill, *The World Crisis 1916-1918* (London: Thornton Butterworth, 1923), II, ap. V, 309-13.

[49]Giulio Douhet, *The Command of the Air* (New York: Coward-McCann, Inc., 1942, tr. Dino Ferrari), 50.

[50]Winston S. Churchill, *Their Finest Hour* (Boston: Houghton Mifflin Company, 1949), 366.

[51]Giulio Douhet, "The War of 19--," as it appears in *The Command of the Air* (New York: Coward-McCann, Inc., 1942, tr. Dino Ferrari), 374-89.

[52]The United States Strategic Bombing Survey, *Summary Report: European War* (Washington: Government Printing Office, 1945), 3.

[53]Albert Speer, *Inside the Third Reich* (New York: The Macmillam Company, 1970), 339.

[54]The United States Strategic Bombing Survey, *Overall Report: European War* (Washington: Government Printing Office, 1945), 71.

[55]The United States Strategic Bombing Survey, *Summary Report: European War* (Washington: Government Printing Office, 1945), 4.

CHAPTER VI

PRECISION BOMBING IN EUROPE

The American bombing was a different breed of cats from the British. While not eschewing the prospect of shattering the will of the Germans to resist, the American attack was aimed primarily at the capabilities of the enemy. The idea was that whereas no one thing will win a war, the absence of some one thing may very well lose it. The American idea was to use the bombs as a sword, not as a club, and to use them not to flail away at the strength of the enemy, say his work force, but to use them as a precision instrument against some vital weakness of the enemy, say his ball bearing industry, aircraft production facilities, oil refineries, or transportation capabilities.

In other words, the American strategy was the vertical approach to bombing, as opposed to the horizontal. The essence of it was to find a target system that met two hard and fast rules. One was that the destruction of the targets would deny the enemy a capability to wage war effectively in some vital element, on land, at sea, or in the air, or possibly all three. And the other was that the targets that had to be destroyed and kept destroyed were ones that could be successfully handled, considering the capabilities and limitations of the force at hand, or planned. If a target system did not satisfy both of these requirements, it had to be rejected.

These rules of target analysis imposed a heavy burden on the planners of air strategy. First, they had to dissect the enemy economy from the point of view of the war effort as a whole, and second, they had to make an accurate estimate of their own bombing capabilities, such as size of force, weather, bombing accuracy and enemy opposition. Otherwise, the errors were liable to be gross in the selection of targets. Douhet did not exaggerate the importance of the task when he said, "it is just here, in grasping these imponderables, in choosing enemy targets, that future commanders of Independent Air Forces will show their ability."[1]

Applying the logic of these principles of target analysis to the problem of selecting targets that would be appropriate for the air offensive against Germany, the first thing the strategic planners had to take into consideration was the fact that the German economy was not only rugged and resilient, but that it also was not under any particular strain at this time. After all, the war was of its own making, and the signal to start the war had not been given until the economy was ready to support it with ample productive capacity and safe stockpiles of strategic materials.

Since then, of course, the conquest of Western Europe had been completed, and the German armies were deep inside Russia when the U.S. entered the war. All the industrial facilities of nearly every European country, including those in occupied Russia, were under Hitler's control, and in the Ukraine he possessed Europe's richest granary. Thus there were no shortages of natural resources, including strategic minerals, that faced Germany at this time, with the possible exception of natural oil and rubber, stockpiled to be sure. In addition to this, the population under German domination or influence exceeded three hundred million, with a labor force of over fifty million, to say nothing of the several million veteran troops, superably equipped, and completely dedicated to the Führer.

The German economy at the time of America's entry into the war was as strong as a rock. It was like any other economy in that there were built into it many safeguards against the effects of bombing. Chief among these was the strong civilian sector of the German economy. Hitler was not satisfied with guns alone, but wanted butter as well, and the German economy was organized to give him just that. He would not bide with any cutbacks in the production of consumer goods and was perfectly content with having industry operate on a single-shift basis. Relatively few women were drawn into the work-force, fewer than in World War One, and the average work week was shorter than it was in England or the United States. To cap it all, Hitler was so confident he had succeeded in Russia that large scale cutbacks in war production were ordered in September 1941.

Looking back at the record, John K. Galbraith expressed astonishment that Germany had lost the war. In a significant article for *Fortune* magazine (December

1945), appropriately entitled "Germany Was Badly Run," Galbraith said, "the simple fact is that Germany should never have lost the war it started; the under-mobilization and overconfidence of the early years help to solve the very real mystery of why it did." However, after America's entry into the war, and the German defeat before Moscow, the German planners began to take a second look at their economy from the point of view of war production, and one of the results was the appointment of Albert Speer to be the Minister of Armaments and War Production. He was given wide powers, and his mission, theoretically at least, was to mobilize the country for war. Total mobilization, however, at least according to the standards of the other belligerents, Russia, England and the United States, was not achieved, at least not at the outset. Women were still not brought into the work force in any large numbers, and there was no sudden transfer of labor from the non-essential to the vital industries. Hitler continued to insist on both guns and butter.

An example of this was recorded by Speer in connection with a segment of the economy essential to both civilian and military production --the steel industry. As late as 1942 the consumer industries in Germany were running only three percent below their peacetime level of consumption of steel. Speer wanted to cut back on the civilian economy and put the economy on a war footing, starting with steel allocations. Hitler went along at first, but after three months changed his mind, and Speer was told that "the fabrication of products for the general supply of the population must be resumed."[2] The fact that the German economy was functioning on a business-as-usual basis meant that there was a great deal of fat in it from the point of view of the war effort. This was good for the Germans, but very bad for the air planners. The fat was a kind of cushion that the Germans could use to offset the effects of strategic bombing.

A modern economy is a tough unit to crack even under the most favorable conditions. There is always a certain amount of slack in it because it is simply not economical to run plants and machinery at full capacity. They wear out too fast. Obviously, this slack is a cushion of safety against the effects of strategic bombing. However, in the German case there was more than the normal cushion in that the Germans could call on the capacity of Western Europe and occupied Russia. The enormous amount of fat in the German economy made the problem of target analysis far greater than would ordinarily have been the case.

Clausewitz quoted Napoleon as saying that war was like "a problem in algebra before which even a Newton might stand aghast."[3] The same thing could be said of the problem that faced the air planners in determining Germany's vulnerability to strategic air attack. Before the breaking point could be expected, how many bombs of what kind and weight would have to be delivered to what targets with what degree of accuracy? The problem of target analysis is one of capabilities. They are of two different kinds, however, since one involves the capabilities of the strategic air force to destroy targets, and the other the capabilities of the defending economy to offset the effects of bombing. It is the old question of the difference between the offense and the defense, which is the stronger form of warfare?

There are many vital factors that have to be taken into consideration in determining the capabilities of a strategic air force. The most obvious of these are its size, range of action, ability to penetrate the enemy's defenses, and the accuracy of its bombing once the targets have been reached. What it all adds up to is bombs on targets. That is, considering everything, the weather, fighter defenses, anti-aircraft fire, how many bombs of what mixtures and weight can be put within the magic circles of the targets every day, week, or month, on an all-year basis? The answer to the question is the definition of bombing capabilities, and from this answer the planners can proceed with the selection of targets. This is the cost factor of strategic bombing. If the cost is too high for any particular target system, that is beyond the capabilities of the strategic air force, the target system has to be ruled out. As Clausewitz said, "a short jump is certainly easier than a long one, but no one wanting to get across a wide ditch would begin by jumping halfway."[4]

Obviously, the British were wrong in their selection of targets for the Royal Air Force. The destruction of three and a half million dwelling units, approximately 20 percent of the total in Germany's principal cities, and the casualty toll of over one million civilians, including three hundred thousand fatalities, were, contrary to the expectations of Churchill and predictions of Lord Cherwell, not enough to break the spirit of the German people, and it did not fatally weaken their capacity to wage war. Thus the British bombing effort must be reckoned a stragegic failure. In other words, the complete destruction of the principal cities of Germany, the sixty-one cities with a population of one hundred thousand or more each, was a target system that was beyond the capabilities of the Royal Air Force. Its selection, therefore, meant that roughly one-fourth of the total tonnage of bombs dropped on Germany was a wasted effort.

The failure of the Royal Air Force to achieve its objective, that is to break the will of the German people by area bombing and night attack, was anticipated by the American planners, which is why they did not throw in their lot with the British, but insisted upon following their own ideas about the strategic air offensive.

These ideas were not new in Air Force thinking but were part of a doctrine that had its origins in a concept that was almost as old as the idea of Air Power itself. This was the idea that daylight bombing of precision targets was within the capabilities of the Air Force, and that by such action it would be possible to accomplish the destruction and dislocation of the enemy's military, industrial and economic system to the point where his capacity for armed resistance would be fatally weakened.

To figure out which targets or target systems the destruction of which was within the capabilities of the American Air Force, and which would bring about a fatal weakening of the ability of the Germans to wage war, was a very difficult problem. Nevertheless, it was one that had to be solved, and fortunately for the Air Force it was a problem that had been under careful study for a period of a decade and a half, or ever since the founding of the Air Corps Tactical School on August 18, 1926. In the old Tactical School the problem had been stated again and again that the primary and principal mission of air power was "the attack on those vital objectives in a nation's economic structure which will tend to paralyze that nation's ability to wage war and thus contribute directly to the attainment of the ultimate objective of war, namely, the disintegration of the hostile will to resist."[5]

Analyzing the economic structure of Germany from the point of view of pinpointing the vital objectives the destruction of which would paralyze the nation's ability to wage war was the task of the tiny unit that General Haywood S. Hansell had established in the intelligence section of the Air Corps in 1939. Starting from scratch, and not without opposition from the War Department's General Staff, the air intelligence specialists began their study of the German economy. The only policy guidance General Hansell gave the air intelligence specialists was that they should look for targets the destruction of which would wreck the German capacity to wage war. This put the spotlight on German industry, not on German morale. It was exactly the opposite of what became the official policy underlying target selection for the British.

The first and most obvious step in analyzing the economy of Germany from the point of view of its destruction was to bring it into focus. Nothing like this had been done in the past, except from a theoretical point of view. As a practical problem, it was a pioneering effort. Ideally, and this was the view of the air force planners, the real target was not the economy of Germany as a whole, but only that segment of it which supported the armed forces. The key question was to find out what was the keystone in the arch of the German economy, the vital segment that, if destroyed, would bring about the collapse of Germany. While this was only a theoretical question in the long years between 1926 and 1939, it became a very practical one after the outbreak of World War II. As a result the tiny unit that General Hansell had established in the intelligence section of the Air Force took on an added significance.

Answering the question meant taking the country apart industry by industry, and plant by plant, not only to determine the feasibility of destruction from the air, but also to make allowances for recuperation after attack. It stood to reason that the Germans would do everything they could to offset the effects of strategic bombing, and the longer the question was studied the more apparent it became that the number of alternatives the Germans would have were almost (but not quite) endless. In other words, it quickly became apparent that the very question itself, a keystone in the arch of the economy, was more of a figure of speech than anything else. Wishful thinking would not make the collapse of the German economy an easy task. It was a tough nut to crack, and no amount of daydreaming could belie that stubborn fact.

The idea of a keystone in the arch of the German economy intrigued everyone who considered the subject in England and the United States. And there were many who did so. In fact, all kinds of new agencies, as well as new parts of established agencies, sprang up to do research on the problem of target analysis. In the United States the Office of Strategic Services (forerunner of the Central Intelligence Agency) and the Board of Economic Warfare were two such, and in England the Air Ministry spent a lot of time making analyses of the German war economy. An Air Ministry study in 1939 took a leaf out of the organismic theory of the state and developed the idea that there were "vital spots in industry as well as in the human body."[6] From this there followed a whole series of "bottle-neck" studies and quick solutions to the problems of target analysis. Another favorite term for the panacea targets was "Achilles' heel." None of them would bear a close examination, however, and eventually they were all exposed as frauds. The only agency that knew what it was doing, and consistently stuck to practicalities from the point of view of bombing objectives, was the small unit in the target information section of Air Intelligence. This was the one that had been activated by General Haywood S. Hansell in 1939 and was closely supervised for the duration of the war by Colonel Malcolm W. Moss.

The Air Intelligence specialists made an important contribution, not only to winning the war, but in another way as well. They worked out a philosophy of air power that has stood the test of time, and which is standing up extremely well in the face of the new challenge posed by the nuclear weapons. However, their work in the field was suddenly brought up short by an unexpected development in the Joint Intelligence Committee. At the time, there was no Air Force representative on the Joint Intelligence Committee, and out of the blue the Committee suddenly challenged all the basic assumptions underlying the selection of targets, saying that industrial and economic information could not be considered military intelligence. General Hansell said that his charge hit at a very vulnerable sector in Air Intelligence since "strategic air intelligence, which centered upon industrial and economic sources of military strength, was a new outgrowth in the field of military intelligence and dealt with dimensions new to warfare."[7]

The military, particularly the older services, do not take kindly to "dimensions new to warfare," and thus the Air Force had to take seriously this challenge to their industrial and economic intelligence. The job of doing so fell to General Muir S. Fairchild, who later took a leading role in organizing the United States Strategic Bombing Survey. On this occasion the staff work done by General Fairchild resulted in the formation of what came to be known as the Committee of Operations Analysis. This was a distinguished group of high-level representatives from the intelligence sections of the Army and Air Force, two officers from Management Control, representatives from the Office of Strategic Services, and from the Board of Economic Warfare, as well as four outstanding civilians.

One of the hardest working members of the Committee was Elihu Root, Jr. of the New York financial firm of Root, Clark, Buckner and Ballantine. Another was the distinguished scholar from the Institute for Advanced Study at Princeton, Dr. Edward Mead Earle. The third civilian member was Thomas W. Lamont of J.P. Morgan and Company. He joined the Committee on January 7, 1943, and participated in the writing of the final report. The fourth member, added at the suggestion of Professor Earle, was Noel Frederick Hall, from the Washington Office of the British Ministry of Economic Warfare, formerly professor of London University and director of Britain's National Institute for Economic and Social Research. The directive establishing the Committee of Operations Analysts was signed by General Arnold on December 9, 1942. It was a very poorly prepared directive, and not only asked a question that was clearly impossible to answer, but asked for it on the basis of an assumption that was directly contrary to basic Air Force thinking.

General Arnold charged the Committee with producing "a report analyzing the rate of progressive deterioration that should be anticipated in the German war effort as a result of the increasing air operations we are prepared to employ against its sustaining sources," and with also producing "as accurate an estimate as can be arrived at as to the date when this deterioration will have progressed to a point to permit a successful invasion of Western Europe."[8]

The first question was clearly an impossible one to answer because no one knew with any degree of certainty what targets would be attacked in the strategic air offensive, much less how successful the attack would be. And as if these were not enough to complicate the problem, General Arnold's second question was stranger still. This was to determine when the attack would have softened up Germany to the point where a successful invasion of Europe would be possible. All Air Force thinking was to the effect that no invasion would be necessary if the air offensive succeeded in accomplishing its objective.

Nevertheless, the Committee of Operations Analysts set to its task with vigor, and it was a difficult task indeed, not only because of the impossible directive, but because the Committee was just that, a committee, not an operating agency with files and personnel of its own. In this respect it was unlike the United States Strategic Bombing Survey, and greatly unlike any of the old or new agencies of the government such as the Office of Naval Intelligence, the Air Intelligence Division, the Office of Strategic Services, the Board of Economic Warfare and the Air Ministry. The result was that the Committee of Operations Analysts could not do any of its own spadework but had to depend on the work of others, not only for the basic factual data, but also for the principles of analysis. Thus to a very great extent the Committee of Operations Analysts was a captive of the very agencies whose work was being questioned, particularly the Target Information Section of the Air Force.

As soon as the organizational work was out of the way, the Committee of Operations Analysts farmed out all its spadework to the new and old-line agencies of the government and depended on them to make the target studies required. By the end of February 1943, nineteen of these studies had been or were nearing completion, and on the 8th of March, the Committee reported its findings to General Arnold. The Committee begged off answering the two unanswerable questions General Arnold had posed in his directive and gave good reasons for doing so, namely that there was no factual basis on which to make a determination of the forces that would be made available to carry out the strategic air offensive, and that even if this were made known, the questions would still be unanswerable because there was no way to forecast with any degree of accuracy the success the forces would have in carrying out their mission.

However, the Committee did come to a conclusion that made the whole effort worthwhile as far as Arnold and Fairchild were concerned. This was the finding on the issue of the strategic air Offensive itself, whether it was worthwhile pursuing. After all, this was the issue that had been raised by the Joint Intelligence Committee, and the one that had led to the formation of the Committee of Operations Analysts in the first place. On this crucial point, the Committee was forthright in saying,

> (1) The destruction and continued neutralization of some sixty targets would gravely impair and might paralyze the Western Axis war effort. There are several combinations of targets from among the industries studied which might achieve this result; (2) In view of the ability of adequate and properly utilized air power to impair the industrial sources of the enemy's military strength, only the most vital considerations should be permitted to delay or divert the application of an adequate air striking force to this task.[9]

In addition to this sweeping endorsement of the strategic air offensive as a perfectly valid concept of war, which was the main thing Arnold and Fairchild wanted to establish, the Committee of Operations Analysts submitted their findings with respect to a long list of target systems. The Committee also added a point about the principles of target analysis. This was that the "Eighth Air Force was aware of and in agreement with the principles of target selection set forth" in the report.[10] The principles themselves, as the official Air Force history points out, were those that had been outlined and presented to the Committee of Operations Analysts when it began its work in December 1942. They were largely the result of the work done in the Air Force going back to 1926, sharpened and brought up to date by the specialists in the Air Intelligence Division under the direct supervision of Colonel Malcolm W. Moss. This was the target unit that General Hansell had activated in 1939.

One of the principles of target analysis that particularly intrigued the Committee of Operations Analysts was that of minimum requirements. "It is better," the report said, "to cause a high degree of destruction in a few really essential industries or services than to cause a small degree of destruction in many industries. Results are cumulative and the plan once adopted should be adhered to with relentless determination."[11]

The concept of minimum requirements was not anything new in March of 1943 when the Committee of Operations Analysts called attention to it in the report to General Arnold. Ever since 1939 minimum requirements had been taken into consideration in all the air estimates, and no one knew better than the air force planners that every industry and vital service had a cushion of production capabilities that served to protect the industry or service against the effects of bombing, and therefore that it was far better to effect a high degree of destruction in a few really essential industries or services than to create a small degree of destruction in many industries or services.

The German transportation system is a good case in point. While it was an essential service, of this there can be no doubt, and while it contained many targets exposed to destruction from the air, it was also not a good target system prior to the spring of 1944 because of the factor of minimum requirements. Germany entered the war with a very strong transportation system, one of the best in the world. It had been designed with a large over-capacity in lines, yards, locomotives, cars, and marshalling facilities, all built up to serve civilian requirements as well as military needs. The result was that the Germans had a capacity to move all kinds of traffic, military and civilian, essential and non-essential, far beyond their minimum requirements. It was estimated that the safety factor in this vital sector of the economy was above thirty percent. In other words, the German transportation system, vital as it was to the war effort, could absorb losses up to thirty percent without undue hardship. This put it light years away from any damage that could have been done to it until the final eighteen months of the war.

Thus the German transportation system in the early years of the war was a perfect example of how the concept of minimum requirements could and did shield a vital segment of the German war economy from the effects of strategic bombing. A premature attack on the German transportation system would have produced negative to minus results. It was not a point of strategic vulnerability for the simple reason that the German transportation system was not an exploitable weakness, until the bombing forces had grown large enough to take on such a formidable adversary or task. This was not until the middle or latter half of 1944. Any effort to peck away at the German transportation system with the limited bombing capabilities available prior to 1944 would have been the height of folly.

The concept of minimum requirements was not always as clear cut and as absolute as it was in the case of the German transportation system. In some of the other vital segments of German industry the concept of minimum requirements was affected by other factors such as stockpiles, imports, substitutes and unused capacity, including stand-by facilities, or even a device as simple as putting the industry on a triple-shift basis. In other words, while minimum requirements was a constant factor in target selection, it was also one of many variables, and the one that came to the fore in one industry was not necessarily the one that would prove decisive in another. Oddly enough, the German aircraft industry is a good example of how excess capacity can be the determining factor in dictating the outcome of what otherwise was a well-planned aerial offensive. The purpose of the attack on the aircraft industry was to break the back of the Luftwaffe by destroying it at its source. What the planners failed to take into consideration was the fact that there was a large factor of excess capacity in the industry and the additional fact that the excess capacity was greater in airframes (the main object of attack) than in engines.

The attack on the industry was made during a fortuitous spell of good weather in the last week of February 1944. The attack lasted from the 20th to the 25th, and in the history of the Air Force it will probably always be known as the "Big Week." During it, nearly four thousand tons of bombs were dropped on aircraft plants in Germany. This tonnage was roughly equivalent to the total tonnage dropped by the Eighth Air Force in the first year of its operations.[12] The results of the air offensive against the aircraft industry were impressive from the point of view of physical damage. Fully seventy-five percent of the buildings in the industry were damaged or destroyed, and it looked as though the assembly lines would have to come to a halt. However, this is not what happened.

There was not a proportionate destruction of vital machine tools, and the assembly lines were soon started up again with seemingly greater efficiency. In fact, as noted by the Bomb Survey analysts, production was higher in March, the month after the attack, than it was in January, the month before. "Part of the explanation," the Bomb Survey said, "was the excess capacity of the airframe industry which, as noted, was considerable. Excess capacity in airframes was considerably greater than in engines."[13]

Another important factor the air intelligence specialists had to take into consideration in connection with every target system was the time lag. To do so, they visualized the German economy as one composed of three horizontal layers of industry. At the bottom were the industries buried deep in the economic process, the basic industries, such as steel, the bombing of which would not have any effect on front-line strength for a year or longer. In the middle were the industries such as general purpose machine tools, the destruction of which would show up on the front lines in less than a year, but not in a matter of weeks, or even a few months. The top layer was composed of those industries that were relatively close to the front lines, such as transportation or electric power. The end-product industries also included those that produced the fighting weapons such as submarines, aircraft, tanks, guns, ammunition and all the other expendable items of warfare. Of course, these were the most important industrial segments from the point of view of the time lag.

Obviously if a time reference was an important factor in picking a target system, say three months before the invasion of Normandy, it would make no sense at all to consider the basic industries from the point of view of selecting an appropriate target system. Thus even if it were suddenly discovered that steel was an important bottleneck, which was not the case in Germany at all, it would still not have made the steel industry an attractive target system on the eve of the invasion of Normandy. This is why so many eyebrows were raised when the American Fortresses suddenly appeared over Czechoslovakia's great Skoda works on April 25, 1945 and proceeded to destroy the plant. Up to this time the great arms and engineering works, the pride of Czechoslovakia, had escaped almost undamaged. Why bomb it now? The war was practically over. German armies were surrendering everywhere, and Hitler was trapped in his bunker in Berlin. Not only that, but American land forces were approaching Prague from the west and Soviet armies from the east. Why pick this time to destroy Czechoslovakia's greatest industrial asset? Its destruction could not possibly affect the outcome of the war one way or another and would complicate the problem of putting Czechoslovakia back on its feet after the war.

The timing of the attack on the Skoda works raised the question of whether bombing had become so ingrained in the minds of the airmen as to lead to such attacks whether they served any military purpose or not. In other words, was the attack on the Skoda works merely another Dresden, bombing for the sake of bombing, or was there a motive beneath the surface? A quip was current in Prague at the time that the attack on the Skoda works at this time was probably "made either to destroy an industrial competitor, or to prevent these great works falling into the hands of the Soviet Union."[14]

Aside from the important factors of timing, capabilities, minimum requirements and the time lag, not to mention imports, synthetics and substitutes, the air intelligence specialists had to take into consideration another important factor in connection with all potential target systems--physical vulnerability. This involved the location of the potential targets, whether they were in range from existing bases, whether there was a high degree of concentration in the potential target system, or alternatively dispersion, but most of all whether the tonnage and types of bombs available were capable of effecting the physical damage required to make the attack worthwhile in terms of production loss. This meant damage not only to the superstructure, but to the vital machinery and equipment inside the superstructure.

The German sub pens are a good example of target system that went wrong because of the factor of physical vulnerability. At the time of the Casablanca

Conference the submarines were put at the top of the list of bombing objectives for what appeared to be a very good reason. This was the altogether certain fact that Germany would win the war if some way could not be found to eliminate the menace of the U-boats. At that time (January 1943), the German submarines were winning the Battle of the Atlantic, and as Churchill said, "never for one moment could we forget that everything happening elsewhere, on land, at sea, or in the air, depended on its outcome."[15] In other wrods, the Battle of the Atlantic had to be won, or the war would be lost.

But it did not follow from this that the German sub pens would make a good target system. Just the opposite turned out to be the case. The twelve or more feet of heavily reinforced concrete that protected the operational bases of the U-boats made the sub pens, the worst possible objectives for air attack. The heaviest bombs available at the time simply bounced off the concrete shelters that had been built to protect Hitler's U-boats. It was a very bad target system. The factor of physical vulnerability should have ruled it out.

Unfortunately, the air planners paid too little attention to the factor of physical vulnerability all during the war. Production losses were assumed when, in fact, they did not exist. It did little or no good to do extensive damage to building structures and leave the vital machinery intact. In fact, a good case can be made for the fact that it would have been better in many instances to have stood the bombers down rather than send them out on pointless missions. The German Armaments Minister, Albert Speer, was a genius at improvisation and won many a round with the air planners as a result of the miracles he wrought in bringing "bombed out" plants back into operation without significant or harmful production losses.

All in all, the air intelligence specialists had their work cut out for them in attempting to analyze the vulnerability of Germany to strategic air attack. To this end, a compilation was made of a number of target systems the destruction of which would accomplish the desired objective. Each of these studies emphasized and highlighted three cardinal points. One was the importance of the industry as a whole to the enemy war machine. Second was a detailed analysis of the minimum requirements of the enemy, taking into consideration the available stockpiles, the unused capacity in the industry, possible substitutes, the practicality of imports, supplies in the pipeline, the economies that could be effected in the civilian economy, the critical point in production losses, as well as the length of time that would be required to rebuild the plants. Finally, the industry was analyzed from the point of view of determining when and to what extent the destruction of the targets would affect front-line strength, taking into consideration the various level of attrition from the highest in an active front to the lowest.

Th official Air Force historians listed a series of questions that were considered by the air intelligence specialists in their target studies:

> (1) What are the minimum requirements of the enemy? In other words, at what point would a shortage impair the front-line military effort of the enemy? (2) What are the production capabilities of the enemy? If every facility in every European country, including occupied U.S.S.R., were utilized to the utmost, what would be the total? (3) How nearly self-sufficient is the enemy? That is, what relationship exists between capacity and minimum requirements, taking into consideration stockpiles and available substitutes? (4) Where are the enemy plants located and what percentage of total capacity is represented by each plant? (5) What are the physical characteristics of the installations themselves? To what extent are the buildings and machinery structurally resistant to high-explosive and incendiary attack? To what extent are they replacable? (6) What is the time lag between the destruction of each plant and the effect on front-line strength? (7) What is the force required to effect the necessary destruction? What, in short, will be the cost?"[16]

When the panacea targets were put under the searchlight of these hard questions some sobering results were bound to surface. Also, when the bubble of an idea burst it was not unnatural or unusual for the proud originator to cry foul and to say in rebuttal that the spoiler was an unimaginative defeatist. In fact, the Air Force historians made note of the fact that some observers at the time were critical of the air estimates because they "seemed to smack of defeatism."[17] However, as later events clearly demonstrated, the error in the air estimates was not one of defeatism, but of failure to give the Germans their due. They turned out to be not only rugged and resilient, but a very ingenious people from the point of view of minimizing the effects of bombing. As the Bomb Survey noted: "the recuperative and defensive powers of Germany were immense; the speed and ingenuity with which they rebuilt and maintained essential war industries in operation clearly surpassed Allied expectations."[18]

Out of all the deliberations on the subject of target analysis a system finally emerged that held great promise. It was small enough in size to be within the capabilities of the very small bombing force that was available in 1943, and if the targets were destroyed the effects would be immediate and possibly catastrophic, affecting the German armaments industry as a whole, with particular reference to the aircraft industry. It was the German ball-bearing industry.

The designation of the ball-bearing industry as the first essential element of the German war economy as a whole to be brought under attack by the Eighth Air Force was a decision of far-reaching importance. It came out of the deliberations of a Planning Team that General Eaker had organized to prepare the plan of operations and to select the bombing objectives for the Eighth's Bomber Command. Chairman and Director of this Planning Team was General Haywood S. Hansell. General Hansell said that all the members of the Planning Team, including the British, accepted with unanimity and enthusiasm the proposal to put ball bearings high on the list of priority target systems. He said that the British were particularly interested in the ball bearings because of a minor incident that had occurred as a side effect of the German bombing of Coventry on November 14, 1940.

A major aircraft engine plant in Coventry had been hit during the German attack, and although the damage was only slight, the bomb blasts had blown out the window glass exposing many trays of ball bearings. A heavy rain had soaked the bearings and rusted them, bringing production in the plant to a standstill, because no reserve of bearings was available. General Hansell recalled the British as saying, "what would have happened if the production capacity of the ball bearing factories themselves had been destroyed or heavily damaged?"[19]

The best part of the ball-bearing industry as a target system was its physical vulnerability. Within the confines of a single city within bombing range, Schweinfurt, there was concentrated one-half of Germany's total production, plus one-half of all the skilled workers. The balance of the productive capacity was also concentrated in five smaller plants at Steyr, Cannstatt, Erkner, Leipzig and Elberfeld. There was every reason to believe that the plants at these six locations could be knocked out and kept knocked out, even with the small force available at the time.

But the high degree of physical vulnerability of the ball-bearing plants was not the only factor that argued in favor of the industry as a target system. Speer himself called the industry "a bottleneck in our efforts to increase armaments production"[20] and lived in fear, as he said later, that this vulnerability, or exploitable weakness, in the war economy, would become the object of bombardment. "In those days," he recalled later, "we anxiously asked ourselves how soon the enemy would realize that he could paralyze the production of thousands of armaments plants merely by destroying five or six relatively small targets."[21]

The ball-bearing industry had other favorable characteristics as a target system. One of these was the fact (or so it was supposed) that there were no stockpiles of ball bearings in Germany, as there were none in England or the United States, and thus a sudden, large-scale shortage of bearings would force a shutdown of the plants producing tanks, airplanes, submarines and other forms of weaponry, just as had been the case with the aircraft engine plant at Coventry. It was also thought that the German civilian economy, fat as it was, could not be relied upon to provide a cushion against the bombing because the highly specialized ball bearings required by the armaments industry were not freely interchangeable with those in the civilian economy. In this respect, the ball-bearing industry was a different breed of cats from say the electric power industry, the transportation industry, or even the basic steel industry. Industrialists in England and the United States also felt that it was highly impractical to think that the Germans could do anything in the way of redesigning their ships, tanks, or airplanes in such a way that they would use fewer ball bearings.

The air planners were not alarmed about the prospects of Germany's being able to obtain ball bearings from outside the country. The plants in France, Italy and Austria could be kept under close aerial surveillance, and if necessary bombed along with the German plants. As for Sweden and Switzerland, the planners were assured by the State Department that enough pressure would be brought on them to practically guarantee that they would not give Germany any meaningful aid. Finally, there was the matter of the time lag between the bombing attacks and the effects on front-line strength. Here again, the prospects looked good. The pipeline between the ball-bearing plants and the armaments industry was not long, a month or six weeks at most, and after that the assembly lines would have to slow down, at last coming to a full stop. Beyond this point, there would be no replacements for battle losses and normal attrition with the result that Germany would be facing a bleak future in all the elements, at sea, in the air and on the ground.

In short, the ball-bearing industry looked like the promised land for the American stratetic offensive in the air. It held out the prospects of dealing the Germans a crushing setback, if not a fatal blow, and all it required to achieve success was the complete destruction of the big plants at Schweinfurt, plus the five other small targets at other locations in Germany.

The initial engagement in the war against the ball bearings, which was a war against the German armaments industry as a whole, was the Battle of Schweinfurt. It began on August 17, 1943, when some two hundred Flying Fortresses struck what the United State Strategic Bombing Survey called "the first great blow aimed at the complete destruction of an entire and essential segment of the German war economy. The segment selected was the anti-friction bearings industry, and the specific target was Schweinfurt."[22] The attack was a huge success from all outward appearances, though the battle was a hard one, and American losses were severe. Eighty direct hits were scored on the ball bearing plants, and a total of over four hundred tons of bombs was dropped. In economic terms, the production loss immediately after the attack was thirty-eight percent. On the other hand, the attack of August 17 was something of a failure from the point of view of the duel between the Flying Fortresses and Germany's air defenses, the German fighters and anti-aircraft batteries. One of the reasons this was so was because the element of surprise was completely lacking in the American attack, and another was the fact that the American planners diverted half of the bombing forces to another target on the day of the mission. These two factors made the job of defending Schweinfurt considerably easier for the German fighters and anti-aircraft gunners.

The reason the element of surprise was lacking when the Fortresses launched the Battle of the Ball Bearings was that the Germans had been expecting the attack for a long time. "As early as September 20, 1942," Speer said,

> I had warned Hitler that the tank production of Friedrichshafen and the ball-bearing facilities in Schweinfurt were crucial to our whole effort. Hitler thereupon ordered increased anti-aircraft protection for these two cities. Actually, as I had early recognized, the war could largely have been decided in 1943 if instead of vast but pointless area bombings the planes had concentrated on the centers of armaments production.[23]

Ball bearings were such a "bottleneck"-- the word in Speer's--in the German efforts to increase armaments production that plans had been made early in 1943 to double the output of bearings. Thus when the American bombers finally came to Schweinfurt it was not only no surprise to the Germans, but they were waiting for them with everything they had in the way of defense against bombers, the cream of their fighters and the best of their ground gunners.[24]

In addition to the disadvantage of having the Germans waiting for them at Schweinfurt, the Flying Fortresses had another obstacle to overcome. This was the mistake the planners made in dividing the bombers on the day of the Schweinfurt attack. This flew in the face of the oldest principle of warfare, that of concentration of force. It also flew in the face of common sense and was the only thing that saved the Germans from what Speer said would have been a "catastrophic blow."[25]

Looking back on the events of August 17, 1943, Speer said,

> In this very first attack the other side committed a crucial mistake. Instead of concentrating on the ball-bearing plants, the sizable force of three hundred seventy-six Flying Fortresses divided up. One hundred and forty-six of the planes successfully attacked an airplane assembly point in Regensburg, but with only minor consequences. Meanwhile, the British air force continued its indiscriminate attacks upon our cities."[26]

Largely because of the combination of these factors --dividing the force and forfeiting the advantage of surprise--but also because of the fact that the bombers were operating without the protection of fighter escorts, the Fortresses had thirty-six of their number shot down over Schweinfurt, most of them by German fighters. This was a loss of eighteen percent, nearly four times what was an acceptable loss rate. In fact, a loss rate this high would be prohibitive in the long run because its cumulative effects would be catastrophic.

The second attack on Schweinfurt came two months later, on October 14, 1943. It was carried out by two hundred twenty-eight bombers, and again they were unescorted by protective fighters after they reached the German border. The result was a massacre. The German fighters had a field day. Wave after wave of enemy fighters attacked the American bombers on their way to Schweinfurt and back again to the border, where the Fortresses came under the protection of their short-range fighters. It was one of the greatest battles in Air Force history, and also one of the most costly. This time sixty-two bombers were lost, twenty-seven percent of the entire force, and one hundred thirty-eight were damaged, some beyond repair. Six hundred crewmen were killed or missing in action. Obviously, such losses could not be sustained, and any attempt to do so would result in having the bomber force wiped out entirely and in short order. Naturally, Goering was jubilant and interrupted a conference Hitler was having with Speer to inform the Führer that the Luftwaffe was once more in command of the skies over the Fatherland. Goering said it was a great victory for Germany and pointed to the pleasant fact that "the countryside was strewn with downed American bombers."[27]

Hitler was elated and could not contain his joy. But Speer could. Getting a first-hand report of the damage to the plants at Schweinfurt, Speer was aghast. Despite the bomber losses on the way in and over the target, the bombing was heavy and accurate. As the Bomb Survey reported,

> Flak over the target was intense, but good visibility enabled the Fortresses to make an accurate bomb run and more than 450 tons of high explosives and incendiaries were dropped in the target area. Heavy damage was inflicted on the major plants, particularly in the ball manufacturing departments. Some 350,000 square feet of buildings were destroyed, and more than 1,000,000 square feet damaged. Ten percent of the machinery was destroyed or damaged and 20 percent of the finished stocks rendered useless.[28]

The August and October missions to Schweinfurt were important building blocks in the very proud tradition of the Air Force that no mission was ever turned back by enemy opposition. They also served to substantiate the feeling Prime Minister Stanley Baldwin expressed in 1932 that "the bomber will always get through."[29] However, considering the bomber losses on both missions, eighteen percent on the first and twenty-seven percent on the second, the Eighth Air Force had no alternative to doing what it did, which was to stand the bombers down from any more deep penetrations of Germany until their long-range fighters appeared in Europe, fighters that could escort and protect the American bombers, not just to the border, but all the way to their targets and back again to their bases in England. This suspension of all deep penetration missions on the part of the American bombers until the long-range fighters appeared on the scene killed the idea, of course, of the Flying Fortresses being flying fortresses. They were not.

But this was not all. Standing down the bombers for the long period that was required to get the long-range fighters -- it was four months before Schweinfurt was attacked again-- meant that the Germans got all the time they needed to get back on their feet from the point of view of ball bearings. They did this, according to the Bomb Survey report, "with great energy and skill."[30] In fact, the success of the counter-measures the Germans took to offset the effects of the attack on this segment of their war economy meant that the Germans, for all practical purposes, won the Battle of the Ball Bearings, or the Battle of Schweinfurt, whichever designation will turn out to be the more appropriate one in the annals of history.

In any case, the steps Albert Speer took to cope with the shortage of ball bearings were precisely those that had been anticipated by the air intelligence specialists several years in advance of the event. For this, the air estimates had been disbelieved because "to some observers they seemed to smack of defeatism."[31] However, when the chips were down the Germans did exactly what had been predicted, and the results, sad to say, were also exactly what had been predicted. In the first place, the pipeline in the ball-bearing industry was good for six to eight weeks, a fortnight beyond what had been predicted. Speer said that after these reserves were used up the sparse production from the plants "was carried daily from the factories to the assembly plants, often in knapsacks."[32] Steps were also taken immediately to import ball bearings from Sweden and Switzerland, steps that Speer said met with indifferent success, and other steps were taken to reduce requirements by substituting slide bearings for ball bearings wherever possible. Some equipment was actually redesigned to substitute other types of bearings. "On certain machines," Speer said, "we managed to conserve 50 percent of the ball bearings."[33]

This was not all. After the second big attack on Schweinfurt on October 14, 1943, all Germany went into high gear to conserve on the use of ball bearings in order to keep the vital armaments industry functioning.

Contrary to the belief of the air intelligence specialists, great economies could and were effected by a program of substitution and redesign in the civilian sector of the economy, and the excessive and often luxurious use of ball bearings in the war economy was considerably reduced. In addition, Speer appointed one of his most vigorous and able associates, General Manager Kessler, as the czar of the bearings industry, with full authority to requisition anything and everything he needed to get the desired results. The results were little short of miraculous. Kessler imposed strict controls on everything connected with the industry, from machinery and machine tools to buildings, plants, raw materials, requirements, personnel and to the use of every single bearing. "Requirements for bearings," the Survey said,

> were drastically reduced by substitution of plain bearings and redesign of armaments. Stocks of finished bearings on hand and in user plants, together with orders and deliveries, were strictly controlled. Finally, vigorous protective measures, including the prompt dispersal from large centers of production to numerous smaller plants, the construction of underground plants, the bombproofing and erection of blast walls around vital machinery and the rapid repair, or replacement of damaged or destroyed machery and other equipment, enabled production to return to adequate levels before the cushion provided by stocks on hand or in the pipeline between producers and users were completely consumed.[34]

Just as General Manager Kessler was getting the production of ball bearings up to an "adequate" level the picture suddenly changed again. This was in Febryary 1944 after the long-range American fighters made their appearance in England (December 1943), and the bombers could have fighters to escort them to targets throughout Germany. "Then," according to Speer (referring to February 1944),

> within four days, Schweinfurt, Steyr and Cannstatt were each subjected to two successive heavy attacks. Then followed raids on Erkner, Schweinfurt and again

> Steyr. After only six weeks our production of bearings (above 6.3 centimeters in diameter) had been reduced to 29 percent of what it had been before the air raids. At the beginning of April 1944, however, the attacks on the ball-bearing industry ceased abruptly. Thus, the Allies threw away success when it was already in their hands. Had they continued the attacks of March and April with the same energy, we would quickly have been at our last gasp.[35]

Of course, at this point in time the attack on the ball-bearing industry had to take a back seat to the great invasion of Normandy. This involved a major decision on the part of the highest authority as to command and control of the heavy bombers just before and immediately after the invasion. Eisenhower had used the heavies in direct support of the surface forces in North Africa and was convinced that this was their porper role. Therefore, he insisted on and got control of the heavy bombers beginning in March and lasting for the next six crucial months. General Eisenhower's Naval Aide, Captain Harry C. Butcher, said that the Supreme Commander felt so strongly about the control of the heavy bombers that he threatened "to go home"[36] if he could not have his way in this matter. Working on the ball-bearing plants, of course, was not exactly what Eisenhower had in mind for the use of the heavies. He wanted them for the kind of work they had done in North Africa, tactical support of the ground troops, not for any esoteric strategic air operations against ball bearings.

The sudden halt in the attack on the ball-bearing plants in April 1944 puzzled Speer. He attributed it to a conviction on the part of the American Staff that the industry was "bombed out" and said that the German General Staff had made a similar mistake with respect to a Soviet synthetic rubber factory that was supposed to have been "bombed out" by the Luftwaffe in the fall of 1943. Speer did not trust the military judgment based on the aerial photographs of the Soviet synthetic rubber factory and said, "I showed these photos to our leading synthetic rubber specialist, Hoffmann, the manager of our plant in Huls, which had undergone much more severe attacks. After pointing out various key sections of the plant which had not been hit, he explained that the plant would be in full production again within a week or two."[37]

The same thing was true of the ball-bearing plants in Germany. They had not been "bombed out" either in the attacks in March and April, and when the attacks abruptly ceased the plants rather promptly bounced back into production. The truth of the matter was that the aerial photographs were misleading. As the Bomb Survey noted,

> Building destruction from all the raids equaled approximately one-half the preraid floor space of the industry, while the equivalent of another half was heavily damaged. The damage to the machine tools was not proportionate to the damage to the buildings. Machine tools destroyed equaled 12 percent of the original inventory and those damaged an additional 30 percent.[38]

This was not enough to stop production either in the short or long run. Production was down in April, of course, but in May it jumped up 25 percent, and the rate continued to climb until September 1944, when it peaked out at 94 percent of the pre-attack rate.

When the Americans returned to the attack on the ball-bearing industry in October 1944 with a large-scale mission against Schweinfurt, it was already too late. The industry had been dispersed and had gone underground to the point where it was no longer an exploitable weakness. The Germans simply won the Battle of the Ball Bearings and could and did boast, "*es ist kein Gerat zuruck gebleiben weil Walzlager fehlten* (no equipment was ever delayed because bearings were lacking).[39]

After the war, Speer was asked what would have been the results if the German ball-bearing industry had been brought under the right kind of air attack? His reply was, "armaments production would have been crucially weakened after two months and after four months would have been brought completely to a standstill."[40] Speer was also asked what the right kind of air attack would have been. Going back to the first attack on August 17, 1943, when the planners divided the bomber force into two parts, one attacking Schweinfurt and the other Regensburg, Speer said,

> This, to be sure would have meant: one: all our ball-bearing factories (in Schweinfurt, Steyr, Erknew, Cannstatt and in France and Italy) had been attacked simultaneously. Two: these attacks had been repeated three or four times, every two weeks, no matter what the pictures of the target area showed. Three: any attempt at rebuilding these factories had been thwarted by further attacks, at two-month intervals.[41]

The failure of the Americans to win a victory in the Battle of the Ball Bearings was the result of the fact that the attacks were interrupted too often and the Germans were given all the time they needed to prevent the industry from collpasing. The attack on the aircraft industry in the famous "Big Week" at the end of February 1944 was also a failure, but for a different reason. It was not because the industry was given a breathing spell, not because the attacks were spread out over too long a period of time, as had been the case with the ball-bearing plants, but simply because there was so much excess capacity in the industry that it could not be pared down to the critical point in one week, not even with four thousand tons of bombs. There was the additional factor, of course, that the wrong segment of the industry was attacked, airframes instead of engines. In an effort to get quicker results, the bombers got practically no results at all. German production of fighters still met, and possibly exceeded, requirements.

If the "Big Week" was a failure in that it did not accomplish its strategic mission, which was the case, it was a huge success from another point of view. Almost as a bonus to the failed attack on production capacity was the fact that in the air fighting that took place during the "Big Week" the back of the German Air Force was broken. The success of the "Big Week" should have been calculated not in the amount of damage to the aircraft industry, but in the number of enemy fighters shot down during the attack. In other words, while the "Big Week" was not a strategical success, it was at the same time an overwhelming tactical victory.

The strange anomaly of the "Big Week," the strategical failure and the tactical success, came about as a result of the new orders that were given to the American fighters during the attack. As the Bomb Survey put it, "whereas previously it had been necessary for fighters to protect bombers, from this time on they were ordered to pursue and destroy enemy fighters as a primary duty."[42] Some two months prior to the issuance of the order to the American fighters to go on the offensive against the German fighters, Goering had issued an order to the German fighter pilots to avoid American fighters the better to concentrate on bombers.[43] The American orders to attack, and the German orders to go on the defensive, complemented each other and at the same time doomed the Luftwaffe. Put fighter pilots on the defensive and they lose the reason for their existence, so great is their disadvantage in combat with other fighters. In this case, it turned out to be a fatal disadvantage.

From this time forward, Luftwaffe Fighter Command was on the defensive in Europe, and the daylight skies belonged to the Eighth Air Force. This did not mean that the German Fighter Command was not still capable of vicious spasms of fighting, but it did mean, as General Hansell said, that "the Stragegic Air Forces with their heavy bombers and long-range escorts could choose their targets almost anywhere in Germany and penetrate the defenses to reach them without incurring intolerable losses."[44] It followed from this that the fate of Germany was a foregone conclusion after the "Big Week." The defeat of the German Air Force was more of a turning point in the war than the invasion of Normandy, the break-out at St. Lo, and crossing of the Rhine, or the battle of the Ruhr. From this point on, Germany's fate was sealed.

The other great turning points in the Second World War were the Battle of Midway in the Pacific War, El Alamein in North Africa and Stalingrad in the Russian War. Midway restored the balance of power in the Pacific and reversed the effects of Pearl Harbor. El Alamein turned the tide of the battle in Africa, cooked Rommel's goose and gave Montgomery the honor of winning the last battle. Stalingrad was not only the worst defeat ever administered to a German field army, but it was here that the Nazis lost the momentum of their advance, never to regain it. From this point on, their armies were pointed in a new direction, away from Moscow and toward Berlin.

That the results of the "Big Week" in February were every bit as far-reaching as the other great battles in World War II was attested to after the war by no less an authority than Albert Speer. Writing on August 9, 1978, Speer said, "the losses inflicted by the American and British air fleets constituted for Germany the greatest lost battle of the war, far exceeding our losses at Stalingrad, in the Winter campaigns in Russia, or during the retreat from France."[45] From the period of the "Big Week" onward, the bombers could choose their targets at will, almost anywhere in Germany, and as the Bomb Survey said, "the German experience suggests that even a first class military power -- rugged and resilient as Germany was -- cannot live long under full-scale and free exploitation of air weapons over the heart of its territory."[46] The net effect of the "Big Week" was that it led to, and made possible, the full domination of the air over Germany, both over the armed forces and its sustaining economy. "Without it," the Bomb Survey said, "attacks on the basic economy of the enemy could not have been delivered in sufficient force and with sufficient freedom to bring effective and lasting results."[47]

The pay-off from the attacks during the "Big Week" did not come immediately. Exploitation of the new balance of power on the continent had to be delayed because of the decision to invade Europe on June 6, 1944. This involved giving control of the heavy bombers to the Supreme Commander, Dwight D. Eisenhower, for a period of six months beginning at the end of March. General Eisenhower's concern, quite naturally, was not with the strategic air offensive as such, but with getting on with the ground war. In other words, the strategic air offensive was put on a back burner for the next six months. All this meant was that the advantage gained in the "Big Week" could not be immediately exploited.

The primary advantage gained in the "Big Week," the victory over German Fighter Command, can be traced back to a flaw in the German economy apparent from the beginning, a shortage of the life-blood of modern warfare, oil, particularly aviation gasoline. It was not for want of ball bearings that the Luftwaffe was beaten, nor for want of good airplanes, but for want of experienced pilots, and that want goes back in the final analysis to a training deficiency brought on by an acute shortage of aviation gasoline. The shortage of gasoline began to put a brake on the training of pilots in Germany as early as 1942. The reason for the shortage was the fact that oil had been a weak point in the German economy from the beginning. Recognizing this, the Germans imposed a cut-back of 90 percent on the civilian use of motor gasoline at the outset of the war, and in 1942 not only canceled a proposed increase in the pilot training program, but ordered a drastic reduction in the number of flying hours for training in order to conserve on aviation fuel. The net result of the cut-back on flying hours in the pilot training programs was that the new pilots were of poorer quality than their predecessors and far poorer than the new pilots being turned out in England and the United States. The shoddy reasoning behind this false economy on the part of the Germans was that numbers were more important than quality, and that their poorly trained pilots could finish their training in the school of hard knocks, where they could be given the finishing touches under the tutelage of hardened veterans.

The trouble with this line of non-reasoning with respect to pilot training was that mistakes made in combat were not always correctable. In other words, the school of hard knocks was not one in which the students could get a make-up examination. There were no second chances in this school, and the first mistake was all too likely to be the last. This was particularly true when the barely trained German pilots had to face up to the highly trained, and often battle hardened, British and American pilots. Poor training was the kiss of death for the Luftwaffe --even if it did save on aviation gasoline.

The Luftwaffe had no shortage of good airplanes at the time it went down in defeat -- in fact it could be said that German Fighter Command was virtually drowning in aircraft--but the missing ingredient was experienced pilots. The planes were useless without good men to fly them. In addition to this was the fact that while the German Air Force was going down in defeat the American strength in fighters was growing rapidly. By March of 1944 the disparity between the two air forces, German and American, was so great that the American bombers were flying selected routes designed with almost a single purpose in mind, and that purpose was, as the Bomb Survey said, "to force the German Air Force to fight."[48] A few simple statistics tell the whole story of the Luftwaffe in this period of defeat. "Whereas," the Bomb Survey noted,

> in 1942 the number of single-engine fighters in operational units, destroyed or damaged in excess of 10 percent, totaled 3,650, it jumped to 10,661 in 1943 and to 16,150 in 1944.[49]

Finally, the once proud Luftwaffe, scourge of the skies of Europe, was reduced to a mere nothingness, or as Shakespeare would say, "sans teeth, sans eyes, sans taste, sans everything."[50]

On D-day in Normandy General Eisenhower told the troops going ashore not to worry, "if you see fighting aircraft over you, they will be ours." The German combat infantrymen made a bitter joke out of their plight on this historic day. A quip was current among them about a foolproof system of aircraft identification. They said that if the planes were of one color, they were American; that if they were of two colors, they were British; and that if you couldn't see them at all, they were German. The Bomb Survey said, "the greatest single achievement of the air attack on Germany was the defeat of the German Air Forces."[51] But this was not all. The Bomb Survey also said that the defeat of the German Air Force came about as a result of a self-inflicted wound, "an over-all shortage of aviation gasoline resulted in the curtailment of flying training as early as 1942 and this decision was reflected in a deterioration of quality of personnel, which was the principal cause of the defeat of the German Air Force."[52]

If there was no excuse for making Training Command the butt of the oil shortage, the shortage itself was not something that could be denied. It was so well known in advance that Germany would face an acute shortage of oil in the event of war that many economists said that Germany would not go to war, and that all the world had to do to stop Hitler was call his bluff.

In the last full year of peace, 1938, Germany consumed seven and a half million tons of oil. Of this amount, only seven percent came from crude production within the Fatherland and another twenty-six percent from synthetic plants. For the balance, Germany depended upon imports, largely from Venezuela and the Middle East. These sources, of course, could not be relied upon in time of war because of the possibility of sea blockade. It looked like the economists were right, that Hitler could not afford to go to war with an oil position as dangerous as this. One way around the difficulties imposed by this impossible position in oil was to accumulate large stockpiles, and the German planners moved energetically in this direction, particularly with respect to aviation gasoline. The plan was to stockpile at least a million and a half tons, but Hitler moved too fast for the planners, and when Poland was invaded only a third of the planned stockpile had been accumulated.

In terms of war requirements, Hitler's stocks of aviation gasoline were adequate for only three months when the war began. The reserve position was even more precarious for motor gasoline and rubber. In these two categories the stockpiles were good for less than three months. In other words, Germany's oil position was tight at the beginning of the war, and the longer it lasted the tighter it got. Aside from capturing enemy stocks and overrunning other sources of supply in Europe, the only hope Hitler had for keeping his airplanes in the air, his tanks and motorized troops moving and his submarines in action at sea, was to increase production of his synthetic plants in Germany. The only way to do this was to enlarge the existing facilities. Decentralizing and scattering the production facilities would provide greater protection against the bombers, of course, but this was a luxury Hitler could not afford. Production was what he needed, and had to have, even at high risk. Thus there was great consternation in Germany when three of the major oil plants were hit on May 12, 1944.

For Albert Speer this date, May 12, was Germany's industrial Pearl Harbor day. He said,

> On that day the technological war was decided. Until then we had managed to produce approximately as many weapons as the armed forces needed, in spite of their considerable losses. But with the attack of nine hundred and thirty-five daylight bombers of the American Eighth Air Force upon several fuel plants in central and eastern Germany, a new era in the air war began. It meant the end of German armaments production.[53]

Up to this time, the doomsayers in Germany had been wrong about the oil economy being so weak that it could not support a war. Precarious as the balance was, it had not grown any worse after the outbreak of war, and in some ways it had improved. The Polish campaign in 1939 had been of such brief duration that it had imposed no real strain on Germany's oil resources, and the same thing was true of the campaign against France and the Low Countries in the following year. In fact, the stocks of gasoline captured in France in 1940 were larger than the German expenditures, leaving the Reich in a stronger position. There were other developments that helped to strengthen Germany's oil position. One of these was the occupation of Rumania in 1940, and the other was the fact that Hungary became an Axis partner in the same year. The imports from Hungary and Rumania totaled about two million tons in 1943. This was mostly in the form of motor gasoline and Diesel fuels.

Long before 1943, and in fact right after their defeat before Moscow in 1941, the Germans realized they were in for a long war and that they would have to speed up the expansion of their hydrogenation plants, as well as begin the construction of new ones out of harm's way in Silesia and the Sudetenland. This is also when they began to make their mistakes in cutting back on the flying time for the Luftwaffe pilots in training. But this was all too little and too late, as some of the Germans themselves knew only too well. The big mistake had been to go to war in the first place, considering the bottleneck in the supplies of oil.

By the Spring of 1944, it was common knowledge among the German industrialists that not only was the war being lost, but that the prospects of Germany's running out of oil, particularly aviation gasoline, would speed up as the armies in the field began to collapse. They were already on the verge of collapse, and the oil shortage promised to get worse, much worse, as a result of the backward movements of the troops in the field. It was only a matter of time before the Rumanian oil fields would be in Russian hands, and the same was true of the Hungarian sources.

The American bombing attacks on seven of Germany's largest hydrogenation plants on May 12 and 28, including the biggest of them all, Leuna, seemed to be the last straw. After a detailed inspection of the Leuna

Works on May 13, Speer warned Hitler that time was running out. The Armaments Minister put it bluntly when he said, "the enemy has struck us at one of our weakest points. If they persist at it this time, we will soon no longer have any fuel production worth mentioning. Our one hope is that the other side has an air force General Staff as scatterbrained as ours."[54]

Speer did not exaggerate when he said that the oil industry, and particularly the hydrogenation plants, represented one of the greatest points of weakness in the German war economy. This was so for several reasons. One was that 80 percent of the production of aviation gasoline came from only ten facilities, and the other was the fact that the chemical industry was closely integrated with the synthetic plants in Germany. Thus a successful attack on the synthetic oil plants, and particularly on such integrated giants as Leuna, would have effects that would go far beyond the loss of petroleum products, important as that was. Bombing the synthetic plants would deal a crippling blow to the munitions and explosives industry in Germany and would severely affect the production of synthetic rubber. From the beginning, Germany had been fighting an ersatz war, and the destruction of the synthetic oil industry would take the erstaz out of the war. No ersatz, no aviation gasoline, no motor gasoline, no explosive, no munitions, no rubber, no war.

The systematic attack on the German oil industry began in May 1944 and continued until March 1945. During this period roughly 225,000 tons of bombs were dropped on the oil targets. This compares to the 11,000 tons of bombs expended in the attack on ball bearings and the 50,000 tons on aircraft factories. In other words, and to put this another way, of the million plus tons of bombs dropped by the United States Strategic Air Forces in Europe the ball bearings got less than one percent, the aircraft industry a little over five percent and the oil targets about 13 percent. Of these three target systems, the ball bearings yielded no results whatsoever, and the attack on the aircraft industry produced negative results, but the attack on the oil targets held the promise of achieving victory through Air Power.

The most serious loss the Germans suffered as a result of the attacks on the oil industry was in the field of aviation gasoline, but there were other losses as well. Rubber was in short supply after the bombing of the synthetic oil plants because of the damage the chemical industry (I.G. Farben) suffered when Leuna, Ludwigshafen and Leverkusen were bombed. The loss of nitrogen put a big dent in the production of military powders and explosives causing a shortage of ammunition. But the loss of aviation gasoline was the most critical item. Its production declined steadily and alarmingly from 130,000 tons per month to 52,000 tons, to 26,000 tons, and virtually to zero by March 1945. The Bomb Survey's oil report said once it became clear that Germany's oil industry was doomed, and that "the Nazis could neither manufacture nor capture any more appreciable supplies of liquid fuel, Germany's defeat became inevitable."[55]

Being inevitable, however, did not mean that the Germans had to recognize it as a fact, and act accordingly. And in fact there were several good reasons the Germans were not forced to accept the fact that they were beaten as they saw their oil industry going down the drain. One was that instead of being highly concentrated, the attack on the industry was spread out over nearly a year, which gave the Germans an opportunity to adjust themselves and their economy to the attacks. In other words, the follow-through was bad. The lack of concentration in the attacks first became clear after the May bombings. General Eisenhower called off the heavies from their strategic bombing in order to put them to work in connection with the invasion of Normandy. Then the oil attacks were resumed. However, after the June attacks there was another pause, and this lasted for nearly a month, while the Germans worked furiously to get the plants back into production. Then pause was again necessitated to let the heavies work on some obstacles holding up the ground forces.

One of General Eisenhower's air planners objected to the diversion of the heavies to do ground support work at this time. General Hansell's point was that the tactical air forces available to Eisenhower were more than enough to do the ground support work, including isolation of the battlefield, and that the heavies should be used on oil targets. In fact, General Hansell said, "Eisenhower's Tactical Air Forces were larger than the entire Luftwaffe, which was fighting on four fronts. They should have been adequate to do General Eisenhower's job."[56] The plain truth of the matter was that General Eisenhower had become something of an addict in the use of heavy bombers to do the work of the

Tactical Air Forces. He thoroughly enjoyed seeing the heavies do this kind of work in North Africa and was more or less convinced that this was their proper role. Sir Arthur Tedder may have been right when he complained that the British Army was also getting too accustomed to calling on the heavies for almost any kind of work. "I am doing my best to get things straight," he said, "but I am sure you will realise that, the Army having been drugged with bombs, it is going to be a difficult process to cure the drug addicts."[57]

Speer was almost hysterical after the June attacks on the synthetic oil plants in the Reich. In a letter to the Führer dated June 30, 1944, the Armaments Minister spoke of the terrible losses and warned of the inevitable consequences. "Our aviation gasoline production," he said,

> was badly hit during May and June. The enemy has succeeded in increasing our losses of aviation gasoline up to 90 percent by June 22. Only through speedy recovery of damaged plants has it been possible to regain partly some of the terrible losses. In spite of this, however, aviation gasoline production is completely insufficient at this time. The already heavy losses in June and the estimated very low production for July and August, on account of increased air attack, will doubtless make us use up the greater part of our reserves in aviation gasoline as well as in other fuels. If it is not possible for us to protect these plants we will be forced to curtail the flow of supplies to the Army in September, which will mean that from that time on there will be a terrible bottleneck which may lead to the most tragic consequences.[58]

Speer took a leaf out of his experience with the ball-bearing industry and appointed Edmund Geilenberg as the czar of the oil industry with full authority to requisition anything and everything needed to restore production in the "bombed out" oil plants. He was not as successful as Kessler had been in the ball-bearing industry, but highly successful nonetheless. Geilenberg soon had a force of well over a quarter of a million engineers, technicians and laborers hard at work repairing damage and restoring production in the oil industry. What he was able to accomplish was nothing short of a miracle, and it took a miracle just to keep Germany in the war.

The Bomb Survey did not make a systematic analysis of the effects of the oil offensive on the front-line strength of the German forces, but they were not ignored either, and the most important of these, it was dutifully noted, was the grounding fo the Luftwaffe. The Bomb Survey was very explicit on this point, saying that the single greatest achievement of the air war was the defeat of the German Air Force,[59] and that the principal cause for the defeat was the shortage of aviation gasoline.[60] The Bomb Survey also quoted General Omar Bradley's description of the effects of the oil shortage on the fighting fronts in the closing months of the war, with particular reference to the Battle of the Bulge. General Bradley said,

> With the defeat of the German gamble in the Ardennes, when the Allies were again in a position to resume the war of movement--in the East as in the West--lack of oil, which the strategic bombing campaign had enforced upon the enemy, told handsomely. The retreat from the Ardennes was an agonizingly slow and costly affair for the enemy. The withdrawal of the six SS Panzer Army begun in daylight on 22 January 1945, was marked mainly by successes of fighter bombers against its tanks and trucks. These successes, however, took place against a background of painfully exiguous oil reserves -- with supply trucks being drained to fill the tanks of fighting vehicles -- and a long pull to the distant loading stations.[61]

General Bradley went on to say,

> When the Allied threat shifted north to the Aachen sector, the enemy was unable to sideslip his 'mobile' formation to meet it in the measure he sought-- again for lack of gasoline. When the allied break-through followed west of the Rhine in February, across the Rhine in March,

> and throughout Germany in April, lack of gasoline in countless local situations was the direct factor behind the destruction or surrender of vast quantities of tanks, guns, trucks and of thousands upon thousands of enemy troops.[62]

The Bomb Survey elaborated on General Bradley's remarks by saying,

> The results of the oil shortage made themselves felt also on the eastern front. The Russian victories in Silesia in February and March 1945 was hastened by the German lack of fuel. At the Baranow bridgehead 1,200 tanks, which had been massed by the Germans to hold the line, were immobilized for lack of gasoline and overrun. On the testimony of Marshal Stalin, bombardment of oil played an important part in the sweeping Russian victories.[63]

The attacks on the oil industry had a number of side effects that caused the Germans a great deal of concern. The chemical industry is a good example. It came close to being shut down as a result of the attacks on the oil industry. The great Leuna Works was important not only as a producer of aviation gasoline, but also as a producer of nitrogen for the munitions and explosives industries, as well as hydrogen for the synthetic rubber industry. Thus when Leuna was bombed a synthetic rubber plant at nearby Schkopau was forced to shut down, making the critical rubber shortage that much worse and also forcing a curtailment in the output of munitions and explosives at other plants. "By February 1945," the Survey said,

> German explosives production, exclusive of extenders, had been reduced drastically to 8,500 tons per month. This decline came at a time when the Germans were fighting on two fronts and consumption was at a peak. Stocks which were ample in mid-1944 evaporated. Thousands of finished shell casings remained unfilled and the Germans were forced on occasion to use as high as 70 percent rock salt to stretch the small supply of explosives.

Supplies of flak ammunition were so short that battery commanders were ordered not to fire at enemy aircraft unless the airplanes were attacking the target the battery was supposed to protect and the commander was sure he could hit the airplanes. The Germans had to give up entirely the manufacture of the powerful explosive hexogen.[64]

The air planners themselves, and perhaps with good reason, were critical of the way the attack on the oil industry worked out. They argued that the timing of the attack was wrong, that the best results would have been obtained if the offensive had been launched in November of 1943 instead of May 1944. In other words, if the attack had been earlier the effects on front-line strength would have been felt that much sooner, or all during the Normandy campaign, including the break-out of St. Lo. The air planners found fault with their own work in connection with another important factor, physical vulnerability, including the vital element of recuperation. "The planning factors used in estimating force requirements," General Hansell said,

> were too low, although this too was primarily due to a lack of air intelligence. More serious was the fact that target plants were more easily repaired than had been anticipated. The enemy worked miracles in terms of repair which meant that attacks had to be repeated again and again. Also, initial destruction of the targets required two to three times as much force as peacetime planning had estimated.[65]

The air planners blamed themselves far too much for the fact that the attacks on the German oil industry were too little and too late to have the maximum effect on front-line strength during and after D-day. This was not their fault. They were not the ones who were responsible for the invasion of North Africa and for the use of the heavy bombers in a purely tactical role. Nor were they the ones who ordered the senseless bombing of the invulnerable sub pens in the vain hope that this would help with the Battle of the Atlantic. Similarly, the air planners were the first to warn that the attacks on the V-1 launching sites was not appropriate work for heavy bombers. Likewise, they did not take the lead in making the decision to further delay the strategic air offensive by putting all the heavy bombers under the control of a ground general for a period of six months, covering the period before, during and immediately after the invasion of Normandy. These decisions were made by surface-minded generals and admirals backed by the highest political authorities, and if they delayed and diluted the effects of the strategic air offensive, which they did, the responsibility does not lie with the air planners. They did themselves and history a disservice at the end of the war by saying *mea culpa* (my fault--I am to blame).

Transportation as a target system was something else again from the oil targets, the aircraft plants and the ball-bearing industry. While transportation always had a high priority as a potential target system, it could not be considered seriously until the bomber force was much larger than it was during the first days of the war. In fact, it was not until the victory of Africa had been won and Eisenhower had returned to England, that transportation began to move to the fore as a serious contender for first place as a bombing system. And when it did, it was within the context of a very special situation, the invasion of Europe to open up the long awaited Second Front.

When General Eisenhower succeeded in winning command control of the heavy bombers for the pre-invasion strikes, while the landings were taking place and during the consolidation period, he quickly found that he had won only half the battle. Equally as important from his point of view was the battle over target selection. This boiled down to a choice between oil and transportation as a target system. Would it be more important to have air supremacy during the critical period of the landings (the oil targets) or would the troops be better served by having the battlefield isolated (the transportation targets)? It was a hard choice to make, and made doubly so because Eisenhower's staff was divided. The airmen rallied behind Spaatz' Oil Plan with its promise of having friendly skies over the troops going ashore, and the surface-minded generals rallied behind Tedder's Transportation Plan, which would give the troops protection against German reinforcements being rushed to the area while the landings were taking place and during the consolidation period immediately following.

Eisenhower wanted both, but this was obviously out of the question. If the heavies were used against oil targets deep in the interior of Germany they could not at the same time be used to pound transportation targets in France and the Low Countries. The question was which use of the heavies would produce the better results from the point of view of the invasion of Normandy. The advocates of Tedder's Transportation Plan argued that the best use of the heavies was to disrupt the local transportation system of France and the Low Countries to the point where the Germans could not move reinforcements into the Normandy area during and immediately following the assault period. On the other hand, the advocates of Spaatz' Oil Plan argued that the destruction of the oil targets would be the better plan because it would practically guarantee command of the air over the invasion beaches.

Spaatz's Oil Plan was a very rational one from the point of view of the principles of war. It had become almost an axiom of warfare by this time that as long as a country maintained control of the air it could never be invaded from its sea approaches. To Spaatz this meant that his primary job was to complete the destruction of the Luftwaffe to the point where it could not put an appearance on D-day. This meant, in turn, that he would have to force the Luftwaffe to fight, and that also meant, again in turn, that his heavy bombers would have to attack sensitive targets deep in the interior of Germany, targets that the Germans could not afford not to defend, such as the oil targets. General Spaatz did not feel that the Germans would be as sensitive about French locomotives and railways as they would be about the vital oil targets deep in the heart of Germany. He thought they would fight to defent the latter, having no choice really, but he was in considerable doubt as to whether the Germans would feel the same way about French locomotives, bridges and rolling stock. Furthermore, as Spaatz said at the time, he was much more concerned about the number of German aircraft that might show up on D-day than he was about the number of railway trains that might put in an appearance.[66]

Spaatz made another point about the Luftwaffe that should have had more impact than it did. This was that the Luftwaffe was presently divided into what amounted to three separate forces, one assigned to the defense of the homeland, one to the German armies in the east. and the third to the German armies in the west. Spaatz made the point that if the attacks on the interior of Germany were broken off, the powerful home defense forces would be speedily transferred to the Normandy area to guard against the expected landings.

Thus, if for no other reason than to pin down the German fighters in central Germany, Spaatz argued that the attacks on the targets in central Germany should be continued and intensified. Spaatz's reasoning was that to do otherwise would be to allow the Germans to take advantage of the principle of mass rather than forcing them to do something they did not want to do, which was to split their forces three ways. The principle of mass or concentration at the decisive point, in this case Normandy, was one that had strong appeal to the surface commanders, and if they could force the Germans to violate the principle, so much the better.

Spaatz' Oil Plan was one of three phases. He proposed that the heavies be used against the oil targets in central Germany for a period of two months prior to the invasion, and that this be followed, three weeks before D-day, by an all-out interdiction campaign against targets in the invasion area. Once the surface forces had secured their lodgment and were well beyond any danger of another Dunkirk, Spaatz proposed that the heavies be released so that they could be returned to their regular work, which, in the language of the Casablanca directive, was to effect "the progressive distruction and dislocation of the German military, industrial and economic system and the undermining of the morale of the German people to the point where their capacity for armed resistance is fatally weakened."[67]

Spaatz's Oil Plan fell on deaf ears as far as the surface-minded admirals and generals, including some of the airmen, were concerned. They felt that the oil targets, or any industrial targets for that matter, were nonsense at this critical stage of the war and that the forthcoming battles on the ground would be decisive in determining the outcome of the war. As to the Luftwaffe, the prevailing feeling among the ground commanders was that the German Air Force had been eliminated as a major factor and that bombing the oil targets to force the few remaining German fighters into combat made about as much sense as flailing a dead horse. On the other hand, the ground commanders argued that no effort should be spared to completely isolate the battlefield and to destroy the enemy on the ground. Their further argument was that anything else was just so much nonsense, including the strategic air offensive. This was war, they insisted, and it was no time to play games.

Eisenhower agreed with his ground commanders, and the decision was made to implemtnt the Transportation Plan. However, the Supreme Commander apparently had some second thoughts himself about the Oil Plan, and this prompted him to open the door an inch or two with respect to Spaatz's oil targets. The result was the bombing of the vital oil targets on May 12 and 28, and the special dispensation that Spaatz was given to bomb oil targets when there was no special requirement for the direct support of the ground forces. The special opening to the oil targets as a secondary system to transportation was finally given by Eisenhower when he too was persuaded that as long as the Germans had planes and pilots they could be dangerous, and that while the Germans probably would not sacrifice their remaining air strength to defend the French railway network, they probably would do so to defend their vital oil resources.

The failure of the Luftwaffe to make its appearance on D-day was the result of the attack on the oil targets. Thus, their destruction assured against the failure of the invasion. However, the consolidations of the Allied forces after the invasion was made possible by the destruction of the targets in another system, the greatly enlarged Transportation Plan that was implemented on the eve of the invasion.

The original Transportation Plan, the one Eisenhower endorsed on March 26, 1944, was not the same one that was finally implemented. Originally, the Transportation Plan was the brainchild of a British scientist in the wartime service of the British Air Ministry, Solly Zuckerman. This is the reason the plan was sometimes called the Zuckerman Plan. The fact that it was adopted and sponsored by Eisenhower's British Deputy, Air Chief Marshal Tedder, and also by the Commander of the Allied Expeditionary Air Froce, Chief Marshal Leigh-Mallory, caused it to be referred to as the Tedder Plan or the Leigh-Mallory Plan. Finally, the plan was also called the Rail Plan.

Solly Zuckerman, later elevated to the peerage as Lord Zuckerman, was a zoologist in civilian life. His specialty was monkeys. After the war, he acknowledged the incongruity of his two careers by entitling his Memoirs, *From Apes to Warlords*. A non-advocate of the Zuckerman Plan commented on the fact that the author's exceptional knowledge of antropoid apes did not necessarily qualify him as an expert on air strategy.

Zuckerman first made a name for himself as a military expert because of a study he made of the heavily fortified island of Pantelleria in the Mediterranean. This was just prior to the invasion of Sicily in 1943. His study not only indicated that the island could be bombed into submission, but precisely how this could be done. However, he got the idea for the Transportation Plan from a study he made of the bombing of the marshaling yards around Rome in July of 1943. This suggested to him that the whole transportation system or network in the area of the Normandy invasion could be knocked out or rendered inoperable by a selective attack on certain targets he called key rail centers in France and Belgium.

Zuckerman's estimate was that these rail targets, or marshaling yards, could be put out of commission by a systematic attack involving not more than 45,000 tons of bombs, which was something less than one-half the tonnage that could be dropped prior to D-day. Specifically, Zuckerman said that "the consistent attacks on the key centers would destroy the marshaling yards themselves, the repair shops, sheds, stations, switches, roundhouses, turntables, signal systems, locomotives and rolling stock." Finally, calling on his scientific studies in anatomy, Zuckerman said that the corpus of "a railway network is like a nervous system, no part of which can suffer damage without effects upon the whole."[68] In other words, Zuckerman practically guaranteed the planners that their "railway desert" in Normandy could be created by D-day from the expenditure of less than 50,000 tons of bombs.

Zuckerman's Transportation Plan turned out to have a number of basic errors in it. In the first place, the excess capacity in the network was far greater than Zuckerman thought it was. The Germans did not need to use more than one-third of the capacity of the French and Belgian railroads to satisfy all their direct military requirements, not two-thirds. In the second place, the Germans proved to be far more skillful in repairing their damaged oil facilities than Zuckerman had estimated. And finally, the rail centers did not turn out to be all that good as targets for the military bombers. Most of the important military traffic, as the Bomb Survey pointed out, used complete trains and did not enter the marshaling yards at all. This was particularly true of troop trains.[69] Therefore, the damage to military traffic was not at all proportionate to the damage to the marshaling yards.

For all these reasons, the Zuckerman Plan was less than a complete success when it was put to the test. Between February and D-day, Allied aircraft dropped 71,000 tons of bombs on the rail centers, nearly half again as many bombs as had been projected as the maximem required, and the final result was that only the civilian rail traffic of the occupied area of France and Belgium was affected. Military traffic moved pretty much as usual. If the Allied forces had depended on this slender reed, the Allied armies might have been driven from the field in defeat. What saved the invasion from almost certain disaster was not the false promises of Solly Zuckerman and the Rail Plan, but the very real results that accrued from an old-fashioned plan to isolate the battlefield by cutting lines of communication. Bridges were the principal targets in this new plan.

Spaatz had always been skeptical of the Transportation Plan, and in a joint target conference on May 6 proposed that a program of destroying bridges be adopted as a more practical method of isolating the battlefield. This was not put forward as a substitute for the Transportation Plan, which was chipping away at the capacity of the local network, but as a supplement to the Transportation Plan. However, the British objected to Spaatz's idea of attacking bridges and called forth a British railway expert who testified that some 1,200 tons of bombs would be required to undermine each one of the bridges over the Seine. Spaatz was skeptical of this too, and the meeting ended with the British and Americans in disagreement.

The following day, May 7, American light bombers began running missions against the bridges across the Seine and found that they could be damaged or put out of use altogether by one or two bombs of the 1,000 pound class. They also found that a low-level attack using skip-bombing techniques against the abutments of the bridges was more effective than dive bombing the bridges themselves. Finally, and to everyone's surprise, it was discovered that only 220 tons of bombs could level almost any bridge, not the 1,200 tons of bombs estimated by the British railway expert. This made a whale of a difference in the cost of the bombing program. Needless to say, this settled the question of whether the destruction of bridges should be included in the plan to isolate the battlefield, and the result was the initiation of a full-blown campaign to interdict the battlefield by means of destroying the bridges on all the approaches to the Normandy area.

This was not all that was accomplished by the vigorous campaign leading to the destruction of the bridges on the approaches to the Normandy area. The bridge-breaking campaign, like the attack on the airfields in the area, also played an important part in the cover and deception plan of the Allies. The single purpose of this plan was to make the Germans believe that the invasion would not be in the Normandy area at all, but in the Pas de Calais, the narrower and more obvious route. To this end, in addition to the Seine bridges, a second line of bridges was brought under attack. These were selected not because they would contribute to the isolation of the Normandy battlefield, but because their destruction would indicate that the Pas de Calais was to be the principal battlefield. Also, the large-scale attacks on the Seine crossings were delayed until just before D-day. However, when they were brought under attack, all nine Seine railroad crossings were destroyed, as well as a dozen highway bridges.

The bridges were only a small part of the whole cover and deception plan, of course, but they played an important part, and the plan itself was a great success. The German Supreme Commander, Marshal Gerd von Rundstedt, kept the bulk of his forces in the Calais area and was completely surprised when he finally realized that the "Calais Invasion" was taking place in Normandy. The result was that the Germans were caught flatfooted. This is what made the difference between victory and defeat on D-day. Because they were taken in by the deception plan in the first place, and because of their enforced immobility in the second place, the Germans could not stop the landings themselves, nor could they do the next best thing, which was to hit the Allies before they could consolidate their forces. Thus the Germans lost the first round in the Battle for Europe.

Surprisingly, Solly Zuckerman's Transportation Plan was not discredited completely because of its failure to live up to expectations. Quite the contrary. The original proponents of the plan--Zuckerman, Leigh-Mallory, Tedder and Eisenhower--hailed the plan for the results it obtained, and some went so far as to say that the invasion would not have been possible without it. And in any case, the bombing of rail centers went on after D-day and continued right up to the end of the war. In fact, on October 25, Marshal Tedder and the Air Ministry moved to revive the Transportation Plan as a high-priority target system, only this time against Germany itself. The theory was that if enough disruption could be caused to the transportation system as a whole, it would not only bring about an immediate and general collapse of the entire economy, but would also adversely affect German fighting strength on all fronts, if not undermine it entirely.[70]

Tedder put a great deal of faith in the new Transportation Plan. He confidently predicted "that if the Allies concentrated their whole aerial effort against the transport system in the western part of Germany they might produce a German collapse within a matter of weeks."[71] This was not a far-fetched idea. Everything in Germany depended on the functioning of the railroads. Without them, industry would have to be shut down, and all the fighting, on the ground, in the air and on the sea, would have to come to an end. The railroads (carrying 72 percent of the traffic in Germany) were the key to everything else, the one essential without which all else would fall apart. If the railroads could be beaten down to the critical point, that is disrupted and destroyed to the point where no one could even keep track of arrivals and departures, Germany would simply descend into anarchy and chaos, militarily and economically. The waterways carrying 25 percent of the bulk tonnage, and the highways accounting for the remaining three percent, would not be enough to keep the country alive. Without the railroads, Germany would simply cease to exist as a going concern.

On October 28, 1944, an agreement was reached that the British and Americans would concentrate on the German transportation system as a high priority target system. To this end, Germany was divided

> into nine zones, of which the approaches to the Ruhr, Frankfurt-Mannheim, and Cologne-Koblenz were to be the three accorded greatest attention. In these zones the heavy bombers would attack marshaling yards, mediums and fighter-bombers would strike railway repair installations and power centers and finish off smaller targets that might escape the heavies within the main rail centers, and fighter-bombers would cut lines and shoot up rail traffic whenever opportunity offered. RAF Bomber Command was to take on the waterways as well as marshaling yards.[72]

The directive to carry out the attacks on the transportation system of Germany was simplicity itself. The orders were to bomb or shoot anything that moved, beginning with the carefully selected targets in the railway network, going on to the vulnerable points in the intricate inland waterway system, and finally including the highways themselves. Germany, in Albert Speer's words, was to be "a country thrown back into the Middle Ages."[73] The strategic attack on the transportation system that got underway in October was a logical extension and outgrowth of the bridge interdiction program and the bombing of the marshaling yards in support of the invasion of Normandy. As the front moved to the German border more and more demands were made on the heavy bombers to supplement the work of the mediums and fighters in support of the ground forces. As a result, the transportation system of Germany came under heavier and heavier attack and was beginning to feel the strain when the decision was made to subject it to a full-scale air attack. Altogether during the war, nearly 600,000 tons of bombs were dropped on the transportation targets of Germany. This was roughly twice the tonnage dropped on the oil targets and more than 25 times the tonnage on the ball-bearing targets. The damage to the transportation system of Germany came very near to being total. How near is difficult to calculate because, as the Bomb Survey said, toward the end, that is after March 3, 1945, "the disorganization was so great that no useful statistics were kept."[74]

However, military (Wehrmacht) traffic, as might be expected, had top priority all during the war, and right up to the end. The last German counter-offensive in the west, which resulted in the Battle of the Bulge, was made possible only because of the priority the military had on the use of the railway network. Even so, the Bomb Survey pointed out, the Ardennes counter-offensive "was probably the last such effort of which the Reichsbahn would have been capable in the west."[75] Nevertheless, the Wehrmacht stumbled on even though, at the end, the uncertainties of rail traffic were so great that they took on comic proportions. At the very end, the Germans took the precaution of sending couriers to accompany troop trains, and this practice was carried over, the Survey noted, to "shipments of tanks and other weapons." The Survey also noted that the task of the couriers was "to get off the train when it was delayed and report where it could be found."[76] That is, of course, if there was anything left of the train, or the couriers.

The magnitude of the attack on the transportation system was such-- 26 percent of the total, second only to the 37 percent in direct support of Army operations --that the Survey gave it special attention and "made a careful examination of the German railway system, beginning as soon as substantial portions were in Allied hands. While certain important records were destroyed or lost during the battle of Germany, enough were located so that together with interrogation of many German railroad officials, it was possible to construct an accurate picture of the decline and collapse of the system."[77]

The basic questions that faced the Survey, of course, were, what did the decline and collapse of the German transportation system do to the German economy as a whole and what did it do to the capabilities of the German fighting forces in the field?

In answer to these questions, the Survey said, "the attack on transportation was the decisive blow that completely disorganized the German economy. It reduced war production in all categories and made it difficult to move what was produced to the front. The attack also limited the tactical mobility of the German Army."[78] The Bomb Survey also said,

> By the beginning of 1945, before the invasion of the homeland itself, Germany was reaching a state of helplessness.

> Her armament production was falling irretrievably, orderliness in effort was disappearing, and total disruption and disintegration were well along. Her armies were still in the field. But with the impending collapse of the supporting economy, the indications are convincing that they would have had to cease fighting-- any effective fighting-- within a few months. Germany was mortally wounded.[79]

Finally, the Bomb Survey said,

> Even if the final military victories that carried the Allied armies across the Rhine and the Oder had not taken place, armament production would have come to a virtual standstill by May. The indications are convincing that the German armies, completely bereft of ammunition and motive power, would have had to cease fighting-- any effective fighting--within a few months. In the actual case--as in most other cases in the history of wars--the collapse occurred before the time when the lack of means had rendered further resistance physically impossible.[80]

It is probably fair to say that between them, oil and transportation, Germany was reduced to a cipher as a military power. Hindsight invariably suggests that there were better ways of attacking oil and transportation than those employed and that there were other target systems that would have produced the same or better results. However, the fact stands out that the German armies were reduced to a state of helplessness by the attacks on oil and transportation and that, when the final surrender came, it simply proved the verity of what the airmen had been saying all along, and what the Bomb Survey did say after the event, that no country, not even one as rugged and resilient as Germany was, can "live long under full-scale and free exploitation of air weapons over the heart of its territory."[81]

FOOTNOTES--CHAPTER VI

[1]Giulio Douhet, *The Command of the Air* (New York: Coward-McCann, Inc., 1942, tr. Dino Ferrari), 59.

[2]Albert Speer, *Inside The Third Reich* (New York: The Macmillan Company, 1970), 265.

[3]Karl von Clausewitz, *On War* (New York: The Modern Library, 1943, tr. O.J. Matthijs Jolles), 575 (Bk. VIII, ch. 3).

[4]Karl von Clausewitz, *On War* (Princeton: Princeton University Press, ed. and tr. Michael Howard and Peter Paret), 598 (Bk. VIII, ch. 4).

[5]Wesley Frank Craven & James Lea Cate, *The Army Air Forces in World War II* (Chicago: University of Chicago Press, 1948), I, 50.

[6]*Ibid.*, II, 351.

[7]Haywood S. Hansell, Jr., *The Air Plan That Defeated Hitler* (Atlanta: Higgins-McArthur/Longino & Porter, 1972, 145.

[8]Wesley Frank Craven & James Lea Cate, *The Army Air Forces in World War II* (Chicago: University of Chicago Press, 1949), II, 353.

[9]*Ibid.*, 355.

[10]*Ibid.*, 356.

[11]*Ibid.*, 355.

[12]Haywood S. Hansell, Jr., *The Air Plan That Defeated Hitler* (Atlanta: Higgins-McArthur/Longino & Porter, Inc., 1972), 183.

[13]The United States Strategic Bombing Survey, *Summary Report: European War* (Washington: Government Printing Office, 1945), 7.

[14]P.M.S. Blackett, *Fear, War, and the Bomb* (New York: McGraw-Hill Book Company, 1948), 26.

[15]Winston S. Churchill, *Closing the Ring* (Boston: Houghton Mifflin Company, 1951), 6.

[16]Wesley Frank Cravens & James Lea Cate, *The Army Air Forces in World War II* (Chicago: University of Chicago Press, 1949), II, 350.

[17]*Ibid.*, 353.

[18]The United States Strategic Bombing Survey, *Overall Report: Eurpoean War* (Washington: Government Printing Office, 1945), 107.

[19]Haywood S. Hansell, Jr., *The Air Plan That Defeated Hitler* (Atlanta: Higgins-McArthur/Longino & Porter, inc., 1972), 160.

[20]Albert Speer, *Inside the Third Reich* (New York: The Macmillan Company, 1970), 339.

[21]*Ibid.*, 340.

[22]The United States Strategic Bombing Survey, *Overall Report: European War* (Washington: Government Printing Office, 1945), 26.

[23]Albert Speer, *Inside the Third Reich* (New York: The Macmillan Company, 1970), 333.

[24]*Ibid.*, 339.

[25]*Ibid.*

[26]*Ibid.*

[27]*Ibid.*, 340.

[28]The United States Strategic Bombing Survey, *Overall Report: European War* (Washington: Government Printing Office, 1945), 26.

[29]George H. Quester, *Detterence before Hiroshima* (New York: John Wiley & Sons, Inc., 1966), 67.

[30]The United States Strategic Bombing Survey, *Over-all Report: European War* (Washington: Government Printing Office, 1945). 26.

[31]Wesley Frank Craven & James Lea Cate, *The Army Air Forces in World War II* (Chicago: University of Chicago Press, 1949), II, 353.

[32]Albert Speer, *Inside the Third Reich* (New York: The Macmillan Company, 1970), 340.

[33]*Ibid.*, 647.

[34]The United States Strategic Bombing Survey, *Over-all Report: European War* (Washington: Government Printing Office, 1945), 29.

[35]Albert Speer, *Inside the Third Reich* (New York: The Macmillan Company, 1970), 340-41.

[36]Harry C. Butcher, *My Three Years with Eisenhower* (New York: Simon and Schuster, 1946), 498.

[37]Albert Speer, *Inside the Third Reich* (New York: The Macmillan Company, 1970), 341.

[38]The United States Strategic Bombing Survey, *Over-all Report: European War* (Washington: Government Printing Office, 1945), 29.

[39]*Ibid.*

[40]Albert Speer, *Inside the Third Reich* (New York: The Macmillan Company, 1970), 339.

[41]*Ibid.*

[42]The United States Strategic Bombing Survey, *Over-all Report: European War* (Washington: Government Printing Office, 1945), 17.

[43]Robert Frank Futrell, *Ideas, Concepts, Doctrines: A History of Basic Thinking in the United States Air Force 1907-1964* (Alabama: Air University, Maxwell Air Force Base, 1974), 79.

[44]Haywood S. Hansell, Jr., *The Air Plan That Defeated Hitler* (Atlanta: Higgins-McArthur/Longino & Porter, Inc., 1972), 183-84.

[45]Ira C. Eaker, "Some Observations on Air Power," in *Air Power and Warfare* (Washington: Government Printing Office, 1978), 354.

[46]The United States Strategic Bombing Survey, *Summary Report: European War* (Washington: Government Printing Office, 1945), 16.

[47]*Ibid.*

[48]The United States Strategic Bombing Survey, *Overall Report: European War* (Washington: Government Printing Office, 1945), 10.

[49]*Ibid.*

[50]"As You Like It," Act II, Scene VII.

[51]The United States Strategic Bombing Survey, *Overall Report: European War* (Washington: Government Printing Office, 1945), 10.

[52]*Ibid.*, 75.

[53]Albert Speer, *Inside the Third Reich* (New York: The Macmillan Company, 1970), 412-13.

[54]*Ibid.*, 413

[55]Haywood S. Hansell, Jr., *The Air Plan that Defeated Hitler* (Atlanta: Higgins-McArthur/Longino & Porter, Inc., 1972), 228.

[56]*Ibid.*, 240-41.

[57]Russell F. Weigley, *Eisenhower's Lieutenants* (Bloomington: Indiana University Press, 1981), 381.

[58]The United States Strategic Bombing Survey, *Overall Report: European War* (Washington: Government Printing Office, 1945), 41-42.

[59]*Ibid.*, 10.

[60]*Ibid.*, 25.

[61]*Ibid.*, 44.

[62]*Ibid.*

[63] *Ibid.*

[64] *Ibid.*, 58.

[65] Haywood S. Hansell, Jr., *The Air Plan That Defeated Hitler* (Atlanta: Higgins-McArthur/Longino & Porter, Inc., 1972), 225-26.

[66] David MacIsaac, *Strategic Bombing in World War II* (New York: Garland Publishing, Inc., 1976), 175.

[67] Winston S. Churchill, *Closing the Ring* (Boston: Houghton Mifflin Company, 1951), 519-20.

[68] Russell F. Weighley, *Eisenhower's Lieutenants* (Bloomington: Indiana University Press, 1981), 59.

[69] Bernard Brodie, *Strategy in the Missile Age* (Princeton: Princeton University Press, 1965), 125.

[70] W.W. Rostow, *Pre-Invasion Bombing Strategy* (Austin: University of Texas Press, 1981), 47.

[71] Russell F. Weigley, *Eisenhower's Lieutenants* (Bloomington: Indiana University Press, 1981), 379.

[72] *Ibid.*, 380-81.

[73] Albert Speer, *Inside the Third Reich* (New York: The Macmillan Company, 1970), 523.

[74] The United States Strategic Bombing Survey, *Summary Report: European War* (Washington: Government Printing Office, 1945), 13.

[75] *Ibid.*, 14.

[76] *Ibid.*

[77] *Ibid.*, 12.

[78] *Ibid.*

[79] *Ibid.*, 16.

[80] The Unites States Strategic Bombing Survey, *Overall Report: European War* (Washington: Government Printing Office, 1945), 58.

[81]The United States Strategic Bombing Survey, *Summary Report, European War* (Washington: Government Printing Office, 1945), 16.

CHAPTER VII

THE FOLLY OF THE SECOND FRONT

General Ira C. Eaker took advantage of a history symposium at the Air Force Academy in Colorado Springs during 1978 to raise the question of whether Germany could have been defeated in World War II without the costly land invasion, that is by virtue of sea and air power alone. In addressing himself to this question, he said that the war against Japan had demonstrated "that by the proper employment of sea and air power, a land invasion of enemy territory can be avoided, saving heavy and unnecessary casualties." As to Europe, he quoted some statistics compiled by Marshal of the Royal Air Force Sir Arthur Harris. These were to the effect that in the last war "only 17 percent of Allied resources were devoted to air power, 33 percent to the sea war and 50 percent to the armies." Sir Arthur continued by saying, "had these resources been distributed between the three services equally, 33-1/3 percent to each, the RAF could have had 1,000 bombers and the 8th Air Force could have had 1,000 bombers and 1,000 long-range fighters in 1942." The conclusion Sir Arthur came to with Eaker's apparent agreement was that in that event "a land invasion of the continent might not have been necessary. All the armies would then have had to do was provide an occupation force in Germany as in Japan."[1]

The speculation that the European War could have been won by sea and air power alone was made with the benefit of hindsight, of course, because the Pacific War was still in progress at the time the decision was made to invade Europe at a certain fixed time and place. Thus the planners did not have before them the example of a great power being defeated by sea and air power alone while the enemy army was still intact.

The decision to launch the Second Front in Normandy was the single most important one made in World War II with the possible exception of the American one on "Europe first." The Normandy decision was made at Teheran in November 1943, and its importance was such that Winston S. Churchill called the Teheran Conference "The Crux."[2] Because of its transcending importance, the question of the Second Front was the most hotly debated one in World War II. Very honest differences of opinion emerged over whether it should be opened up at all, and if so when and where. Tempers flared to such a point on this subject that they almost led to the defection of Russia, and the conclusion of a separate peace with Germany. If this had happened it would have been the third "betrayal treaty" on the part of the Soviet Union, the first being the treaty of Brest-Litovsk that took Russia out of World War I in 1918, and the second the Non-Aggression Pact of 1939 that gave Adolf Hitler the green light to start World War II. The controversy over the Second Front also came perilously close to driving a permanent wedge between England and the United States. There the issue was the difference between the peripheral strategy of the indirect approach, strongly favored by Churchill, and heavily loaded with political considerations, and the frontal attack favored by the Americans, with a jaunty disregard for political considerations.

In other words, the differences of opinion on the Second Front created problems not only between the Western Allies but within the Grand Alliance, problems so serious that it is almost a miracle they were resolved without handing the victory over to Hitler on a silver platter.

Oddly enough, it was not Franklin D. Roosevelt or Premier Joseph Stalin that first broached the subject of a Second Front in Europe, but the British Prime Minister Winston Churchill. He did this quite inadvertently in a communication to Stalin on July 7, 1941, in which he referred to the very heavy air attacks on Germany and said, "thus we hope to force Hitler to bring back some of his air power to the West and gradually take some of the strain of you."[3] In reply, Stalin made it clear that while the very heavy air attacks were very welcome, they were not at all what he had in mind. He then proceeded to tell Churchill that the way England could really take the strain off the Soviet Union was to charge right back on the continent and open up a Second Front "against Hitler in the West -- Northern France, and in the North -- the Arctic." Stalin added the thought that this front "would be popular with the British Army, as well as with the whole population of Southern England."[4]

Stalin did not reveal his sources or explain why a return to the continent would be so well received in England at this time, and in fact the timing could not have been worse, coming as it did on the heels of the worst string of military defeats in English history, in Africa, in Greece and on Crete. Nevertheless and needless to say, Churchill was not impressed with the Soviet dictator's logic and quickly informed him that a Second Front in 1941 was not possible, no matter how much the Russians needed help, and no matter how anxious England was to give them all the help she could. However, this did not settle the matter. Stalin would not take no for an answer, and in one particularly insulting message told the Prime Minister that the Germans themselves were conviced "that no second front exists in the West, and that none will exist. Germans consider it quite possible to smash their enemies singly: first Russia, then the English."[5]

This was the beginning of the bitter dispute over the Second Front. It took a turn for the better, however, from Stalin's point of view, on December 7, 1941, when America became a full belligerent as a result of the Japanese attack on Pearl Harbor. The Soviet dictator then found that he had an ally in the White House, a very powerful ally that could be used against the stubborn Churchill, if that became necessary. This became apparent on April 11, 1942, when President Roosevelt made his personal views known to Stalin and proposed that he send Foreign Minister Vaycheslav Molotov and a reliable general to Washington in the immediate future to discuss "a very important military proposal involving the utilization of our armed forces in a manner to relieve your critical western front."[6]

When Molotov left Washington he carried with him a communique practically guranteeing the Russians a second front in 1942. This was in the form of a press release written by Molotov and concurred in by President Roosevelt saying that at the Conference a "full understanding was reached with regard to the urgent tasks of creating a Second Front in Europe in 1942." The only trouble was that Churchill refused to take the communique seriously, and his opinion was final because, as he said, the whole operation would have to be from British bases, and the forces employed "would have to be almost entirely British. The Navy, the air, two-thirds of the troops and such landing-craft as were available must be provided by us."[7]

Churchill agreed with Roosevelt's known desire for action on the part of the ground forces in 1942 but strongly disagreed that it should take place in Western Europe. His first choice was an invasion of North Africa, supported by an operation against Norway. Churchill quoted Hitler as having said that Norway was "the zone of destiny in this War,"[8] and for that reason proposed that it would be better "to lay out right claw on French North Africa, tear with our left at the North Cape and wait a year without risking our teeth upon the German fortified front across the Channel."[9] The only reason Churchill abandoned the Norway plan was because he could not get anyone to take him seriously, and as he said, he did not want to "queer the pitch"[10] for the North African operation by being too stubborn about the Norway plan. However, he never really gave up on the idea of invading Norway and of beginning "to roll the map of Hitler's Europe down from the top."[11] Norway was the invasion route Hitler feared the most-- the real straight-line approach to the heart of Germany.

As a result of Churchill's strong stand against the cross-Channel operation, the invasion of North Africa was launched on November 8, 1942. What this amounted to was that England got a second front in Africa at Rommel's expense, and Russia was denied a second front in Europe. It is small wonder that Stalin was a disillusioned and disappointed man -- no "Europe in 1942," in spite of the Roosevelt promise.

By the end of November 1942, the invasion of North Africa was attended with enough success to convince President Roosevelt and Prime Minister Churchill that a conference of the Big Three was necessary to determine what should be done after North Africa was cleared. Stalin proved unwilling to attend, so plans went ahead without him. The result was the Casablanca Conference held from January 14 to January 26, 1943

The Casablanca Conference was an important one, and the decisions far-reaching. Perhaps the most important of these was the one regarding the invasion of Sicily. On this were was a very sharp difference of opinion. The Americans were opposed to a further involvement in the Mediterranean on the grounds that it would delay the cross-Channel operation, which they believed was the only one that could be decisive in the war. "Every diversion or side issue from the main plot," Marshall said, "acts as a suction pump"[12] that will draw men and material from the main effort. However, he was overruled, and the decision was made to invade Sicily in July 1943. This, and the decision to undertake the Italian campaign the following September, ruled out any Second Front in 1943. The suction pump took care of that. As an American intelligence officer at Casablanca said, "we came, we listened and we were conquered."[13]

The decision to invade Sicily, important as it was, was not the only major decision to be made at Casablanca. There was also the announcement that electrified the world of "unconditional surrender," and the volte face on the part of Churchill with respect to the Flying Fortresses. He quoted General Eaker as sayint that this reversal on his part "saved the Fortress bombers from abandonment by the United States at the moment when they were about to come into their own."[14]

Churchill's decision to support Eaker's daylight bombing and thus to achieve an "around the clock" bombing capability against the Germans, meant that there would not be one, but two air offensives against the enemy, the one at night by the British, and the other in the daytime by the Americans. Together, the day and night assault was called the Combined Bomber Offensive, but since the philosophies of the two air forces were totally different, working out an operational order that would fit both became quite a chore. The result was a giant straddle, and a great victory for the drafting team. The order, dated February 4, 1943, was called the Casablanca Directive and said to the two air forces, "your primary object will be the progressive destruction and dislocation of the German military, industrial and economic system, and the undermining of the morale of the German people to a point where their capacity for armed resistance is fatally weakened."[15]

The famous Casablanca Directive was a godsend to the British and American airmen. To the British it said, win the war by continuing to bomb the great cities of Germany until the "*morale*" of the German people is fatally weakened. To the Americans, the same directive said, continue the destruction of the industrial and economic systems of Germany until the "*capacity*" of the Germans is fatally weakened. And to the airmen of both countries the famous directive said, your capacity as a war-winner has been officially recognized. Now you have a "mission" of your own, and it is an independent mission. The airmen made the most of this directive to "fatally weaken" Germany. As General Hansell said, the directive gave the airmen "a coordinated air objective on a level at least as high as the objectives of ground warfare, if not higher."[16]

But there was a Catch 22 to the war-winning Casablanca Directive. The airmen found this out on May 18, 1943, when the Casablanca Directive was amended in connection with the decision to invade France in the summer of 1944. The amendment consisted of a single sentence that was added to it after the words "fatally wakened." The sentence read, "this is construed as meaning so weakened as to permit initiation of final combined operations on the Continent."[17]

This put the airmen back at square one and made it clear that the basic plan was to defeat Germany on the ground. Out of the window went the war-winning version of the Casablanca Directive, and back in place was the old theory that only ground armies can finally win wars. More importantly, it meant that the one and only purpose of air power was to prepare the way for an invasion of the continent by the ground forces, and any idea of bombing Germany into submission could be forgotten. In other words, the airmen lost what they had counted on the most and what was implicit and explicit in all their planning, that they would have a period of six months after the bomber force reached full capacity to bomb Germany to the point where an invasion would be unnecessary, other than to occupy a defeated country. This is what "independence" meant to the airmen. It is also what they thought they had finally won at Casablanca. And it was what was taken away from them by the May 18 amendment to the famous Directive.

The airmen were the first to concede that air power was as yet too untried to rely upon it for final victory. In fact, General Hansell said in so many words, "victory through air power was pure theory. It had never been accomplished and it could hardly be expected that the United States would voluntarily pin its future existence on an untried theory."[18] On the other hand, to do exactly the opposite did not make much sense to the airmen either. In other words, while they granted that air power was as yet too untried to rely upon it exclusively for victory, it seemed to the airmen to be equally unwise to rule out victory through air power simply because it had not been accomplished in the past. That was a rather wooden approach to strategy, or at least so it seemed to the airmen. Nevertheless, this was the decision that was made, and it was implemented to the letter when General Eisenhower became Supreme Commander of the invasion forces. The folly of it was underscored by Churchill when he observed that "there were more British and American aircrew casualties than there were killed and wounded in the great operation of crossing the Channel."[19]

Before the Casablanca Conference ended on January 27, 1943, provisions were made to continue the heaviest possible air offensive against Germany, and at the same time to speed up the concentration of forces in the United Kingdom for an invasion of the continent as soon as Germany had been weakened to the required extent. Provisions were also made for an emergency return to the continent in the event that Germany suddenly cracked up under the avalanche of bombs or for any other reason, including the death of the Führer, a landslide of defections in the Balkans, or even in case of popular uprising in Germany itself. A highly qualified and respected British staff officer, General Sir Frederick Morgan, was put in charge of this planning, and he worked out Cases A, B and C, to cover any kind of a breakdown in Germany. In case of C, a general collapse, a million men were to be thrown on the continent, and an airborne division was to be dropped into Berlin. However, for all these plans and agreements, Casablanca did not achieve a meeting of the minds between the British and the Americans. Their differences were deep and fundamental. They were the differences between a strategy of attrition (the British view) and one of frontal attack (the American view). No compromise was possible, and the basic issues were not resolved until the Big Three met at Teheran in November 1934, and then Churchill was bludgeoned into accepting the American view.

Churchill's misgivings about a premature crossing of the Channel was rooted in the mass slaughters of the World War I experience and on the certain knowledge that Hitler's strength on the ground in Western Europe was very formidable. It did not make sense to

Churchill to fight the war where the enemy was strongest, but where he was weakest. For this reason, he believed the airmen should be supported and given an opportunity to bring Germany down with their bombs, and for this reason also he believed in his Mediterranean-Balkans strategy, particularly getting behind the Germans in the Balkans, as well as in the descent from Norway.

On the other hand, the Americans were hell bent on what they called the "straight line" approach to peace. As Roosevelt explained it to Churchill, "I always think of my early geometry-- 'a straight line is the shortest distance between two points'"[20] and said that he saw no reason to question this truism in the present circumstances. Therefore, with the strong support of Marshall and Eisenhower, the American President decided to stare Churchill down at Teheran and to team up with Stalin to bury the idea of anything other than the "straight line" approach to the defeat of Germany. This meant tearing up all the plans for a Mediterranean-Balkans strategy, giving up the descent from Norway and choking off any possibility that the planned strategic air offensive would make an invasion of the continent necessary.

There were several "good" reasons the United States teamed up with Russia at the expense of Churchill at Teheran. First of all, Roosevelt and the military leaders were strongly convinced of the validity of the "straight line" approach to the defeat of Germany. None of Churchill's arguments were convincing to the Americans, and in fact he was suspect in their eyes because they believed his strategy was rooted in a desire to preserve the British Empire in the post-war world. There was also a strong belief that Russian help would be needed to defeat Japan in the Pacific War, that is when the home islands were invaded, as they would have to be to win the war. Finally, there was the certain knowledge that Russia held the key to making the World Organization a success in the post-war world, in that with her cooperation and support it could be a viable instrument for peace, but without Russian support the instrument was bound to be a failure. Therefore, President Roosevelt put all his eggs in the Russian basket at Teheran and decided that Churchill would simply have to make the best of it. Finally, there was the matter of the Combined Bomber Offensive. This was an easy thing to write off because it was only an untried theory at best, and it was clear that the time for playing games was over. Arnold was no problem on this score because he was inclined to defer to Marshall in order to be sure of his support in connection with the battle over the post-war organization of the Armed Forces.

Under these circumstances, what happened at Teheran was a foregone conclusion and is best described by the victim of the plot, by Churchil. "There I sat," he said, "with the great Russian bear on one side of me, with paws outstretched and on the other side the great American buffalo, and between the two sat the poor little English donkey." This forced the Prime Minister to realize, probably for the first time, "what a small nation Britain was." Nevertheless, Churchill never gave up the struggle and insisted to the end that England's poor little donkey was the only one "who knew the right way home."[21]

Roosevelt opened the Teheran Conference by saying "that the Russians, the British and the Americans are sitting round the table for the first time as members of the same family, with the single purpose of winning the war."[22] He continued by reviewing and outlining the objectives of the war in the Pacific and then addressed himself to the war in Europe. He said the question was whether the Anglo-American armies should concentrate on the invasion of France in 1944, or whether it would be better for them to exploit the opportuniites open to them in the Mediterranean? He continued by explaining that these were competing attractions because of the shortage of landing-craft.

"Many plans," he continued

> had been mooted--increasing the strength of our attack on Italy, the Balkans, the Aegean, Turkey and so forth. The most important task of the Conference would be to decide which of these to adopt. The governing object would be for the Anglo-American armies to draw the greatest weight off the Soviet forces.[23]

In other words, President Roosevelt was appealing to Stalin to be the final judge between the competing strategies that had been mooted between the British and the Americans.

Stalin did not hesitate to hand down a decision. He began by ridiculing the whole idea of the Mediterranean-Balkans strategy, including the capture of Rome, which he called a mere diversion, and continued by saying that if the Conference had been called to discuss military matters, there was only one that was worthy of discussion, and this was the invasion of northern France, supported by an assault on the coast of south France. As to the latter, Stalin said, if this could be carried out two or three months before the invasion of northern France, "so much the better, but if not, it might coincide with it, and, if it could not coincide, it would still help if it took place a little after it. The assault on the south of France as a supporting operation would be definitely helpful."[24] On the subject of the invasion of northern France, Stalin said there were only two matters to be decided. One was the date, and he made it clear that "this should be in May and no later," and the other was the appointment of a Commander-in-Chief, which Stalin said should be made "before the Conference ended, or at least within a week thereafter."[25]

This has to be one of the most bizarre chapters in the history of American diplomacy, letting the Soviet dictator become the arbiter of the world's fate by casting the deciding vote on the use of Anglo-American forces in the greatest war in history. It is no wonder that Churchill called the Teheran Conference The Crux.[26] It was a Crux in more ways than one.

After Churchill was forced to gulp down the Normandy decision at the expense of the Mediterranean-Balkans strategy, the descent from Norway, and the strategic air offensive, the Soviet dictator sought the Prime Minister out to impress on him the fact that there was no turning back on the Second Front. Stalin reminded Churchill that Russia always had the alternative of making a separate peace with the Germans. Churchill recalled this by quoting Stalin as saying,

> he must warn me that the Red Army was depending on the success of our invasion of northern France. If there were no operations in May 1944, then the Red Army would think that there would be no operations at all that year. The weather would be bad and there would be transport difficulties. If the operations did not

> take place, he did not want the Red Army to be disappointed. Disappointment could only create bad feeling. If there was no big change in the European War in 1944, it would be very difficult for the Russians to carry on. They were war-weary. He felt that a feeling of isolation might develop in the Red Army.[27]

It did not take much imagination to figure out what Stalin was talking about when he said that the war-weary Russians would not be able to carry on if there was no big change in the European war in 1944. The big change Stalin had in mind was the Second Front. In 1939 a feeling of "isolation" had developed in the Red Army, and the result was the "betrayal treaty," with Germany, a treaty that led to the outbreak of World War II. Stalin did not want the Western Allies to forget that a separate peace was always possible, and indeed probable if there was another double-cross on the Second Front.

The idea of a separate peace was not anything new with Stalin. Like Clausewitz, he never forgot the primacy of the political objective in war, and shortly after the battle of Stalingrad took the lead in putting out peace feelers to the Germans, only to be rebuffed by Hitler who demanded a satellite Ukraine plus economic concessions in other parts of the Soviet Union. The Russians, on the other hand, insisted upon their frontiers of 1941.[28] The talks were resumed in June of 1943 at an even higher level, and according to the British historian, B. H. Liddell Hart,

> In June, Molotov met Ribbentrop at Kirovgrad, which was then within the German lines, for a discussion about the possibilities of ending the war. According to German officers who attended as technical advisers, Ribbentrop proposed as a condition of peace that Russia's future frontier should run along the Dnieper, while Molotov would not consider anything less than the restoration of her original frontier; the discussion became hung up on the difficulty of bridging such a gap, and was broken off after a report that it had leaked out to the Western Powers.

> The issue was then referred back to the judgment of battle.[30]

The judgment of battle was against the Germans, and following the Russian victory at Kursk in July, Stalin let the Germans take the initiative in the peace talks. In early September he was still willing to listen to anything the Germans had to say, and it was at just about this time, one month earlier in fact, that Robert E. Sherwood reported that the atmosphere was "alarmingly reminiscent of that which preceded the Molotov-Ribbentrop peace of August 1939."[30] This was only three months before the Teheran Conference, and thus the "peace scare" at Teheran should not have come as any surprise to the Allies. General Albert C. Wedemeyer was probably right when he said that Stalin "was simply holding the idea over our heads and blackmailing us."[31]

It says something about the lack of common sense on the part of the Americans that they could be blackmailed by the threat of a separate peace at Teheran. Why was this such a nightmare? A separate peace between Germany and Russia at the time of Teheran would have guaranteed a balance of power on the continent after the war instead of a one-power domination and might have been the best possible way the war could have ended. Of course, it might have played havoc with the idea of "unconditional surrender," but the loss of that specious doctrine would have been a blessing and certainly not a tragedy. There never has been any real peace in the world without a balance of power in Europe, that is unless the unrest of the Cold War can be called a real peace.

In any case, after all the major decisions had been taken at Teheran, and the poor little English donkye was left to find his own way home, Stalin, relishing the "May and no later"[32] triumph, was highly cooperative about supporting the Normandy landings with an offensive of his own to prevent the Germans from transferring any forces from the East to repel the invasion. He said he fully realized that the "danger period" for this type of operation "would be at the time of deployment from the landings," and to this end he pledged "a large-scale Russian offensive in May."[33] But this is not the way it worked out, of course. The Allied invasion of Western Europe began on June 6, 1944, and the main Russian attack did not get underway until June 23. Liddell Hart said that by the time the Russians launched their offensive "the Anglo-American invasion of Normandy was well established. This, together with the Allied advance beyond Rome, ensured that the Germans were hard pressed everywhere before the Russians struck."[34]

In other words, the Russians did not support the Western Allies in the "danger period" of their landings, but used the landings instead for support of the Russian summer offensive. This set the pattern for the future and made it possible for the Soviets to get a piggy-back ride on the offensives in the West. Otherwise, they would have been held far back in the East, at least within their 1941 frontiers.

The Russian drive that was launched on the back of the Normandy landings advanced 460 miles to Warsaw at the Vistula. The distance covered was approximately to the line Molotov had demanded in the peace talks with Ribbentrop a year earlier. After reaching the line of the Vistula in August 1944, the Red Army did not resume its advance until January 1945. This time the drive was from the Vistula to the Oder, and again it was helped, or rather made possible, by a serious setback the Germans suffered in the West. This was the disastrous failure of Hitler's audacious gamble in the Ardennes.

The truth of the matter is that Hitler refused to take the Russians seriously. Though his generals assured him that he was mistaken in this, the Führer persisted in the notion that the Russians were incapable of mounting a major offensive on their own, and argued that even if this was not the case it would not make any difference because it was easier "to give up territory in the east than in the west where the Ruhr was directly threatened."[35] In fact, at the time of Teheran, six months before the Normandy invasion, German strength in the West "totalled no fewer than 2,440,000 men compared with the figure of 2,800,000 in Russia."[36] In addition, Hitler maintained a substantial reserve of "more than twenty-five divisions"[37] in Germany itself. In 1942, when the vital offensive against Stalingrad was launched it came within 300 yards of being successful and failed only because of a shortage of two or three divisions that Hitler had refused to release from the defenses in the West.[38]

The plain fact of the matter is that Hitler underestimated the Russians and took exaggerated counsel of his fears in the West. At a briefing conference on December 20, 1943, Hitler expressed the firm conviction that the war would be decided in France, not in Russia.[39] In other words, the last thing Hitler was willing to do was weaken the defenses in the West in order to meet the needs of the East. And it is only against this background that the Ardennes offensive makes any sense at all.

Hitler called Field-Marshal Gerd von Rundstedt out of retirement to put the Ardennes offensive together. It was a bold plan designed to catch the Allies by surprise, which it did, and its objectives were to split the Western Allies, seize Antwerp, and thus turn the tide of battle in the West. To carry out this plan, Hitler seriously weakened the Eastern Front and threw in "nearly the whole of the German Strategic reserve."[40] Nothing was held back in this win-the-war offensive even though the worried generals on the Eastern Front warned the Führer that the huge gamble left the Soviets with a superiority in the East of "eleven to one in infantry, seven to one in tanks, and twenty to one in guns."[41]

The first time Albert Speer got word of the Ardennes offensive was on October 12, 1944, when the Führer told him he needed his help in organizing a great win-the-war offensive. "For that," Hitler said, "you must organize a special corps of German construction workers, one sufficiently motorized to be able to carry out all types of bridge building even if rail transportation should be halted."[42]

In the course of organizing the construction workers and shifting industrial priorities Speer had a conference with one of the German industrial chiefs, Albert Vogler. However, all Vogler wanted to talk about was the bombings in the Ruhr and "when are we going to call it quits?" It was only with some difficulty that Speer got the conversation back to the business at hand, and when he did so, the first question he got from Vogler was, where is the operation going to take place? Not getting a straight answer, the industrialist answered his own question, "of course, it will be on the eastern front. Nobody could be so crazy as to strip the East in order to hold back the enemy in the West."[43]

Be that as it may, Hitler's offensive in the Ardennes began on December 16, 1944. Under cover of darkness and bad weather, which grounded the Allied air forces, von Rundstedt struck the blow that Hitler hoped would be a repetition of the successful strategy of 1940. However, history did not repeat itself, and the initial breakthrough was not powerful enough to carry the Germans through to their objectives. As soon as the skies cleared, Eisenhower's overwhelming superiority in the air was more than enough to turn the Germans back, and by Christmas Eve it was apparent even to Hitler that he had lost his last gamble. Von Rundstedt said, "I wanted to stop the offensive at an early stage, when it was plain that it could not attain its aim, but Hitler furiously insisted that it must go on. It was Stalingrad No. 2."[44] Speer was even more outspoken. He said that Hitler played his "last card" in the Ardennes offensive and that its failure meant "the war was over." Speer also said, "what followed was only the occupation of Germany, delayed somewhat by a confused and impotent resistance."[45]

With the opening of 1945, after the Battle of the Bulge had run its course and the American First Army had crossed the Rhine at Remagen on March 8, General Eisenhower, acting entirely on his own, sent a message to Stalin advising him that Berlin had lost its purpose as a military objective and telling him that the American armies would advance along the Erfurt-Leipzig-Dresden axis, not in the direction of Berlin. He asked Stalin for his comments. The Soviet dictator assured Eisenhower that his plans meshed with those of the Soviet High Command, and with respect to Berlin Stalin added the thought that it has "lost its former strategic significance," with the result that the Red Army would move only "secondary forces in the direction of Berlin."[46] As to the timing, Stalin told Eisenhower that the Soviet advance would get underway about the second half of May.

Stalin had no sooner sent this message to Eisenhower than he called his top generals to Moscow and ordered them to make their major objective Berlin, and to speed up their preparations so that the drive could be underway by April 16. It is obvious that Stalin wanted to forestall any possibility of Eisenhower's changing his mind about the strategic importance of Berlin, and it is also obvious that the decision to allow the Russians to "liberate" Berlin was Eisenhower's. In fact, in his Memoirs, General Eisenhower said, *"I decided"* that Berlin was not the logical nor the most desirable objective, and *"I decided"* to invade southern Bavaria, western Austria and northern Italy to prevent the Germans from establishing a National Redoubt.[47]

A slightly different version of the "liberation" of Berlin is in the Memoirs of Nikita Khrushchev. Recounting the very "good relations" Stalin always had with Eisenhower and the even "better ones" with Roosevelt, Khrushchev said,

> Stalin appealed to Eisenhower in a letter to hold back his armies; Stalin told Eisenhower that according to his agreement with Roosevelt and in view of the amount of blood our people had shed, our troops deserved to enter Berlin before the Western Allies. Eisenhower then held his troops back and halted their offensive, thus allowing our troops to take Berlin. If he hadn't done this, Berlin would have been occupied by the Americans before we reached it, in which case, as Stalin said, the question of Germany might have been decided differently and our own position might have turned out quite a bit worse. This was the sort of chivalrous generosity Eisenhower demonstrated. He was true to Roosevelt's word.[48]

World War II did not have to end this way and did not have to be fought this way. If the Western Allies had said no to Stalin at Teheran and if the original air-war plan had been implemented in letter and spirit, there would have been no need for a Second Front, and the war could have ended with a balance of power in Europe instead of a one-power domination. This would have been far better for the history of Europe, for American history and possibly for the history of the world. Some of the broader details of how this might have worked out can be ascertained by examining what did happen in the light of "what might have been" had the war followed a different course, beginning with a no to Stalin at Teheran.

The original air-war plan has been rightly called "one of the most remarkable documents in the history of American air power,"[49] and it was remarkable in several respects. It accurately predicted the exact number of air groups that would be required in the war against Germany, and also accurately predicted the exact time it would take to get these airplanes and their combat crews in position for the attack. Also, it made an accurate prediction of the time it would take for the strategic air offensive to produce the desired results. More specifically, the original plan projected a force requirement of 239 combat groups, and at peak strength the Air Force had 243 such groups. The original plan predicted that the force would reach effective strength by the middle of April 1944, and it was just about this time that it reached such strength. Finally, the original plan predicted that after a period of six months of ceaseless pounding from the air Germany would be reduced to the state of helplessness that was actually achieved in May 1945, when Germany surrendered.

What would the airmen have done differently if the Western Allies had said no to Stalin at Teheran? Using as a starting point the Teheran Conference of November 28 to December 1, 1943, it is probably only fair to say that there would have been no immediate difference between what actually took place and what might have been under the new scenario. The airmen had already learned their bitter lesson at Schweinfurt on August 27 and October 14, 1943, that their Flying Fortresses were not flying fortresses, and that deep penetrations of Germany could not be undertaken until the bombers could be escorted all the way to their targets and back again. But just as the Teheran Conference was coming to an end the long-range fighters began arriving in England, and the door that had been closed since October 14 began to swing open again. The result was the resumption of the strategic air offensive. But the question is would it have been any different?

The answer is probably no, at least not at the outset. The battle for air supremacy had to be won or there would be no strategic air offensive, and the only way this could be done was by destroying the German Air Force, particularly Fighter Command. This left the planners no choice and forced them to concentrate on the destruction of the German aircraft industry, as they did in the famous Big Week. However, if the planners had not been boxed-in by the May deadline for the invasion of the continent they might have chosen a different target system, and a better one at that, aircraft engine plants instead of the final assembly facilities. After the event, Field-Marshal Erhard Milch said,

> It is true that air attacks on the airframe factories did hold up our production, but they did not reduce it. I have never understood why you did not attack the engine factories in 1944. If you had attacked them straightway, it would have been far worse. You needn't have attacked the airframe factories at all.[50]

As for the abortive attack on the highly vulnerable ball-bearing industry in Germany, would it have gone any differently if the Western Allies had said no to Stalin at Teheran?

There is good reason to believe it would have. British and American airmen saw eye-to-eye on the subject of ball bearings as a short-cut to the destruction of the German armaments industry. This is what led to the premature attack on the ball-bearing industry, before the long-range fighters were available in 1944, but the ball-bearing targets were attacked again in the Big Week. The airmen also came back to the ball bearings in the March-April period, but then, as Speer said, the attacks "ceased abruptly."[51] This coincided with the transfer of the control of the bombers to the ground forces in April and the attack did not pick up again until October 1944, after control of the heavies was switched back to the airmen.

It almost goes without saying that if the airmen had retained control of their bombers they would have pushed the attack relentlessly after the Big Week, and if they had done so, as Albert Speer said, it would have paralyzed "the production of thousands of armaments plants."[52] As it was, and as Speer also said, "the Allies threw away success when it was already in their hands. Had they continued the attacks of March and April with the same energy, we would quickly have been at our last gasp."[53]

In other words, if the bombers had been allowed to bomb strategically instead of tactically, German armaments production might have come to a full stop by the end of April 1944.

The story of the oil industry is much the same thing. Germany was starved for oil right from the beginning, particularly aviation fuel, and the industry was highly vulnerable to air attack because 80 percent of the aviation fuel came from only 10 facilities. In addition, the chemical industry of Germany was closely integrated with the large hydrogenation plants. Kill one, and the other dies along with it. With the loss of the synthetic plants, Germany could not produce aviation fuel, munitions, explosives, rubber and a host of other essentials to carry on a war.

However, the industry did not come under serious attack until May 1944, and then the campaign against it was stretched out over nearly a full year, instead of being compressed into a period of only a few months. The reason for this was because there was a prior demand on the part of the ground forces for the heavy bombers. In fact, only 13 percent of the tonnage of bombs dropped in the war went on the oil targets. The attack on the oil targets could have begun much earlier than it did and could have been completed by June or July, except for necessary policing. The paralysis of the Germany economy and the lack of mobility on the part of the armed forces that came about when Germany ran out of oil could have been brought about nearly a year earlier if it had not been for the folly of the Second Front.

Transportation was the obvious target system once the bomber force had reached peak strength. It was an ideal system for a Combined Bomber Offensive and also for all types of bombers, heavy, medium and light. Transportation was the key to everything else in Germany, to the movement of coal and raw materials to the factories, to the movement of food to the cities and to the movement of all the armed forces, land, sea and air. And the railroads were the key to the system as a whole. The statistics tell the story. Of all the traffic that moved in Germany, military and civilian, 72 percent was on the railroads, 25 percent on the waterways, and three percent on the airways. Without the railroads, Germany would not be a going concern, militarily or economically. However, it was not until the end of October 1944 that the transportation system of Germany as a whole was brought under heavy, serious and sustained attack. Within five months, Germany was a helpless wreck. As the Bomb Survey said,

> The attack on transportation was the decisive blow that completely disorganized the German economy. It reduced war production in all categories and made it difficult to move what was produced to the front. The attack also limited the tactical mobility of the German army.[54]

Yet of the total tonnage of bombs dropped over the whole period of the war, only 26 percent were on transportation targets. Germany could have been reduced to "a state of helplessness"[55] by the July to August period of 1944 had it not been for the decision at Teheran and the folly of the Second Front.

The Bomb Survey pinpointed the date of the German collapse as the beginning of 1945, which would put it shortly after the failure of the Ardennes offensive and before the final military victories that carried the Allied armies across the Rhine and the Oder had taken place. In fact, it was just at about this time that the Germans quit keeping any records as to the arrivals and departures of the railroad trains in the Fatherland.[56] What the Survey said was:

> By the beginning of 1945, before the invasion of the homeland itself, Germany was reaching a state of helplessness. Her armament production was failing irretrievably, orderliness in effort was disappearing and total disruption and disintegration were well along. Her armies were still in the field. But with the impending collapse of the supporting economy, the indications are convincing that they would have had to cease fighting -- any effective fighting--within a few months. Germany was mortally wounded.[57]

General Hansell took a more constructive view of the collapse of Germany when he said:

> All experts agree that by March of 1945, interior Germany was a shambles, on the verge of total collapse. By that time Germany was no longer able to carry out the essential functions of corporate life, to say nothing of carrying on a major war. Both her industrial capacity to support the war and her national economic structure were shattered. Thus, the strategic objectives of air warfare as presented in the plans were actually attained.[58]

This was not all General Hansell had to say about the original air plans and the folly of the Second Front. He also said that while the strategic objectives of the original air plans were achieved, the attainment of these objectives:

> was seriously out of phase with the timing intended in the plans. Had the strategic air effort not been so often diverted to tactical targets the mortal wounding of Germany would probably have been attained through air warfare by June of 1944, rather than March of 1945. That would have been in keeping with the concept of the original strategic air plan. Unfortunately, the strategic air plans were not closely followed and their effect was consequently diluted. In my considered opinion, this dilution had a tragic effect in prolonging the war with attendant loss of life.[59]

There is no question about whether the airmen could make a shambles out of Germany and do it within a period of six months or less. This was done in the way it had been promised in the six months following the time when control of the bombers was switched from the ground forces back to the airmen. But by then it was too late. The case of the airmen had been predicated on the assumption that when the strategic air forces reached a planned peak, the airmen would be given six months to bomb Germany into submission "*before* the invasion, if an invasion should prove necessary."[60] Instead, the invasion took place while the bombers were under the control of the ground forces. Nevertheless, when this control was lifted the airmen proceeded to do what they had promised, and within six months Germany was mortally wounded. By this time it was already too late. The toll exacted by the ground invasion had been paid in full. The heavy price that had to be paid for this violation of the rule of common sense can never be recovered, for the price was in the blood of nearly half a million American casualties in the Second Front that never should have been. If ever there was a case of the cart being put before the horse, this was it.

In other words, if the timing of the Second Front had been in accordance with common sense, as it was in the original air plan, a Second Front might not have been necessary other than to provide the occupation force for a defeated country. This is precisely what happened in the case of Japan later in the year, and it was precisely what had been anticipated at Casablanca when General Morgan had been given the task of making detailed plans for an emergency return to the continent. Thus of all the blunders of World War II, the timing of the Second Front was probably the worst, and to say this is to put the finger of blame where it belongs, on the "May and no later"[61] decision at Teheran.

Actually, there was nothing to be gained by the decision at Teheran except the obvious, that it gave Stalin a free hand with respect to East Europe. Hitler was heading for certain defeat in the war and was on the defensive in all the elements, land, sea and air. He was on the retreat in Russia, had been evicted from Africa, was being defeated in the Battle of the Atlantic and was on the verge of having his Luftwaffe wiped out. Which made the "danger" of a separate peace a spurious reason for appeasing Stalin, and the other arguments turned out to be equally as fallacious. Russia's help was not needed to defeat Japan, and no amount of appeasement was enough to gain Stalin's support for the World Organization in the post-war world. In other words, there was nothing to be gained by the "May and no later" decision except the defeat of Churchill. It would have been much better if the Americans had kept all their options open at Teheran, the better to switch from one plan to another, as Churchill said, "as and where the cat jumps."[62]

The only interest that was served at Teheran was Stalin's, and it was served well. Of the participants, it turned out that Churchill was the only one who really knew what was going on. As he said later, "The Soviet menace, to my eyes, had already replaced the Nazi foe."[63] In 1944, as the Russian armies were sweeping across the face of East Europe liberating all the ancient capitals, Churchill said, "Russia was the Deliverer and Communism the gospel she brought."[64] It was at Teheran and not at Yalta or Potsdam that Russia won the war.

At Teheran, there was another factor that should have been considered (and was not) in connection with the "May and no later" decision. This was the possibility (probability?) that Germany would suffer a breakdown on the home front, as was the case in World War I, and that Germany would collapse as a result. By the time of Teheran, fear, defeatism and war-weariness were widespread in Germany. The Bomb Survey estimated that by the beginning of 1944 (which was right at the time of the Teheran Conference, and only seven months before the July 20 attempt on Hitler's life) "three-fourths of the German people regarded the war as lost. Bombing aided greatly in producing conviction of Allied superiority both through severity of raids and through the unchecked passage overhead of fleets of Allied planes."[65] The attempt on the life of the Führer on July 20 was only the tip of the iceberg. Legions wanted Hitler dead so Germany could be saved. Discontent was rapidly approaching the point it reached in Japan when the Emperor bowed to the unconditional surrender terms. And in a way this is not surprising. In Germany, seven and a half million people were rendered homeless as a result of the bombing, and there were over a million civilian casualties.[66] In fact, there were more civilian casualties in Germany than there were deaths on the part of British and American soldiers in World War II.[67]

The plain fact of the matter is that at the time of the Teheran Conference, Germany was about to explode, and if ever the time was right for a shocker such as the atomic bomb, this was it. Germany was probably more vulnerable to an atomic shock at this time than

Japan was at the time of Hiroshima and Nagasaki. Then why was it that the bomb did not figure at all in the planning for the defeat of Germany, that it was not a factor at Casablanca (January 1943), at Teheran (December 1943), and at the time of the Normandy invasion (June 1944)?

The airmen were confident that a non-atomic air offensive was all that was needed to bomb Germany into submission, and if the atomic bomb had figured in their planning they would have been twice as certain, as indeed would any other reasonably minded person. Then why was the atomic bomb not taken into consideration in connection with the explosive issue of the Second Front, not at Casablanca, not at Teheran, and not at the time of the actual landings?

As early as 1940, two respected British scientists, Otto R. Frisch and Rudolph E. Peierls, predicted that "an atomic bomb might quickly be built."[68] Largely on the strength of this prediction, England and the United States pooled their efforts to build a bomb on October 11, 1941, and on December 2, 1942, got their big break when Enrico Fermi attained a self-sustaining nuclear chain reaction in his laboratory at Stagg Field, Chicago. This was one year to the day *before* the final session of the Teheran Conference.

Fermi's spectacular breakthrough turned out to be the halfway point in the development of the bomb. After that, it was all downhill, and the only question was when the bomb would be produced, not whether a bomb would be made. Oddly enough, at almost the exact time that Fermi made his breakthrough the Germans decided to scuttle their atomic project. Hitler lost interest in the subject, and when he referred to the German atomic energy program, according to Albert Speer, he did so with a sneer, asking not about nuclear physics but about "Jewish physics."[69]

In October 1943, one month before the beginning of the Teheran Conference, one of President Roosevelt's principal advisers, Vannever Bush, told the President there was a "good chance"[70] the atomic bomb would be ready for use in 1945. Which makes it stranger still that the use of the atomic bomb against Germany was not considered at Teheran because, as Herbert Feis has pointed out, "the prime incentive of the effort to make an atomic bomb had been to assure and hasten the defeat of Germany,"[71] not Japan.

As the date for the invasion of Normandy approached, the Director of the Manhattan District Project, General Leslie R. Groves, confirmed Vannever Bush's estimate and became even more specific, "telling Marshall that current expectations were that several bombs of one type would be available between March and June 1945, but if further work on this type did not come up to this mark, he was confident that another and more powerful type would be ready by the first of August."[72] Then on March 15, 1945, *three months before the invasion of Normandy*, Secretary of War Henry L. Stimson met with President Roosevelt for what turned out to be the last time, and he did so in order to suppress any lingering doubt"[73] the President might have about the reality of the bomb. Finally, in early April, Stimson paid a visit to the Oak Ridge plant, after which he confided to his Diary, that "success is ninety-nine percent assured, yet only by the first actual war trial of the weapon can the actual certainty be fixed.

Thus it seems clear that the atomic bomb was a reality, and a reality in plenty of time to have been a factor in the European war for which it was intended. But it was not a factor, and the million or so American lives the bomb saved in the Pacific war were not saved in the European war. As it was the Second Front took its grisly toll of lives and took them needlessly.

If there had been no Second Front, only a bomber offensive for six months at full strength, Germany would have been bombed into submission "by June of 1944,"[75] according to General Hansell, or at least "by the beginning of 1945,"[76] according to the estimate of the Bomb Survey. However, if they were both wrong, and if the war went on until August 1945, the atomic bombs could have been used against Berlin and Hamburg just as easily as they were used against Hiroshima and Nagasaki. Surely an additional three months of war beyond May would have been a small price to pay for the advantage of not having a Second Front.

And there was another disadvantage of the Second Front that should not be forgotten. This was that Germany was bled white by the fighting on two fronts, so much so that her armies failed in doing what otherwise would have been an easy task, which was to hold the Russians far back in the East. The Russian drive to the Vistula on June 23 was made possible by the Allied landings in France on June 6, 1944, and the Russian drive to the Oder on January 12, 1945 was made possible by the German collapse in the Ardennes on December 25. Finally, the dash from the Oder to Berlin on April 14 was also made possible by the collapse of the Rhine and the fall of the Ruhr in February and March, plus the chivalrous generosity that Eisenhower displayed when he held his troops back to allow the Russians to take Berlin.

On the other hand, if the original air plan had been followed, and if the Russians had been held far back in the East, as would have been the case in a one-front war, the peace that would have followed would have been far more compatible with the legitimate war aims of the democracies than was actually the case. East Europe would have been free instead of being behind the Iron Curtain; Germany would have been united; and the Russians would have been confined to their legitimate and natural frontiers. In other words, instead of a Cold War and the nuclear arms race, peace might have broken out in the world.

To put this another way, if the air strategy had prevailed the war would have ended with the Russians in no position to dominate either East or West Europe, and there would have been a balance of power on the continent. While this would not have attained the Mackinder ideal of a balanced globe of human beings, "and happy, because balanced and thus free,"[77] the prospects of a happy globe of human beings would have been considerably enhanced. In other words, if the strategy of common sense had prevailed in the European War instead of the folly of the Second Front, the final result would have been much better for the world as a whole.

FOOTNOTES--CHAPTER VII

[1]Ira C. Eaker, "Some Observations on Air Power," in *Air Power and Warfare* (Washington: Government Printing Office, 1979), 356

[2]Winston S Churchill, *Closing the Ring* (Boston: Houghton Mifflin Company, 1951), 375.

[3]Winston S. Churchill, *The Grand Alliance* (Boston: Houghton Mifflin Company, 1950), 381.

[4]*Ibid.*, 383

[5]*Ibid.*, 456

[6]Mark A. Stoler, *The Politics of the Second Front* (London: Greenwood Press, 1977), 42.

[7]Winston S. Churchill, *The Hinge of Fate* (Boston: Houghton Mifflin Company, 1950, 323.

[8]*Ibid.*, 112.

[9]*Ibid.*, 323-24.

[10]*Ibid.*, 325.

[11]*Ibid.*, 350.

[12]Forrest C. Pogue, *George C. Marshall: Organizer of Victory* (New York: The Viking Press, 1973), 22.

[13]Mark A. Stoler, *The Politics of the Second Front* (London: Greenwood Press, 1977), 77.

[14]Winston S. Churchill, *The Hinge of Fate* (Boston: Houghton Mifflin Company, 1950), 680.

[15]Winston S. Churchill, *Closing the Ring* (Boston: Houghton Mifflin Company, 1951), 519-20.

[16]Haywood S. Hansell, Jr., *The Air Plan That Defeated Hitler* (Atlanta: Higgins-McArthur/Longino & Porter, 1972), 197.

[17]*Ibid.*, 168.

[18]*Ibid.*, 75.

[19]Winston S. Churchill, *Closing the Ring* (Boston: Houghton Mifflin Company, 1951), 531

[20]Winston S. Churchill, *Triumph and Tragedy* (Boston: Houghton Mifflin Company, 1953), 66.

[21]Mark A. Stoler, *The Politics of the Second Front* (London: Greenwood Press, 1977), 150.

[22]Winston S. Churchill, *Closing the Ring* (Boston: Houghton Mifflin Company, 1951), 347.

[23]*Ibid.*, 349.

[24]*Ibid.*, 369.

[25]*Ibid.*

[26]*Ibid.*, 351.

[27]*Ibid.*, 379-80.

[28]Vojtech Mastny, "Stalin and the Prospects of a Separate Peace in World War II," *The American Historical Review*, v. 77, no. 5 (December 172), 1376.

[29]B.H. Liddell Hart, *History of the Second World War* (New York: G.P. Putnam's Sons, 1970), 48.

[30]Vojtech Mastny, "Stalin and the Prospects of a Separate Peace in World War II," *The American Historical Review*, v. 77, no. 5 (December 1972), 1386.

[31]*Ibid.*, 1384.

[32]Winston S. Churchill, *Closing the Ring* (Boston: Houghton Mifflin Company, 1951), 369.

[33]*Ibid.*, 383.

[34]B.H. Liddell Hart, *History of the Second World War* (New York: G.P. Putnam's Sons, 1970), 578.

[35]Albert Seaton, *The Russo-German War 1941-45* (New York: Praeger Publishers, 1971), 528.

[36] *Ibid.*, 395-96.

[37] *Ibid.*, 527.

[38] *Ibid.*, 270.

[39] *Ibid.*, 395-96.

[40] *Ibid.*, 527.

[41] *Ibid.*

[42] Albert Speer, *Inside the Third Reich* (New York: The Macmillan Company, 1970), 492.

[43] *Ibid.*

[44] B.H. Liddell Hart, *History of the Second World War* (New York: G.P. Putnam's Sons, 1970), 659.

[45] Albert Speer, *Inside the Third Reich* (New York: The Macmillan Company, 1970), 497.

[46] Winston S. Churchill, *Triumph and Tragedy* (Boston: Houghton Mifflin Company, 1953), 460.

[47] Dwight D. Eisenhower, *Crusade in Europe* (New York: Doubleday and Company, 1949), 396-97.

[48] Nikita Khrushchev, *Khrushchev Remembers* (Boston: Little, Brown and Company, 1970), 221.

[49] Eugene A. Emme, "The American Dimension," *Air Power and Warfare* (Washington: Government Printing Office, 1979), 78.

[50] The United States Strategic Bombing Survey, *Overall Report: European War* (Washington: Government Printing Office, 1945), 22.

[51] Albert Speer, *Inside the Third Reich* (New York: The Macmillan Company, 1970), 340-41

[52] *Ibid.*, 340.

[53] *Ibid.*, 341.

[54]The United States Strategic Bombing Survey, *Summary Report: European War* (Washington: Government Printing Office, 1945), 12.

[55]*Ibid.*, 16.

[56]*Ibid.*, 13.

[57]*Ibid.*, 16.

[58]Haywood S. Hansell, Jr., *The Air Plan That Defeated Hitler* (Atlanta: Higgins-McArthur/Longino & Porter, Inc., 1972), 251-52.

[59]*Ibid.*, 252.

[60]*Ibid.*, 78.

[61]Winston S. Churchill, *Closing the Ring* (Boston: Houghton Mifflin Company, 1951), 369.

[62]*Ibid.*, 162.

[63]Winston S. Churchill, *Triumph and Tragedy* (Boston: Houghton Mifflin Company, 1953), 569.

[64]*Ibid.*, 208.

[65]The United States Strategic Bombing Survey, *Overall Report: European War* (Washington: Government Printing Office, 1945), 96.

[66]*Ibid.*, 95.

[67]Quincy Wright, *A Study of War* (Chicago: The University of Chicago Press, 1965), 1542.

[68]Martin J. Sherwin, "The Atomic Bomb and the Origins of the Cold War," *The American Historical Review* (October 1973), v. 78, no. 4, p. 948.

[69]Albert Speer, *Inside the Third Reich* (New York: The Macmillan Company, 1970), 273.

[70]Herbert Feis, *The Atomic Bomb and the End of World War II* (Princeton: Princeton University Press, 1971), 28.

[71]*Ibid.*

[72] *Ibid.*

[73] Henry L. Stimson and McGeorge Bundy, *On Active Service in Peace and War* (New York: Harper & Brothers, 1947), 616.

[74] Herbert Feis, *The Atomic Bomb and the End of World War II* (Princeton: Princeton University Press, 1971), 30.

[75] Haywood S. Hansell, Jr., *The Air Plan That Defeated Hitler* (Atlanta: Higgins-McArthur/Longino & Porter, 1972), 252.

[76] The United States Strategic Bombing Survey, *Summary Report; European War* (Washington: Government Printing Office, 1945), 16.

[77] Halford J. Mackinder, "The Round World and the Winning of the Peace," *Foreign affairs* (July 1943), v. 21, no. 4, p. 605.

CHAPTER VIII

JAPANESE TACTICAL AIR POWER

To men of vision it was perfectly obvious by the end of World War I that a new era of military thinking and planning had been born. The potentialities of the airplane, not as a new "weapon" of war, but as a new *method* of waging war, were clearly recognized. Yet these men were usually ignored, often ridiculed and occasionally shot. Even in Japan, Italy and Germany, where the art of war was so sedulously developed, this vision was lost upon the highest military authorities. It was only in the United States and England, and to a greater extent in the latter rather than the former, that the idea of military air power was allowed to germinate and grow.

But the fact that the idea of air power was allowed to germinate and grow did not mean that it had been transformed into a reality by the time England realized that war was upon her in the fall of 1939, nor did it mean that air power was a reality in the United States when this country accepted the fact of the matter of war in the winter of 1941. However, what it did mean was that in both countries there were men capable of transforming their ideas into such tremendous and overwhelming realities that they were to have a decisive influence in determining the outcome of World War II and the course of civilization--at least until the outbreak of a World War III.

The nearest the Japanese came to displaying real air power in World War II was in the attack on Pearl Harbor. However, the Pearl Harbor attack was more in the nature of a raid than a battle. Even so, the Japanese virtually wiped out both the Pacific Fleet and the American overseas air arm; yet could not follow through on the attack and destroy the facilities in the United States that could and did reconstruct the fleet and the air forces that eventually crushed Japan.

Air power in the hands of the Japanese on December 7, 1941, would have sped them across the undefended Pacific and enabled them to overwhelm the pitiful six hundred combat planes that were left, and then to destroy the factories that were to produce the fleet and the air power that in four short but eventful years forced unconditional surrender upon an Emperor who was still undisputed ruler of half a billion people, a land area of nearly three million square miles and an undefeated army of considerable size and proven ability. The Japanese finished the war with the battle or raid idea connected with aircraft. The strategic concept of military air power consisting of balanced air forces that could operate independently of ground and sea forces to achieve decisive results in war (not battles or raids) remained foreign to the Japanese to the end.

The confrontation at Pearl Harbor was over the Monroe Doctrine, not the American Monroe Doctrine, but the Japanese. Japan refused to be the butt of the double standard of morality that condoned the building of empires by the British and the Americans, but said no to the Japanese. If America was for the Americans, as James Monroe said it was, it only seemed logical to the Japanese that Asia should be for the Asiatics, and if this meant war, then so be it; wars were the way empires were won, and the Japanese were not afraid of going to war, not even if it meant war with the United States and Great Britain as well, if that hard choice had to be made. At stake was the principle of equality as a Great Power, a principle so cherished by the Americans that on the day of their independence they proclaimed it to be their right "to assume among the Powers of the earth, the separate and equal station to which the Laws of Nature and of Nature's God entitle them."[1] What was right for the Americans was certainly right for the Japanese.

Even as late as the attack on Pearl Harbor, not all the Japanese by any means were convinced that war with the United States was necessary or desirable, much less inevitable. One such was the Admiral of the Combined Imperial Fleet, Isoroku Yamamoto, who had first-hand knowledge of the United States, having studied briefly at Harvard and having also been a naval attache in Washington. To him, it was preposterous to think that a small country like Japan could defeat a country the size of the United States with ten times the productive capacity. As early as 1940 Yamamoto had publicly expressed these views, and privately he had warned the Premier of Japan, Prince Fumimaro Konoye,

> If you tell me that it is necessary that we fight, then in the first six months to a year of war against the U.S. and England, I will run wild, and I will show you an uninterrupted succession of victories; I must also tell you that, should the war be prolonged for two or three years, I have no confidence in our ultimate victory.[2]

At the time of the Yamamoto warning, Japan was agonizing over the dilemma with which she was faced. With the fall of France and the invasion of Russia, the democracies were fighting for survival, and had their backs to the wall. Their preoccupation with Hitler left the Japanese free, or so it seemed, to strike out boldly in the Far East. Was this the moment for which Japan had been waiting all these years, or was it only an illusion? Was this the time to strike out for the "Co-Prosperity Sphere in Greater East Asia," or would that only bring on a war with the United States and England? If so, was the prize worth the gamble, taking into consideration Yamamoto's warning?

While the Japanese were on the horns of this dilemma, Franklin D. Roosevelt was on the horns of a dilemma of his own in the White House. His problem was that he was President of a divided country. Personally, he was firmly convinced that it was one world, and that if the totalitarian states (the forces of evil) were not stopped soon, they would dominate the world. However, the problem was to unite the divided country on this issue so that the great weight of the United States could be thrown in on the side of freedom. In this connection, he could visualize Japan as a backdoor entrance to the global conflict, and some of his closest friends agreed with him.

On the day after the German invasion of Russia, which was on June 22, 1941, one of the President's key Cabinet members, Harold Ickes, presented a formal Memorandum to the President advising him of a wonderful way to beat the isolationists in the United States and at the same time get the full weight of this country thrown into the war. Oil was the key to Ickes' plan.

"To embargo oil to Japan," he said,

> would be as popular a move in all parts of the country as you could make. There might develop from the embargoing of oil to Japan such a situation as would make it, not only possible but easy, to get into this war in an effective way. And if we should thus indirectly be brought in, we would avoid the criticism that we had gone in as an ally of communistic Russia.[3]

Everything Harold Ickes said made sense to the President. He honestly believed that the freedom of mankind was at stake in the war against Germany, and that unless the full power of the United States was thrown into the balance the war would be lost. On the other hand, no one knew better than the President that the country was painfully divided on the subject of war or peace, and that unless some way could be found to unite the country it was useless to talk about throwing the full weight of the United States in the cause for freedom. The solution might be just the one that Ickes recommended.

If imposing an oil embargo on Japan backed the country into a corner, as Ickes believed it would and forced Japan into some kind of "dastardly deed" against the United States, as the President thought likely, then it would be "not only possible, but easy, to get into this war in an effective way," as Ickes said. Moreover, Ickes was also right when he said that coming into the war by the back door of Japan would silence all of those who otherwise would have been critical of seeing the country get into a war "as an ally of Communistic Russia."

The Japanese played right into Roosevelt's hand at this time by moving strong elements of their armed forces into southern Indochina, thus posing a threat to the whole of the South Pacific. Roosevelt's answer to this new "provocation" was swift and stunning in its implications. He immediately issued an order freezing all Japanese assets in the United States, and when England and the Netherlands followed suit, the quarantine of Japan was complete. It meant that Japan was cut off from all trade with the western powers, including the large imports of oil without which Japan could not continue to exist, once her scanty reserves had been used up. These were so short that even with strict wartime controls Japan's stockpile would be exhausted in less than a year.

In other words, the executive order putting Japan under quarantine was tantamount to a declaration of war on Japan. As the oil gauge dropped, Japan realized that her only alternatives were to do Washington's bidding, which meant giving up her ill-gotten gains over the past fifty years and forgetting about the establishment of a New Order in Asia, or to strike out on her own in the direction of the plentiful oil supplies in the Dutch East Indies. This was the moment for which President Roosevelt had been waiting. He was leading from great strength (or so he thought), and the Japanese had their backs to the wall. Roosevelt reasoned (wrongly) that the United States had nothing to lose by forcing the issue with Japan. If the Japanese surrendered forthwith, that is backed down, this would be a complete vindication of American policy in the Far East, embracing the territorial integrity of China and the Open Door, and if Japan opted for war, then so be it, the back door of Japan would be the opportunity the United States had been waiting for to get into the larger global conflict "in an effective way," as Ickes put it.

President Roosevelt was living in a fool's paradise all during the period from the inception of the embargo on July 25 to the attack on Pearl Harbor on December 7. He was playing a cat and mouse game with Japan, with the United States being the cat, and little did he dream that the mouse could turn into a cat with very sharp claws. Of all the millions in the United States who refused to take the Japanese seriously, none exceeded the President. In fact, the President embarked upon his collision course with Japan very casually, almost lightheartedly as was evident in the meeting with Winston Churchill on August 9, 1941.

At this meeting between Roosevelt and Churchill, the Prime Minister had two objectives--one was to draw the United States closer to a war with Germany, and the other was to get help in the Far East. He had no difficulty on either score. As for the Far East, the President volunteered that he would be the policeman in the area and said that he could hold the Japanese in check at least temporarily. Sumner Wells enlarged on this and said that the United States would repel any Japanese thrust in the area whether it was directed "against China, against the Soviet Union or against British dominions or British colonies, or the colonies of the Netherlands."[4]

Churchill pressed the point with Roosevelt about holding back the Japanese in the southwest Pacific because England needed time to strengthen her position there, particularly at Singapore. In fact, Churchill went so far as to show Roosevelt two messages he had drafted to warn Japan against any further encroachments in the area, one a message to come from Washington, the other from London. Churchill said the President was very agreeable to the idea, and had already drafted a note that he would hand to the Japanese Ambassador when he returned to Washington. Churchill suggested adding a final sentence to the note, and to his surprise, Roosevelt accepted it. The sentence read, "Any further encroachment by Japan in the Southwest Pacific would produce a situation in which the United States Government would be compelled to take countermeasures, even though these might lead to war between the United States and Japan."[5]

However, everything was not as rosy as it seemed. Roosevelt had second thoughts on his way home and instead of delivering the harsh memorandum, took a more moderate course, obviously playing for time. Thus on seeing the Japanese Ambassador, Roosevelt rebuked him mildly and suggested that, if Japan would mend her ways and halt her expansionist activities in the area, it might be possible for him to have a meeting with the Japanese Prime Minister, possibly in Juneau, Alaska, "around the middle of October."[6]

On the subject of the European war, Churchill was more successful with Roosevelt at Argentina. Not only did he get him to agree that the primary aim of the war was "the final destruction of the Nazi tyranny"[7]-- which raises the question of the propriety of a neutral having any legitimate war aims -- but also to agree that in the post-war world the "guilty nations" must be disarmed, and the peace maintained by "strong united British and American armaments both by sea and air for a long indefinite period."[8] On balance, Churchill was proud (and rightly so) of the outcome of his first personal meeting with the American President, and of their "Atlantic Charter," the first draft of which was in Churchill's own hand. Churchill said, the profound and far-reaching importance of this Joint Declaration was apparent. The fact alone of the United States, still technically neutral, joining with a belligerent Power in making such a declaration was astonishing."[9]

After the Atlantic Conference in August 1941, the negotiations between the United States and Japan continued fitfully without either side showing any real disposition to compromise. The hard liners in Japan were determined to press forward along expansionist lines, and they were opposed by a moderate group, including the Emperor, who were anxious to avoid war with the United States, if at all possible. The only question in the minds of the Japanese was whether the position of the United States was so wooden, particularly with respect to the oil embargo, that Japan had no alternative to striking out for herself?

In the United States, there was also a problem. The country was still divided on the issue of war or peace. The grass-roots strength of the anti-war group can be measured by the fact that as late as mid-August 1941 the Congress approved of the extension of the draft by only one vote. On the other hand, the President sincerely believed that the cause of human freedom was at stake in the war and that it was his duty to lead the country into the war. This is where the dispute with Japan came in. To the President, the Far East was only a sideshow to the main event, which was in Europe, and the main importance of the sideshow was that it might serve as a back door through which the larger and more important conflict could be entered. This was certainly Harold Ickes' view, and it was one that was shared by the Secretary of War, Henry L. Stimson. The issue at stake came to a head, as did nearly everything else, at a meeting of the so-called War Cabinet at the White House on Tuesday, November 25, 1941. At the meeting, Secretary Stimson quoted the President as saying he thought the Japanese were likely to attack perhaps by the following Mondey, December 1, and the question was, "what we should do?"[10]

The question was a difficult one for several reasons. In the first place, there was the question of the power of the President: could he take this country into the war on his own authority without waiting for the Congress to declare war? In addition, there was the political mood of the country to consider: would the people support or impeach a President who acted on his own authority in this great crisis? And finally, there was the question of where the Japanese would strike: would it be against the British and Dutch possessions in the Far East, or would those of the United States be included? -- There were no answers to these questions, and without answers, it was difficult to proceed.

In the President's mind the danger was clear and present that the Japanese would strike at the British and Dutch possessions and also possibly at Thailand, but that they would carefully avoid those of the United States, thus making it difficult, if not impossible, for him to take the necessary action. In this eventuality, it was clear to the President and the War Cabinet that the long view of the national security was far from reassuring. What if Japan were allowed to run wild in the Pacific, Germany conquered Russia, and England was invaded, where would that leave the United States? The answer was all too clear. America would be left alone to face an overwhelming combination of aggressive powers. The chance for survival would be slim, almost to the vanishing point.

Which brought the subject back to the original question, "what we should do?" To Stimson, the answer was obvious. He, like Ickes, had long been urging the President to act more forcefully against all the agressors, and on this occasion went the President one better by saying the question was not what should be done in the crises at hand, but only "how we should maneuver them into a position of firing the first shot without allowing too much danger to ourselves?"[11]

In other words, if the Japanese would not do so on their own, Stimson wanted to provoke them into firing the first shot, and was confident that it would be interpreted as an attack on the United States, even though the actual victim might be the British or the Dutch, thus justifying the entry of the United States into the war. This was also Ickes' line of reasoning, that Japan should be provoked into taking some overt act, or doing some dastardly deed that "would make it, not only possible but easy, to get into the war in an effective way."[12]

The Japanese are a very practical people, and as such had been planning for a long time against just such a contingency as this, that the diplomatic negotiations would fail, and they would be forced to make a decision between adject surrender, as they saw it, and a widening war, certainly one against the British and Dutch, and possibly one against the United States as well.

On November 26, the day after the meeting of the War Cabinet, Secretary of State Cordell Hull gave the Japanese their final answer. It required them, if they wanted to restore normal diplomatic relations between the two countries, to turn the clock back almost to the beginning of the 20th Century, at least to 1931, as far as her conquests were concerned and to renounce their treaty of alliance with Germany and Italy. Hull told the Japanese that if they would accept these terms not only would normal relations between the two countries be restored, but it would be to their economic advantage also, because in that event the United States would unfreeze the Japanese funds and allow a certain amount of normal trade to take place, including trade in oil.

As far as the Japanesd were concerned, Hull added insult to injury when he told them of the cash benefits of accepting the final U.S. offer. The Japanese construed the cash benefits in Hull's trade offer as something in the nature of a bribe, or the sort of inducement that would be extended to a servant, or to the representatives of a colonial power. Hull didn't care. The next day he informed Stimson that he was no longer concerned with the Japanese problem. "I have washed my hands of it," Hull said, "and it is now in the hands of you and Knox--the Army and the Navy."[13]

Ten days before Hull handed the Japanese the note outlining the final position of the United States in the matter of restoring mormal relations, the Pearl Harbor Carrier Strike Force (*Kido Butai*) gathered at the mouth of the Inland Sea, and by twos and threes took off under orders to head on separate courses to Eterofu Island in the Kuriles. There the fleet was to lay at anchor in Hitokappu Bay until the Japanese were convinced that all hopes of reaching any kind of understanding with the United States were lost. Then the ships were to proceed to Pearl Harbor for the Sunday-morning attack on the United States Pacific Fleet.

This was Yamamoto's plan all along. It rested on three untested assumptions. One was that the Japanese fleet could get to Pearl Harbor undetected; the second was that aerial torpedoes would be effective in a body of water as shallow as Pearl Harbor; and the third was that the American fleet would be in Pearl Harbor on the day of the attack. Of these three assumptions, the most tenuous, and by far the most important one, was that the task force could get to Pearl Harbor without being detected. The aerial torpedoes were also a problem, but not an insurmountable one. Three Italian battleships had been sunk at Taranto in November 1940, and the feat was accomplished by torpedoes launched from airplanes in a water depth of about 75 feet. This clearly indicated that the shallow waters of Pearl Harbor, 30 to 40 feet, would not protect the battleships from aerial torpedo attack. In any case, this was merely a technical problem, and one that Yamamoto was confident the Japanese could solve. However, as to the third problem, whether the American fleet would be in the harbor when the Japanese arrived, this was just a chance that Yamamoto had to take, but knowing the Americans as well as he did, Yamamoto figured the odds were on his side, particularly if the attack came without warning on a sleepy Sunday morning.

Leaving the problem of the torpedoes to the Japanese technicians who were confident of getting results to the point of guaranteeing them within the time allotted and trusting to luck about catching the American fleet in harbor on a Sunday morning, Yamamoto concentrated on the problem of how to get the Japanese task force to Pearl Harbor without having it detected. There were three possible routes to the target area, and each had certain advantages and disadvantages.

The advantages of the central and southern routes, the one straight across the Pacific via the Midway Islands, and the other up from the naval base in the Marshall Islands, were that they were the shortest routes that could be taken and that the weather this time of year could be counted on as being favorable. The disadvantage of these routes was that they were the obvious ones, and therefore to take either one would be to forfeit any chance of surprising the enemy. The third choice was the northern route, down from the Kuriles. The disadvantages were that it was a much longer route, which involved refueling at sea, and also involved the worst possible weather conditions. In fact, a study Yamamoto made revealed that no ships of any nationality had traversed this route in November and December over the past ten years. This is the reason Yamamoto opted for the dangerous northern approach, rough seas or no rough seas, refuelling or no refuelling. Surprise was the one advantage Yamamoto had to have.

Having made the decision to strike from the north in the event of war, Yamamoto ordered a Japanese merchantman, the *Taiyo-maru* to traverse the exact course he had planned for the Strike Force to take the following month. This was to test out the hypothesis of the attack from the Kuriles. Before dawn on November 1, the *Taiyu-maru* approached Oahu, and Yamamoto was satisfied that his plans were sound. Thus it was that Yamamoto was not at a loss as to what to do when the negotiations with the United States broke down on November 26.

The carrier task force that had slipped out of the Inland Sea on November 16 was a formidable one. Its commander was Admiral Chuichi Nagumo, and it consisted of two fast battleships, two heavy cruisers, eight destroyers and an assortment of oilers and supply ships. The last carrier to leave port was the *Kaga*. She was loaded with the final modified torpedoes; the earlier ships had had only thirty of the one hundred thirty required in the attack on Pearl Harbor. All ships were put under complete radio silence and even had their communications system rigged to enforce the order. The ships left behind had set up and maintained a large volume of radio traffic to mislead enemy listeners.

Admiral Nagumo's orders were not absolute in any sense. He was told to proceed according to plan, but to await further orders before actually mounting the attack on Pearl Harbor. Yamamoto wanted to leave the door wide open for a cancellation of the attack in case there was a favorable turn in the negotiations. In that case, Nagumo was ordered to proceed to a point of latitude 42 degrees north by longitude 170 degrees east, where he would maintain his fleet in a state of readiness until further instructions. Meanwhile the negotiations showed no signs of improving and actually came to a standstill on November 26 when Hull presented his all-or-nothing note. While this was being typed, Nagumo's task force in Hitokappu Bay was making last-minute preparations to begin its long voyage south, the long voyage that would make history and change (at least temporarily) the face and nature of things in the Pacific.

What prompted Hull to send the uncompromising note of November 26 was the receipt of intelligence in the White House that a Japanese task force of about fifty ships, including troop transports, had been sighted heading southward from Shanghai, apparently in the direction of Indochina. To President Roosevelt this represented an act of bad faith on the part of the Japanese, who were supposed to be negotiating on a settlement of all the issues between the two countries. As a result, the President gave his approval to the stiffer U.S. policy embodied in the Hull note. Roosevelt categorically refused to pay any attention to Stimson's proposal that the United States should bomb the southbound Japanese expedition with B-17s from the Philippines.

On the other hand, the President listened to Stimson when he told him that the Japanese reaction to the Hull note might be explosive and that he had better protect himself by putting the commanders in the field on notice that war might break out at any moment. Roosevelt thought this was a good idea, and Stimson immediately got the ball rolling with General Douglas MacArthur in the Philippines, who was put on notice that hostile enemy action was "possible at any moment," and that he had better take all the necessary precautions. A similar message was sent to General Walter C. Short, commander of the Army Department in Hawaii.[14] Stimson also saw to it that the Navy commanders in the field were put on the alert. To Admiral Thomas C. Hart in the Philippines and to Admiral Husband E. Kimmel in Hawaii went the message that "an aggressive move by Japan is expected in the next few days."[15]

With Hawaii and the Philippines presumably on a full war-alert, negotiations dragged on for the next ten days, with both sides playing for time, the Japanese for time to get everything in place for the trial of battle, and the Americans for time to strengthen their outposts in the Pacific.

On December 6 a detailed British Admiralty report gave Washington the intelligence that "a Japanese fleet of about thirty-five transports, eight cruisers and twelve destroyers was moving from Indochina across the Gulf of Siam. Other Japanese fleets were also at sea on other tasks."[16] Taking cognizance of these Japanese troop movements, Secretary of Navy Frank Knox, at a regularly scheduled briefing conference, asked the assembled Admirals whether this meant that Japan was about to make war on the United States? Admiral Richard Kelly Turner spoke for the Staff when he said, "No, Mr. Secretary. They are going to hit the British. They are not ready for us yet."[17] This was the day before Pearl Harbor.

Admiral Turner's complacency about the security of the American outposts in the Pacific was fully shared by the American commanders in the field, none of whom took the orders they had received in late November seriously enough to put their stations on a full war-alert. In fact, the naval commander in Hawaii, Admiral Kimmel, was so relaxed about the imminence of war that, as late as December 2, he made a joke out of the fact that for the past two weeks naval communications intelligence had lost track of six of Japan's ten aircraft carriers. Not having been able to pick up any radio traffic from these six carriers over a period of two weeks, the Office of Naval Intelligence assured Admiral Kimmel that the carriers were probably in "home waters" handling their routine communications by some means other than radio. Admiral Kimmel was undisturbed by this strange development and refused to read anything ominous into it. He dismissed the incident by jokingly asking the intelligence officer doing the briefing, "do you mean to say that they could be rounding Diamond Head this minute and you wouldn't know?"[18]

This was about the size of it. The six carriers that naval communications intelligence carried as "missing" were the ones in Admiral Nagumo's Pearl Harbor Strike Force. Even as Kimmel was making light of them, the carriers were on their way to Pearl Harbor. In fact, on the very day Admiral Kimmel made his joking remark about Diamond Head, December 2, Admiral Nagumo received the coded message for which he had been waiting. The message was from Yamamoto, and it was an action one. The signal flashed to Admiral Nagumo was, "Climb Mount Niitaka." This meant that Nagumo should proceed with the attack as planned.

Mount Niitaka on Formosa was the highest peak in the Japanese Empire.[19]

In his blindness to what was going on all about him, Admiral Kimmel was not unlike the other field commanders in the Pacific, none of whom was the slightest bit disturbed by the critical war-warnings of late November. The Army's general in Hawaii was so obtuse, in fact, that he made a bad situation worse by proceeding to line up his planes in neat rows, wing-tip to wing-tip, thus making them sitting ducks for the Japanese attack. He thought that the Army's job was to guard against sabotage and to defend the beaches. War from the air was not in General Short's mind when he made his dispositions for a Japanese attack. Admiral Kimmel was just as parochial in his thinking as General Short. He expected a submarine attack, if there was any attack at all, and the last thing he was prepared for was a mighty blow from the air. The result was the greatest debacle in the history of American arms. Not only was the fleet destroyed without ever getting out to sea, but the air force was also destroyed while it was still on the ground.

The only saving grace was the absence of the big aircraft carriers the day the Japanese struck. They were not at Pearl Harbor on the day of infamy because of a stroke of good luck. As a sop to the war-warning, Admiral Kimmel had sent them to sea to carry additional aircraft for the defense of Midway and Wake. Otherwise, they would probably have shared the fate of the other ships at Pearl Harbor.

Of all the inept commanders of the American forces in the Pacific at the time of Pearl Harbor the most culpable of all was probably General Douglas MacArthur in the Philippines. He began getting the news that Pearl Harbor had been attacked just as the last of the Japanese planes were winging away from the devastation they had wrought at that unhappy base. Yet nine hours after the word reached him MacArthur was a witness to the destruction of his own air force on the ground. While it is still a moot question as to why this happened, whether it was MacArthur's fault or one of his subordinates, what is not moot is that MacArthur was the commanding officer, and as such was responsible and answerable for what happened.

The American debacle at Pearl Harbor and in the Philippines vindicated the soundness of Yamamoto's original plan and proved that he knew what he was talking about when he said, "in the first six months to a year of war against the U.S. and England, I will run wild, and I will show you an uninterrupted succession of victories."[20] The victories came off, as Roberta Wohlstetter said, "like a string of Japanese firecrackers,"[21] and Yamamoto began to emerge as a man of military genius. He had not been wrong on a single point, and what he had promised he had delivered.[22] Everything had gone according to plan and without a slip. Pearl Harbor had proved to be an easy nut to crack, much easier than had been anticipated, as were all the other objectives in the original plan, including the Philippines, Guam, Hong Kong, Malay, Singapore, Burma and the Dutch East Indies. In addition, the *Prince of Wales* and the *Repulse* had been sunk, and Japan had gained the mastery she had sought in the whole area of her ambition.

Japan was supreme in all three elements, land, sea, and air, and her defenses against counterattack were also secure. Now she had China closed off from outside help through both the Pacific and the Indian Ocean approaches, and with her island outposts, including Guam and Wake, Japan had a picket line across the Western Pacific, a series of island fortresses from which she could guard all the approaches to the Empire.

The Japanese defensive perimeter looked like it was impregnable from the Pacific and the Indian Ocean, and of course the land border in the north was no problem because Russia was fighting for her life against the Germans. The Japanese defensive perimeter extended in a wide arc from the Kurile Islands in the north down through Wake and the Marshall Islands, around the Gilberts and the Bismark Archipelago, then through northern New Guinea, Java, Sumatra and finally up to and including Burma. The only thing that remained to be done was to make it so strong that it would not even be challenged by the former colonial powers.

The vast area that Yamamoto conquered covered more than a million square miles of land in southeast Asia, and practically the entire western half of Pacific Asia. It increased the Emperor's subjects by more than a hundred fifty million people and included more than half a million whites, as well as hundred fifty thousand prisoners of war. The marvel is that Yamamoto accomplished this at all, much less in six months. After all, it had taken the old colonial powers several centuries to acquire this much land and to have this much influence in Asia. In fact, looked at from this point of view, Yamamoto eclipsed Hitler as a conqueror and was probably more successful than any man since Napoleon. Yamamoto was also different from Napoleon and Hitler as a military strategist. His sphere was not ground warfare, the principle of mass that Napoleon perfected as he ousted an old system of warfare, nor was it the ground-air blitzkrieg that Hitler used so successfully in the opening period of World War II. Yamamoto's sphere was that of the amphibious blitzkrieg, the blitzkrieg that employed all three elements, land, sea and air and employed them simultaneously. This is what made Yamamoto stand out.

In this respect, Yamamoto was almost an original in his approach to war. He was an original in some other ways as well, not all of them good. Instead of a one-front war, Yamamoto planned a war on six fronts all at the same time, and instead of a policy of concentration, he followed one of scatterization. The essence of his strategy was a centrifugal offensive with blows at almost every point on the compass from widely separated bases, and all at the same time. It was a system that was enough to make a Hitler or a Napoleon go mad. However, not only did it work, but it worked well, and worked despite the fact that mischance and bad luck intervened at critical moments in the implementation of the plan.

One stroke of good luck that Yamamoto had was in connection with, or more accurately because of, his decision to strike out in all directions on D-day. This required the movement of nearly half a million men and hundreds of vessels at the same time, not only troop transports, but warships of all kinds and varieties. There was no way to hide the sea-lift of this magnitude, and no efforts were made to do so. As a result, the intelligence eyes of the world were all focused in the wrong direction during the critical period from November 26 to December 7.

Instead of worrying about where the six carriers were, the carriers that were to destroy the fleet and the air arm at Pearl Harbor, the intelligence eyes of the world were on the massive movement south. This obscured the fact that the danger was in the north, not in the south. It also led to the false conclusion that Pearl Harbor would not be the target, that certainly the British and Dutch possessions would be attacked, and possibly the Philippines, but not Pearl Harbor. Viewed in this light, the complexity of the Yamamoto plan was one of its most important advantages. The movement south in full view of all the intelligence services served to camouflage the attack on Pearl Harbor, and thus acted as the best cover and deception plan that could have been devised-- even though it was all an accident, not planned that way at all.

The razzle-dazzle of Yamamoto's D-day was not without its disadvantages, of course. The greatest of these was that something might go wrong which would give advance warning of the attack on Pearl Harbor. Without the element of surprise, the Pearl Harbor attack had little or no chance for success and might have meant that the Japanese Strike Force would have been completely wiped out by the greatly superior American force supported by land-based air. This almost happened, and it is still somewhat of a mystery why it did not. The original Japanese plan called for a predawn attack on Pearl Harbor, but at the last moment the first strike was delayed by about two hours in order to avoid the hazard of having the flyers taking off in darkness. The decision was taken too late, however, to warn all the other commanders, and thus the invasion of the Malay Peninsula began two hours ahead of the strike at Pearl Harbor. If the British had relayed the message to Pearl Harbor promptly, the Japanese might have been robbed of the vital element of surprise in the Pearl Harbor attack.

Another close call Yamamoto had with his razzle-dazzle had to do with the Japanese attack on the Philippines. Yamamoto's plan was to bomb the airfields of the Philippines at the same time Pearl Harobr was brought under attack, but a heavy fog descended on Formosa at just the time the planes were scheduled to take off, and this forced a delay of several hours. The delay was serious. It meant that the aviators would not achieve surprise in the attack on the Philippines because MacArthur would be warned that the Pearl Harbor attack was underway, and if so, the airfields in Formosa would be bombed before the planes could take off. In fact, when the American bombers did not appear over the Japanese airfields in Formosa, no one was more surprised than Yamamoto, and the Japanese aviators were delighted when they got to the Phillippines to find that the American airplanes were all neatly lined up on the airfields while the pilots were indoors having a noon-day meal. Not even the bad luck of the fog on Formosa was enough to spoil Yamamoto's fun on this the first day of the war. However, the first day was not the whole war for the Japanese.

The only exception to this rule of luck for the Japanese was Wake Island. Held by a small force of Marines, less than five hundred, Wake was bombed on the first day of the war and had all but four of its antiquated airplanes destroyed, but the Japanese made no attempt at an actual landing until December 11. To repulse it, the Marines followed the example of the patriots at Bunker Hill and held their fire until the invasion flotilla was at point-blank range. The accuracy of their fire convinced the Japanese that they had taken on more than they had bargained for, and the result was that the fleet withdrew leaving the island to the Marines. This prompted the Navy to make a pro forma inquiry as to what the Marines on Wake needed in the way of help. Major James Devereux, among other things, facetiously replied, "send us more Japs." The reply of the Marine commander at Wake became part of the American folklore. It put him right up with the Army chaplain at Pearl Harbor ("Praise the Lord and Pass the Ammunition"), and with General McAuliffe, who turned down a demand that he surrender during the Battle of the Bulge with a one word reply--"Nuts." The Japs, however, would not take nuts for an answer at Wake.

On December 23 the Japanese came back to Wake Island, and this time they came in great force. The invasion force included not only six heavy cruisers and two of the aircraft carriers that had been at Pearl Harbor, but a detachment of some two thousand tough Japanese Marines. It was enough. Though the Marines fought stubbornly and gallantly, they had to give up their island, but their courage and valor inspired a country that needed something like this to lift the sagging spirits of the people. Wake was the only bright spot in these somber and defeatful days.

The string of successes that Yamamoto promised and delivered during the first six months of the war was made possible by tactical air power in the hands of very skillful Army and Navy commanders. This was the case at Pearl Harbor, in the Philippines and at Singapore, the three key victories that made all the rest possible. In other words, the upsurge in the Far East that fundamentally changed the strategic balance was an upsurge of tactical air power.

Tactical air power was not only the key to the Japanese victory at the American Pearl Harbor on the island of Oahu, but since it was the only force the Japanese employed on the day of infamy, it is hardly open to question that tactical air power was the decisive factor. The same thing can almost be said about the Pearl Harbor in the Philippines, though to a lesser extent, of course. The only hope MacArthur had of defending the Philippines was in the strength of his air force, and when that was destroyed on the ground, he was in a hopeless position. The third Pearl Harbor, the British variety, was at Singapore, and there the case was a little different though upon closer examination it is clear the principle was the same. England lost Singapore because of the tactical air power in the hands of the Japanese and because the defenders of the British base did not have any tactical air power worthy of the name.

Singapore was more than just another British naval base in the Far East. It was designed to be the key to the whole of the southwest Pacific and became a symbol of western power in the Far East.

The Japanese looked at Singapore as an insult and a challenge. The construction of Singapore as a naval base began in 1924. This was a time of trouble between the United States and Japan. In the first place, on December 13, 1921, the British, under strong American pressure, politely declined to renew the alliance with Japan. Then in 1924, Congress passed the Oriental Exclusion Act, and the fat was in the fire. The announcement by the British that they intended to build a new naval base at Singapore adequate for a battle fleet only served to add fuel to the flames of the Japanese anger and convinced them that the white people of the world were joining forces against them. The Japanese were also convinced that their only salvation was in making themselves strong enough to protect their own vital interests in this part of the world.

While the Singapore naval base was under construction, there was a great debate in England over the best means of defending it. Lined up in the debate were the First Sea Lords and the Air Staff. The battleship men still believed in battleships, not in airplanes and believed that a naval base should be protected by naval power, not by air power or land power. Consequently, the admirals argued for the emplacement of big guns at Singapore, guns that could protect the base from the danger posed by enemy warships. On the other hand, the Air Staff argued that the air defense of the base was of paramount importance and that a neglect of this feature would leave the base wide open to attack from the air and also to attack from the land, that is from the Malay Peninsula.

The admirals won the debate, and the result was a Singapore that was practically invulnerable to attack from the sea but was highly vulnerable not only to tactical air power, but to an attack from the land behind. The trouble was all the big guns were pointed in the wrong direction--out to sea-- and not in the direction from which the danger came, that is from the air and from the land approaches to the great naval base.

When England had her back to the wall after the fall of France, and even after some of the pressure had been taken off by the German invasion of Russia, another great debate broke out in England, and again over Singapore. This time the question was one of priorities, whether Singapore's defenses should have a lower or higher priority over those of Egypt and the Middle East? Churchill felt strongly that the defense of Singapore had a lower priority than that of Egypt and the Middle East and refused to budge from that position: "Our priorities during 1941," he said,

> stood: first, the defence of the Island, including the threat of invasion and the U-boat war; secondly, the struggle in the Middle East and Mediterranean, thirdly, after June, supplies to Soviet Russia; and, last of all, resistance to a Japanese assault. It was, however, always understood that if Japan invaded

> Australia or New Zealand, the Middle East should be sacrificed to the defence of our own kith and kin.[23]

Some of the critics of Churchill's policy have charged (quite wrongly) that his judgment was flawed because his eyes were focused on North Africa, and because he could not rest until British arms had cleared the enemy from that part of the world. In fact, one of Churchill's critics, Liddell Hart, has charged that it was Churchill's preoccupation with the exploits of Erwin Rommel that really brought about the disaster in the Far East. "Thus," Hart said, "Rommell indirectly produced the fall of Singapore--and as much by the personal impression he made on a personality-minded Prime Minister as by his potential threat to the Nile Valley and the Suez Canal."[24]

Be that as it may, Singapore was certainly starved to feed the build-up in Africa in 1941, and this meant that Malaya as a whole was starved. There was not a single tank in Malaya, and of the minimum number of airplanes required for the defense of Singapore, some five hundred aircraft, only one hundred fifty-eight were available. Not only was the size of the force totally inadequate, but the quality of the aircraft was sub-standard. All the defending fighters were of antiquated design, slow and cumbersome, and none of them was a match for the very high-quality airplanes in the hands of the Japanese. The Japanese also had five times as many airplanes as the hapless defenders of Malaya and Singapore.

LiddellHart and the other critics of Churchill's policy are quick to point out that, while he was starving Malaya and Singapore to mount a premature offensive in North Africa, he was also sending some six hundred modern aircraft to Russia and carrying on a nightly bombardment of Germany with long-range aircraft, not one of which ever found its way to the defense of Malaya and Singapore. It was almost as if Churchill had written off the defense of Singapore and had resigned himself to the loss of the great base if Japan should enter the war.

This was not necessarily the case at all, merely a matter of priorities, and after the Atlantic Conference with Roosevelt in August, Churchill had some second thoughts about the priorities themselves. Accordingly, on August 25 he proposed that the Admiralty immediately dispatch a squadron to the Indian Ocean, even though he realized the number would have to be small. Churchill's idea was that the Japanese would recognize this for what it was, a "deterrent"[25] force, and that a larger number of ships would follow as the British built up a formidable Far Eastern Fleet at Singapore, one that could act in conjunction with the American Fleet on Pearl Harbor.

The Admiralty did not take kindly to Churchill's idea to build up a Pacific Fleet because it would have to be at the expense of the Atlantic, of course, and there was always the danger that the *Tirpitz* would break out and become a menace in the Atlantic. Nevertheless, the Prime Minister's orders had to be obeyed, and the result was that England's newest battleship, the *Prince of Wales*, and the battle cruiser *Repulse* with four destroyers were sent to Singapore in 1941.

The appearance of the *Prince of Wales* and the *Repulse* in the Pacific was only a down-payment on what Churchill had in mind. This was the creation of a British Far Eastern Fleet that by the spring of 1942 would consist of seven capital ships, one first-class carrier, ten cruisers and twenty-four destroyers. This would give the United States and England an edge over Japan of sixteen capital ships to nine, but the Japanese would still have the advantage of ten aircraft carriers over four. On the overall, Churchill was confident that this imbalance against Japan would preserve the peace in the Far East for a long time.

But the Prime Minister had reckoned without taking into consideration the vital factor of time in his equation. Ever since July 25, when the oil embargo had gone into effect, and particularly after the Atlantic Conference, the Japanese were aware that they would have to go to war with England and the United States if they were not to face the ruin of their ambitions in Asia. The mystery is why they waited so long to strike. Time was not on their side, and Churchill was simply living in a dream world when he thought the Japanese would wait another six months for the British Fleet to get organized and on station at Singapore, before they struck. It may be a reflection on the Japanese acumen that they did not strike four months before they did.

Nonetheless, Churchill calmly went ahead with his plans, and on December 2 the *Prince of Wales* and the *Repulse* put in their appearance at this new battle station, Singapore. Oddly enough, and by coincidence, this was the very date that Yamamoto sent his final message to Admiral Nagumo confirming the order to attack Pearl Harbor on December 7, "Climb Mount Niitaka."

Two and a half months before this, on October 24, Admiral Sir Thomas Phillips--nicknamed "Tom Thumb" because he was an inch shorter than Napoleon and had to stand on a box to see over the bridge of a ship -- was named commander of the new British Far Eastern Fleet. He arrived in Manila on December 5 to see if he could work out some joint operational exercises with the small American fleet stationed there under Admiral Thomas C. Hart. While Admiral Phillips was in Manila, the news came in that a large Japanese seaborne expedition had been seen entering the Gulf of Siam, destination unknown but clearly not friendly. Admiral Phillips took his departure immediately and arrived back in Singapore on the morning of December 7.

On arriving in Singapore, Admiral Phillips learned that the Japanese were disembarking troops at Kota Bharu in northern Malaya and that landings were also being made at Singora and Patani. He felt that duty required him to proceed to the area as quickly as possible with the force available. Before sailing, however, Admiral Phillips took steps to assure that air cover would be available, particularly in the strike zone to the north. What Phillips found out, to his dismay, was that the Japanese had already overrun all the airfields in northern Malaya and that, therefore, no fighter protection would be possible. Nevertheless, Admiral Phillips set sail with the *Prince of Wales* and the *Repulse*, determined to do what he could with the force he had.

At first, Admiral Phillips' daring sortie was protected by rain and clouds, but this kind of luck was too good to last, and when the weather changed, as it did on the evening of the ninth, he was immediately spotted by the Japanese lookouts. Prudently, Admiral Phillips reversed his course and headed south for Singapore.

Admiral Phillips' change of direction fooled the Japanese, and the bombers failed to discover his whereabouts. However, he did not hold to his new course long because he received another radio message from Singapore, this time to the effect that an enemy landing was reported in progress at Kuantam, a point much closer to Singapore than Kota Bharu and Signore. Thus he changed his course again and proceeded to Kuantan.

When Phillips arrived at Kuantan, he found that no Japanese landing was in progress. But while he was milling around in the harbor, a strange meeting was going on in London, where the question was, what should be done with the *Prince of Wales* and the *Repulse*, now that England was at war with Japan. For his part, Churchill said,

> I thought myself they should go across the Pacific to join what was left of the American fleet. It would be a proud gesture at this moment, and would knit the English-speaking world together. We had already cordially agreed to the American Navy Department withdrawing their capital ships from the Atlantic. Thus in a few months there might be a fleet in being on the west coast of America capable of fighting a decisive sea battle if need be. The existence of such a fleet and of such a fact would be the best possible shield to our brothers in Australasia. We were all much attracted to this line of thought. But as the hour was late we decided to sleep on it and settle the next morning what to do with the *Prince of Wales* and the *Repulse*. Within a couple of hours they were at the bottom of the sea.[26]

The loss of the *Prince of Wales* and the *Repulse* on December 10 was another staggering blow to the Allies. The Pacific Fleet had already been destroyed at Pearl Harbor, of course, and Japanese troops were landing in the Philippines after MacArthur's air force had been destroyed on the ground. Now the Malta or Gibraltar of the British Far East was about to fall following the loss of the two great capital ships.

"This stroke," Liddell Hart said,

> settled the fate of Malaya-Singapore. The Japanese were able to continue their landing unchecked and establish air bases ashore. The superiority of their air force over the meagre British air strength in Malaya was decisive in crumbling the resistance of the British troops and enabling their own troops to push down the Malay Peninsula and force the backdoor into Singapore. Its fall was the consequence of earlier oversight and misjudgment--mainly in London.[27]

One of the most interesting might-have-beens of history is what might have happened if Admiral Phillips had been more respectful of the tactical aviation of the Japanese and had not lost his fleet in such a capricious manner. In other words, what would have happened if, instead of joining the marine life at the bottom of the sea, the *Prince of Wales* and the *Repulse* had joined what was left of the American Paciic Fleet?

The chances are that if the American and Dutch fleets had immediately been augmented by the *Prince of Wales* and the *Repulse*, the battle of the Java Sea might have taken a different turn, and certainly the fall of the Indies might have been delayed, if not averted altogether. Also, the setbacks the Japanese took in the battles of the Coral Sea and Midway not only might have been much more serious, but possibly disastrous. To put this another way, if the two big British ships had been available to the Americans at the end of 1941 and the beginning of 1942 the war might have been shortened by a year, possibly two. At the same time, Churchill was probably guilty of the understatement of 1941 when he observed, "the efficiency of the Japanese in air warfare was at this time greatly underestimated both by ourselves and by the Americans."[28]

FOOTNOTES--CHAPTER VII

[1]The Declaration of Independence, in *Documents of American History* (New York: Appleton-Century-Crofts, 1958, ed. Henry Steele Commager), 100.

[2]Roberta Wohlstetter, *Pearl Harbor: Warning and Decision* (Stanford: Stanford University Press, 1963), 350.

[3]John Toland, *The Rising Sun* (New York: Random House, 1970), 108.

[4]Charles Callan Tansill, "Japanese-American Relations: 1921-1941," in *Perpetual War for Perpetual Peace*, ed. Harry Elmer Barnes (Caldwell: The Caston Printers, 1953), 303.

[5]Winston S. Churchill, *The Grand Alliance* (Boston: Houghton Mifflin Company. 1950), 440.

[6]John Toland, *The Rising Sun* (New York: Random House, 1970), 117.

[7]Winston S. Churchill, *The Grand Alliance* (Boston: Houghton Mifflin Company, 1950), 435.

[9]*Ibid.*, 441.

[9]*Ibid.*, 444.

[10]John Toland, *The Rising Sun* (New York, Random House, 170), 180.

[11]*Ibid.*

[12]*Ibid.*, 108.

[13]*Ibid.*, 219.

[14]*Ibid.*, 220.

[15]*Ibid.*

[16]Winston S. Churchill, *The Grand Alliance* (Boston: Houghton Mifflin Company, 1950), 601

[17]John Toland, *The Rising Sun* (New York: Random House, 1970), 243.

[18]*Ibid.*, 233.

[19]*Ibid.*, 223.

[20]Roberta Wohlstetter, *Pearl Harbor: Warning and Decision* (Stanford: Stanford University Press, 1962), 350.

[21]*Ibid.*, 341.

[22]*Ibid.*, 350.

[23]Winston S. Churchill, *The Grand Alliance* (Boston: Houghton Mifflin Company, 1950), 587-88.

[24]B.H. Liddell Hart, *History of the Second World War* (New York: G.P. Putnam's Sons, 1970), 182.

[25]Winston S. Churchill, *The Grand Alliance* (Boston: Houghton Mifflin Company, 1950), 590.

[26]*Ibid.*, 616.

[27]B.H. Liddell Hart, *History of the Second World War* (New York: G.P. Putnam's Sons, 1970), 227.

[28]Winston S. Churchill, *The Grand Alliance* (Boston: Houghton Mifflin Company, 1950), 619.

CHAPTER IX

"SLAUGHTER BOMBING"

Up to the time of the Coral Sea and Midway Yamamoto was flawless in the execution of his prewar plans, but then he began to slip and stumble. This was also the time when the one great error in the original planning began to surface. It was an error that Yamamoto had called attention to well before the war started, namely, that if the war should "be prolonged for two or three years, I have no confidence in our ultimate victory."[1] What he meant was that if tactical air power failed to produce a quick victory, Japan would lose the war. However, the early victories seemed to vindicate the Japanese concept of air power.

The Japanese may have been less than comprehensive in their thinking about air power, in that they limited it to the tactical role, and ignored strategic air power altogether, but in their favor it must be said that in their thinking on the subject they were clear and unified. There were no Billy Mitchells, no Douhets, and no Hugh Trenchards in Japan to becloud the issue. To the Japanese, the airplane was just another weapon with which to wage war. It was certainly not a new weapon system and not a new method of waging war. As a result, the Japanese developed two air forces, one to serve the needs of the Army, and the other those of the Navy. They had nothing in common beyond their blind allegiance to a divine Emperor. The airplanes they flew were different, their training was different, and there were no joint exercises to give them any experience working together. This non-cooperation between the two Services, and the non-integration of their air forces, did not augur well for a war that demanded joint operations. In fact, two-thirds of the fleet air forces were lost when they were required to perform functions for which they had not been trained, that is when they were required to work from land in support of ground forces at Guadalcanal and in the defense of Rabaul.[2]

The non-cooperation and intense rivalry between the Army and Navy air services even spilled over into the field of procurement. The Army and Navy suppliers fought each other tooth and nail over the available raw materials for aircraft production and bid against each other for obtaining the engineering resources that were spread far too thinly. In addition, the two Services fought each other over intelligence procurement and interpretation. Radar is a good example. Rather than pool their meager resources, the two Services built and operated separate stations, often side by side, thus duplicating information gathering and the interpretation services. It was a case of never the twain shall meet, and the rivalry was very harmful to the development of aviation in Japan.

On the other hand, the Japanese were among the world leaders in the development of naval aviation. As early as 1919 the world's first aircraft carrier, *Hosho*, was under construction in a Japanese shipyard. It was followed by the *Akagi*, the *Kaga*, and the smaller *Ryujo*. Later, other aircraft carriers were added, and in 1941 the First Air Flotilla was formed. This was Admiral Nagumo's Pearl Harbor Strike Force of six carriers. Admiral Yamamoto, Chief of Engineering at the Imperial Japanese Naval Air Forces Headquarters in 1931, had been the prime mover in the construction and design program. He revolutionized the thinking in this area by laying down the rule, "fit the aircraft carrier to the fighter plane, not the plane to the carrier."[3]

Between 1930 and 1935, this same logic was applied to two other developments, aerial torpedoes and dive-bombing tactics. By 1936, doctrine was beginning to catch up with the technological developments, and the Japanese were starting to think in terms of a decisive air battle taking place before the engagement of the surface fleets. In other words, Japanese air force thinking in the Navy was very gradually coming round to the revolutionary thinking of Douhet about supporting and striking forces, with the surface battleships being in the supporting rather than the striking role. The idea was only in the formative stage and had not been crystallized, but more and more was being heard about it in Japanese naval circles.

While the Japanese staff officers (some of them) were playing with these heretical thoughts about the relationship between sea and air power, the service chiefs in the Army and Navy were sparing no efforts in the training of aircraft pilots and crews. In fact, the Japanese were sticklers for training. Before the war began, their standards were so high that flyers for the Combined Fleet got over eight hundred hours of flying experience before they were assigned to a tactical unit, and the average Army and Navy pilots got from three to six hundred hours. A cadet in basic and advanced training in the United States at the same time got only about two hundred hours of training. Not only that. A very large percentage of pre-war pilots in Japan had front-line combat experience before the war began. They got this from the war against China in 1937, and in the border skirmishes with the Russians. All in all, the Japanese pilots were among the best in the world by the time World War II began. They also flew airplanes that were as good as any in the world, and superior to most. That is why Pearl Harbor was such a success and why the conquest of the southern region was brought about with so few losses.[4]

After the battles of the Coral Sea and Midway, Japanese fortunes began to wane, so much so that in the fighting over the Solomon Islands, the Bismark Islands, and New Guinea, the Japanese lost over ten thousand planes.[5] While the planes could be replaced from new productions, the pilots could not unless drastic cuts were made in their training programs. These were mandated anyway by the acute shortage of oil. So the net result was that the pilot training programs were cut back, and the Japanese had to resign themselves to having pilots of very poor quality for the rest of the war.

In fact, by the end of the war the training standards were so low in the Japanese air forces that the new pilots hardly had any training at all. The replacement pilots were being turned out with less than one hundred hours of flying time and were immediately thrown into combat. In fact, over half the new pilots had to join the fray with less than a hundred hours of training. After all, a suicide pilot did not need much training, since he was limited to one combat mission, and already knew how to die.

While the Japanese air forces were declining, but before they disappeared from the skies altogether, the Japanese were losing another kind of battle. This was an engineering and production battle, a battle that had both qualitative and quantitative aspects. Whereas at the beginning of the war the Japanese had superior planes on all the fighting fronts of the war and had them in larger numbers than any of their enemies, they could not count on this to last for very long, and it didn't. Not only did the Americans catch up with them qualitatively in very short order and surpass them before too much time had elapsed, but they also quickly closed the gap in the number of aircraft available on the fighting fronts. And they did this while their training programs were turning out pilots and crewmen that were every bit as good as the Japanese, if not far better.

The story is in the statistics. While the Japanese were producing 60,000 planes between 1942 and 1945, the Americans were turning out 260,000. Not only were the Japanese beaten by better than four to one in the production process, but the new American planes were of superior quality. The Japanese were never able to close the gap, not in the quality of the aircraft, not in the number of airplanes, and not in the quality of the pilots.

Yamamoto's flood tide set in at Pearl Harbor and reached its zenith six months later. After that, the ebb set in, and with it came the inevitable defeat of Japan, completely swamped, as Yamamoto knew it would be, by the ten to one superiority of the United States in production capacity. Japan's final act of desperation was the conversion of substantial parts of her air forces to Kamikaze tactics. However, this was also a failure. The suicide pilots were so poorly trained that most of them missed their targets altogether, and when they scored hits the damage was usually ineffectual because the explosives were too small to do any real harm. It was customary to use trainer-aircraft in the Kamikaze attacks, and hand grenades were often heaped around the pilot to beef up the explosive charge. In this respect, and except for their limited success at Okinawa, the Kamikazes were like the German rockets, great for morale value at home, but not much good on the battlefields. By May 1945 the Kamikaze attacks had all but ceased altogether and were of no significance. Okinawa was the last big splurge.

A year before the Japanese began the conversion of parts of their air forces to Kamikaze tactics, that is by the middle or later part of 1943, the American Navy became superior to the Japanese in every respect, in capital ships, in submarines, and in the air. This is all the more remarkable because the naval losses were heavy during the first six months of the war when Yamamoto was running wild. It was made possible, in no small part, by the prodigious accomplishments of American industry as well as by the losses inflicted on the enemy at the battles of the Coral Sea and Midway. Not only did American naval power surge ahead of the Japanese, but its program of construction (exceeding that of all other nations in speed and size) soon made the United States the mightiest sea and air power in the world.

With its new-found superiority in sea and air power, it was now possible for the United States to take the offensive against Japan. This took the form of a two-pronged attack, one aimed at Buna on the north coast of New Guinea, and the other at Guadalcanal in the southern Solomons. This was a test of the Japanese outer perimeter, and it was earlier than the Japanese had expected. It was also successful. By the end of 1942, both Buna and Guadalcanal had been taken, and the strategic plans of the Japanese had been seriously upset. Not only that. In the fierce fighting for the Solomons and New Guinea the Japanese lost the bulk of their fully trained naval air pilots, as well as a portion of their best Army airmen. The Japanese never fully recovered from these losses, and as the Bomb Survey said, "the initiative had passed to the United States."[6]

The advantage gained as a result of the penetration of Japan's outer defense perimeter was pushed with skill and determination. The advance followed two main paths. One was up the north coast of New Guinea to the Philippines (MacArthur's Command), and the other was across the central Pacific to the home islands themselves (Nimitz's Command). In both cases, the range of the advance was determined by the combat radius of fighter aircraft. When the carriers furnished the main air support, they were not withdrawn until the airfields ashore were prepared and operational. In some cases the entire forward movement was by air. In all cases, the forward movements were timed to be mutually supporting, "even though," as the Bomb Survey noted, "concentration on one line might have been more rapid."[7]

The two approaches to Japan converged, as had been planned, when the line of the Philippines was reached. This was in the summer of 1944. By this time, MacArthur had completed his conquest of New Guinea, and the way was open for the return to the Philippines. By this time also, the American invasion of the Marianas was a *fait accompli*, and the Japanese were in headlong retreat from the area, flying toward Okinawa in the Ryukyu Islands south of Japan. Their losses were 402 aircraft to the Americans 27. This is why the battle became known as "The Great Mariana Turkey Shoot." The Navy's gifted historian, Samuel Eliot Morison, said that the Battle of the Philippine Sea was "even more important than the subsequent Battle of Leyte Gulf"[8] in bringing about the ultimate defeat of Japan.

Up to this point the war in the Pacific had been dominated by the air element. It had been the decisive factor in all the engagements, whether they were on the part of the ground or the naval forces. In Guadalcanal, victory or defeat depended on who controlled Henderson Field, and at Coral Sea and Midway, it was not the battleships that did the fighting, but the airplanes. The surface ships never came in sight of each other. They traded blows at ranges of a hundred miles or more instead of the twenty miles or less that had been the extreme limit of fire in the old battleship days.

In other words, World War II was the witness to the end of the battleship as the core element of sea power. The fighting in World War II, as Liddell Hart said, "showed the hollowness of the battleship bogey and exposed the folly of the faith that had been placed in such out-of-date monsters. Their only important value in World War II was for shore bombardment--a role for which, ironically, they had in previous generations been considered unsuitable, and too vulnerable."[9]

With all this emphasis on airplanes, their role in the Pacific War, up to and including the Battle of the Philippine Sea, was purely tactical, not strategic in any sense. This changed, however, as a result of the Battle of the Philippine Sea because with the conquest of Saipan, Tinian and Guam, the home islands of Japan came within range of the B-29s, the new bombers that were coming on the scene. This put a new face on the war in the Pacific because it meant that Japan's inner line of defenses had been penetrated.

It is no small wonder that the loss of the Marianas caused General Tojo's government to fall. The highly vulnerable and almost defenseless cities of Japan were now exposed to destruction from the air. This was a new strategic dimension. Now Japan itself was a part of the battlefield, her cities a new battleground. It was time for the Japanese to do some rethinking about the war as a whole.

The B-29 was not a new airplane in any sense of the word. It had been in the planning stage ever since 1939, and the decision to go ahead with its production had been made before the airplane was airborne. The production order was signed on May 17, 1941. Boeing was simply told "to begin manufacture when ready."[10]

General Kenneth B. Wolfe was designated by Henry H. Arnold to be the project officer for the B-29s in April 1943. Production plans called for 150 B-29s to be produced in the early part of 1944, and General Wolfe's job was to organize, equip and train the units for combat. Various plans were bruited about for the employment of the B-29s, and the one Arnold favored was the India-China plan. This called for basing the bombers on airfields in the Calcutta area of India and for staging them through forward airfields in the Chengtu area of China for strikes against Japan.

The principal disadvantage of this plan was its logistics. The B-29s had to fly seven missions from India to China just to build up enough gasoline to carry out one combat mission against Japan, and the ferrying missions were over the hump of the highest mountains in the world. Another disadvantage of the plan was that the bombers could not reach the heart of the home islands from Chengtu but only the outer edges of Japan. The pilots spent so much of their energy carrying gasoline over the hump from India to China that they referred to themselves as "a goddam trucking outfit."[11] Their commander at this time, Curtis E. LeMay, agreed with his men. He said that the whole plan "was founded on an utterly absurd logistic basis."[12]

Since this was clear from the outset, and since it was also clear that the B-29s could not hurt Japan from the Chengtu bases, the question naturally arises as to why this operation was undertaken in the first place? The most obvious answer is the one given by Curtis LeMay--the pressure of public opinion. The public had been delighted with Jimmy Doolittle's daring raid of April 19, 1942, when he surprised the Japanese by bombing their home islands, including Tokyo, from the deck of the aircraft carrier *Hornet*, and demanded more of the same. LeMay said that the "over the hump" bombing of Japan was just another stunt like the Doolittle raid. "I've never been able to shake off the idea that General Arnold himself never believed that this would work. It didn't work. No one could have made it work." Nevertheless, it was undertaken and carried on, LeMay said, because "our entire Nation howled like a pack of wolves for an attack on the Japanese homeland. The high command yielded. The instrument wasn't ready, the people weren't ready, nothing was ready. Folks were given an impossible task to perform."[13]

Another reason the B-29s were sent to bomb Japan from "over the hump" was because of President Roosevelt's great concern about keeping China in the war. Chiang Kai-shek was always needling Roosevelt to give China as much aid as was being sent to Russia, and had him convinced that the war could not be won unless China was saved. It was also apparent to both Roosevelt and Chiang that the bombing of Japan from China bases would do much to elevate the morale and spirit of the Chinese people. Therefore, Roosevelt paid no heed to the advice of his military advisers, and kept promising the Generalissimo that come what may the United States would soon be bombing Japan from China bases. President Roosevelt was also lavish in his other promises to Chiang and refused to heed the advice of others that the war could actually be lost by trying to do too much to come to the rescue of the heroic and downtrodden people of China.

Roosevelt's concern about the Chinese, and his mistaken notion that the war could not be won without first saving China, was a view that never ceased to amaze Churchill, among others. It was also an idea, possibly an *idee fixe*, as the French would say, that on more than one occasion threatened to disrupt serious considerations of strategy. One such was the Cairo Conference of November 22-26, 1943, a conference that was called at the insistence of Churchill so that England and the United States could get their ducks in a row before sitting down with Stalin at Teheran. It was an important Conference because the work *not* done there is what left it up to Stalin to be the arbiter between the United States and England at Teheran. In other words, the non-work at the Cairo Conference set it up for Stalin to make the crucial decision to open up the Second Front in Europe *before* rather than *after* the strategic air offensive was launched.

Roosevelt agreed with Churchill that a conference was needed to iron out any differences between the two countries before sitting down with Stalin at Teheran, but then the American President proceeded to sabotage the Cairo Conference by insisting on inviting Chiang Kai-shek to the Conference. "What we had apprehended from Chiang Kai-shek's presence," Churchill said,

> now in fact occurred. The talks of the British and American Staffs were sadly distracted by the Chinese story, which was lengthy, complicated and minor. Moreover, as will be seen, the President, who took an exaggerated view of the Indian-Chinese sphere, was soon closeted in long conferences with the Generalissimo. All hope of persuading Chiang and his wife to go and see the Pyramids and enjoy themselves till we returned from Teheran fell to the ground, with the result that Chinese business occupied first instead of last place at Cairo.[14]

One of the promises (among others) Roosevelt made to Chiang Kai-shek at Cairo was that the B-29 bombers would soon be arriving in China, not later than April 1944, and that soon after that they would be bombing the home islands of Japan, at least by the first of June. The President also made it clear that he was not satisfied with these dates and wanted the program and schedules speeded up. Remembering the glorious publicity that had resulted from the Doolittle raid of 1942, the President wanted the headlines again, and wanted them as soon as possible. Thus on October 15th Roosevelt had sent a message to Marshall, saying,

> I am still pretty thoroughly disgusted with the India-China matters. The last straw was the report from Arnold that he

> could not get the B-29s operating out of China until March or April of next year. Everything seems to go wrong. But the worst thing is that we are falling down on our promises every single time. We have not fulfilled one of them yet. I do not see why we have to use B-29s. We have several other types of bombing planes.[15]

The President apparently just forgot that the "other types of bombing planes" did not have the range to bomb Japan from existing bases, and the B-29s only barely.

The arrangement that eventually prevailed was the one that had been worked out for the initial deployment of the B-29s to India and for their command from the Washington Headquarters. This was approved of by the White House in conferences on February 11 and 19, and on April 4 the Twentieth Air Force was constituted and activated. Subsequently, on April 10, the Joint Chiefs gave their approval to the plan, and this constituted the formal charter under which the Twentieth Air Force operated with Arnold as its first commander.[16]

The terms of the charter were not unprecedented. On January 21, 1943, the Casablanca Directive had put the Combined Bomber Offensive under the direct control of the Combined Chiefs of Staff, with Air Chief Marshal Sir Charles Portal, Chief of Staff of the Royal Air Force, as the executive agent. Now Arnold had a similar position in the war against Japan. As the executive agent of the Joint Chiefs of Staff, he would be responsible for carrying out strategic operations, and carrying them out under directives which he, as a member of the Joint Chiefs of Staff, would help to frame. It was a responsibility that was more or less thrust upon General Arnold by the dedication of his staff and by the certain knowledge that unless the strategic operations were carried out under a single command, there would be no strategic air operations, since the B-29s would fall under scattered theater commanders.

The Pacific did not have a single commander. Instead, it was divided up in a number of commands, and for the most part they were warring commands. There was one in the Central Pacific, another in Southwest

Pacific, and still a third in the China-Burma-India Theater. If they were to be effective at all the B-29s needed a single commander, and it was for this reason General Arnold said, "I could do nothing but retain command of the B-29s myself-- something I did not want to do."[17] The command was passed to Arnold not only because of the President's keen interest in getting the bombing of Japan started without further delay, but also because of some outstanding staff work done by one of General Arnold's key officers, General Haywood S. Hansell, Jr.

General Hansell's principal accomplishment in this area was to bring Admiral Ernest J. King into line with the Air Force thinking on the subject of a unified command for strategic air operations against Japan. This was no small achievement. The Navy considered the conflict in the Pacific to be the "Navy's war," and preferred to win it all by themselves without interference from the Army or the Air Force. What the Air Force was asking Admiral King to do was to open up the Pacific to an independent command, the strategic air force, that would operate in the Pacific but would not be under the Navy, but at the same time would look to the Navy for logistical support. This violated the sacred principle of unity of command in a theater of war and made a mockery of the idea that the conflict in the Pacific was the "Navy's war." At the same time, as long as Admiral King felt that way the Air Force could forget about the idea of obtaining Joint Chiefs of Staff approval for their strategic operations in the Pacific, that is unless they were put under the Navy.

General Hansell, in his personal Memoirs of the war, recalled the manner in which he approached Admiral King on this touchy subject and the surprising results. First off, General Hansell said he obtained General Arnold's permission to discuss the subject with Admiral King, and the next step was to corner the Admiral when he was in the company of General Arnold. What followed, according to General Hansell, was this,

> I asked Admiral King if I might have a word with him. I described briefly the problem of concerted command and control of the long-range bombers which would be attacking common targets in Japan but would be operating from bases under the

> command of several separate theatre commanders. I suggested a similarity with the problems attendant on control of the U.S. Fleet, whose command was centralized under him as Commander in Chief, U.S. Fleet, as well as Navy member of the Joint Chiefs of Staff. Would it not be sensible to concentrate the very long-range bombers in a strategic air force under General Arnold, Commanding General of the Army Air Force?[18]

General Hansell continued by saying,

> Under this arrangement, the B-29s would actually fall under the control of the Joint Chiefs of Staff, with General Arnold serving not only as Commanding General, but also as executive agent for the Joint Chiefs. The Joint Chiefs would provide unified strategic objectives. Like fleet units of the Navy, logistic support could be provided through directives to appropriate area and theater commanders. Admiral King reflected for a moment and then said, "I could find such an arrangement acceptable."[19]

This was all that was needed to get the job done. General Marshall did not need any convincing as he had already seen the light, and as a result the Twentieth Air Force was established. The way was now open for trial by battle. It was the opportunity the airmen had been denied in Europe. There the ground invasion had been launched before the strategic air offensive even got off the ground. Here it was different. The B-29s were to go into action ahead of the ground forces. The only question was whether the softening up process could be carried to the point where no ground invasion, beyond the occupation of a defeated country, would be necessary. If so, the world would have an answer to the long-debated question of whether Air Power was capable of producing decisive results in war.

The preliminary answer to this question was a resounding no. The answer came from the performance of the XXth Bomber Command in India, a command that had four outstanding leaders during its short existence, including Curtis LeMay.

The XXth Bomber Command ran its first combat mission on June 5, 1944, against the railway facilities at Bangkok and its last on March 31, 1945, against oil storage targets at Singapore. Two of the most spectacular missions were run on the night of August 10-11 against Palembang in Sumatra and Nagasaki in Kyushu. These targets were three thousand miles apart. The dual mission was considered a tactical success but a strategic failure. The B-29s reached their distant targets, but the bombing was so poor that no real damage was done at either place.

Altogether, during the brief period of its existence the XXth ran 49 combat missions and dropped 11,477 tons of bombs. Considering the original complement of 150 B-29s, this would average out as about two combat sorties per plane per month, a very low average. Not only that. The official Air Force historians said,

> Only a small fraction of this effort was directed against industrial targets within the Inner Empire. Some 5,200 tons, roughly 45 percent of the total load, were carried by planes flying out of China bases, and of that weight more than half were expended in the PAC-AID strikes or against other non-industrial targets.[20]

PAC-AID was the code name for other Pacific operations.

The diversion of the B-29s from their strategic bombing to other operations was not an unexpected development. Following the pattern of the European Theater, where upwards of a third of the so-called strategic bombing effort was on targets primarily in support of Army campaigns, the Joint Chiefs also included a Catch 22 in the basic charter for the Twentieth Air Force. This was the provision, should strategic or tactical emergencies arise requiring the use of the B-29 forces for purposes other than the missions assigned to them by the Joint Chiefs, Theater Commanders are authorized to use the B-29 forces, immediately informing the Joint Chiefs of such action."[21]

On the whole, and from the point of view of purely military considerations, which is never a very safe point of view under any circumstances, the decision to put the first contingent of B-29s in India turned out to be a mistake. The B-29s could not hit Tokyo or any other industrial targets in Honshu from the Chengtu bases; in fact the Inner Empire was almost as invulnerable from the Chengtu bases as it had been before. Also, there was no element of surprise in the attack from these bases as had been the case with the Doolittle raiders in 1942. Long before the first combat mission the presence of the B-29s in India was known to the Japanese, and the work on the Chengtu airfields was also known. Added to this was the impossible logistic support for the B-29s, and the poor training of crews unfamiliar with the new airplane, and uncertain about the best techniques of flying it, or bombing. In short, the strategic bombing of Japan from the bases in India was both a failure and a mistake.

The United States Strategic Bombing Survey passed the same judgment on the bombing of Japan from the India bases. "The overall results achieved," the Bomb Survey said,

> did not warrant the diversion of effort entailed, and the aviation gasoline and supplies used by the B-29s might have been more profitably allocated to an expansion of the tactical and antishipping operations of the Fourteenth Air Force in China. The necessary training and combat experience with B-29s provided by this operation might have been secured through attacks on "Outer Zone" targets from bases more easily supplied.[22]

The Commander of the Fourteenth Air Force in China was General Claire Chennault. He was also Chiang Kai-shek's Chief of Staff for air. Not only did he agree with everything the Bomb Survey had to say about the B-29s in India, but added a thought of his own. This was that the B-29s in India, as far as the war in China was concerned, were "a liability rather than an asset."[23] In fact, in August General Chennault demanded that the B-29s be "withdrawn from China" altogether unless it was possible to use them in a "counter-air campaign."[24] The considered judgment of the Air Force historians was that the bombing of Japan from India "was not successful" and "achieved no significant results."[25]

Under the circumstances, the marvel is that the Joint Chiefs did not reconsider their position with respect to the Twentieth Air Force and do what the Army and Navy, particularly the latter, had wanted to do all the time, that is parcel out the B-29s to the various Theater Commanders. However, the Air Force luck held, and the Twentieth was not demolished, thus giving the B-29s another chance in the war against Japan.

The other chance for the B-29s came from their bases in the Marianas, when they were finally won. The first inkling that something was afoot in this area came on June 6, 1944, when General Kenneth B. Wolfe, Commanding General of the XXth Bomber Command, got an urgent message from Arnold to mount an attack on Japan proper in order to relieve pressure on General Chennault's forward airfields and to support an "important operation"[26] in the Pacific. Arnold and Wolfe argued back and forth about the date of the attack, whether it would be on the fifteenth of June or the twentieth, and about the size of the force, whether it would be with 50 B-29s or 25. The earlier date and the higher figure were agreed upon in the usual compromise with higher authority, and on June 15th the B-29s made their first appearance over Japan proper.

This was the first time since the Doolittle raid of April 18, 1942, that American bombers had appeared over Japan proper. However, this time the target was not Tokyo, but the Imperial Iron and Steel Works at Yawata in Kyushu, the most important single objective in Japan's vital steel industry. The day Yawata was hit the Fourth U.S. Marine Division swarmed ashore at Saipan. This, then, the invasion of the Marianas, was the "important operation" that General Arnold had referred to in his urgent message to General Wolfe.

From their bases in the Marianas, the B-29s went to work on Japan proper. At first, they were under the command of General Haywood S. Hansell, Jr., the officer who had been largely responsible for winning over the Army and Navy to the Air Force point of view in connection with the establishment of the Twentieth Air Force. He was also looked upon as one of General Arnold's proteges, and his job, obviously, was the most important command assignment in the Air Force.

General Hansell's tenure as commander of the Twentieth Air Force was from August 28, 1944, to January

20, 1945. Then he was succeeded by Curtis E. LeMay, who commanded until July 16. On that date, which was after the victory in Europe, there was a major reorganization of the strategic air forces in the Pacific, and in the reorganization, Carl Spaatz was named Commander-in-Chief. U.S. Army Strategic Air Forces, Pacific, with headquarters on Guam. The two principal combat commanders under him were Nathan F. Twining and James H. Doolittle. The latter became commander of the Eighth Air Force, newly equipped with B-29s and moved to Okinawa, and the former took over as the new commander of the Twentieth Air Force (as the XXIst Bomber Command was reconstituted). On relieving Curtis LeMay, General Twining said, "taking over the outfit from Curtis LeMay is about like taking over the Notre Dame football team from Knute Rockne."[27] Even so, General Twining's first major job was to do something a little bit different. It was to deliver the atomic bombs.

The B-29s in the Pacific began their operations under commanders who were strongly influenced by the concept of Air Power that had been developed before the war, and which was finally brought to fruition in the European Theater. This was that victory could be won by Air Power through the destruction of vital targets by precision daylight bombing. That this was a valid concept was attested to by the Bomb Survey after the war when it said, "the German experience suggests that even a first class military power--rugged and resilient as Germany was -- cannot live long under full-scale and free exploitation of air weapons over the heart of its territory."[28] Japan was nowhere near the equal of Germany from the point of view of being rugged and resilient, and the airmen were confident that she could be defeated by Air Power.

By August of 1944, when General Hansell took over the XXIst Bomber Command, the crews were being trained for radar bombing by night. The first thing he did was to reverse this and to reinstate the idea that the bombing would be by daylight against precision targets. As General Hansell said, the crews were "openly hostile to me and everything I wanted to do,"[29] but he was determined on his course and turned the training round to emphasize daylight operations against precision targets. This training continued for the entire period General Hansell was with the XXIst Bomber Command. In fact, the XXIst was more of a training than a bomber command while General Hansell held the reins.

General LeMay was having the same experience with the XXth Bomber Command in India. He had been sent there as a trouble shooter when operations lagged under General Wolfe, and the first thing he did was to introduce the idea of Lead Crew bombing, and his famous combat box for protection. When these failed to solve the problem, as fail they must because the problem was not operational but logistical, the base was phased out and LeMay was transferred to the Marianas in a similar role. The question was whether the "magic" that had failed LeMay in India would produce results in the Marianas. Meanwhile, in the Marianas, General Hansell was making progress with his training program and was actually getting the crews accustomed to combat conditions. The bombing still left a great deal to be desired, however, and the non-combat losses were higher than they should have been, but on the overall progress was being made.

The main emphasis in the bombing from the Marianas, once it got underway for real, was on the Japanese aircraft industry. The reason for this was the certain knowledge that control of the air had been and would continue to be the key factor in all operations in the air, at sea, and on the ground. Not only had this been the primary lesson learned thus far in the Pacific, but it was reinforced by the European experience. As the Bomb Survey said, "the greatest single achievement of the air attack on Germany was the defeat of the German Air Forces."[30]

Therefore, for General Hansell's opening blow against Japan the target selected was the Musashino plant in the northwest part of Tokyo. This was a good target because about one-third of all the engines for Japanese combat aircraft were produced there. However, the results of the bombing were not encouraging. Bad weather conditions prevailed over the target, and most of the bombing had to be done by radar. As a result, only one percent of the building area and two percent of the machinery were damaged, and casulaties were light. The only thing General Hansell could take satisfaction in was the fact that he had bombed the toughest target in Japan, and in doing so he had not been forced to take heavy combat losses. In fact, there was only one combat loss in this mission, and that was to a Kamikaze pilot who flew his plane into the tail of a B-29.

The mission to Tokyo on November 24, 1944, set the pattern for the next few weeks. For the most part, the bombing was against aircraft targets, and the tactics were according to the standard doctrine of the Air Force, bombing from high altitude in daylight and using precision techniques. However, the record was still spotty. Bad weather over the targets interfered with the accuracy of the bombing, and the extremely strong winds at high altitudes made navigation and formation flying a problem. The truth of the matter was that the program fell far short of the high expectations General Hansell had set for the B-29s, and he began to have some second thoughts about the high-level tactics and the precision bombing techniques.

While General Hansell was in this self-critical mood, he received a message from General Lauris Norstad asking him to launch a full-scale incendiary attack on Nagoya. Norstad had replaced Hansell as Chief-of-Staff for the Twentieth Air Force when General Hansell was assigned to the Marianas. Therefore, he was uncertain about the Nagoya idea, was it Norstad's or Arnold's? In any case, General Hansell was upset about the idea of going off in an entirely new direction just when he seemed to be making some progress in solving the problems of high altitude, precision bombing and protested the abrupt change. He queried Arnold directly as to whether this was a policy change and explained that he had only "with great difficulty implanted the principle that our mission is the destruction of primary targets by sustained and determined attacks using precision bombing methods both visual and radar."[31] Norstad replied immediately in a conciliatory message that the Nagoya directive did not constitute a change in policy and that the aircraft industry remained the primary objective, but insisted that the test fire-raid mission be undertaken as a "special requirement resulting from the necessity of future planning.[32]

The test run was made on January 3, and it was largely a failure. There were 97 B-29s in the mission, and only 57 reached the target area. They bombed through cloud cover from high altitudes and started some fires, but there was no holocaust. The Nagoya fire department had the blaze under control in a reasonable time, and it was certainly not another Hamburg. In fact, the Nagoya attack strongly suggested that fire raids against Japanese cities would *not* be any more effective than the incendiary attacks on German cities by the Royal Air Force. And if there was one lesson that was learned from the European experience it was that wars are not won by urban area attacks, even on the scale of a thousand bombers or more.

Accordingly, General Hansell resumed his attacks on the Japanese aircraft industry. High winds, however, continued to break up the formations, and bad weather persisted in playing havoc with visual bombing. But on January 19, 1945, everything seemed to come together for General Hansell. Oddly enough, the date of his greatest triumph was the day before General Hansell was relieved of his command and posted to another assignment in the B-29 training program, while General LeMay took over the XXIst Bomber Command.

General Hansell's last mission of the war was run against an airframe and engine complex at Akashi about twelve miles west of Kobe. Akashi was an important plant of the Kawasaki Corporation, the third largest aircraft company in Japan. The mission was a complete success. No bombers were lost, and the bombing was so accurate that there was a 90 percent production loss in both airframes and engines, a loss that was never recovered. However, even more important was the fact that this loss accelerated the move to disperse the aircraft industry in Japan, and to protect it by going underground. This had almost as much to do with the loss of production in the country as the bombing itself.

General Hansell said that the success of the Akashi mission proved that

> the extensive training program was paying off. It was clear to all, and most especially to the combat crews, that the XXIst Bomber Command was capable of destroying selected targets when weather conditions made visual bombing possible. It was an immense first step. The next would be the achievement of acceptable accuracy in radar bombing of selected targets. That, too, was on the way and

was attained later in the war.[33]

In more detail, General Hansell said that at Akashi he dropped the bombing altitude about 5,000 feet below previous levels. "This decision was made," General Hansell went on to say, "In order to improve bombing accuracy. It reduced the problem caused by very high winds and extreme turbulence encountered above 30,000 feet, and it reduced errors which tend to be proportional to bombing altitude."[34] Finally, General Hansell said, this was only a

> first step in a planned schedule of altitude reductions by successive steps to improve bombing, based upon the discovery that the enemy fighter attacks, though bitter and reckless, were not as effective against the B-29 as German fighter attacks had been against B-17s and B-24s. I had planned to continue this step-by-step reduction in altitude to improve bombing accuracy until we found a level at which further reduction incurred too many losses from enemy fighters. The mission of 19 January was a magnificent success.[35]

General LeMay inherited a going concern when he took over General Hansell's XXIst Bomber Command. As General Arnold said, it was a well organized, well trained and efficient organization.[36] As a result, General LeMay made no immediate change in anything, not in the training program, not in the way the missions were run, and not in the concept of bombing. Instead, the old and established pattern was followed to the letter.

No one rocked the boat in the XXIst Bomber Command until February. Then toward the end of the month Nordstad in Washington renewed the suggestion that an incendiary attack, this time on Kobe, be tried. Obviously, he had not been satisfied with the earlier fire-raid on Nagoya and wanted another test of incendiary tactics against the "flimsy" Japanese cities. LeMay responded with a fire-bombing mission against Kobe on February 3. The bombing was from about 25,000 feet, and it was a very large-scale attack. Considerable damage was done, but the results were far from being catastrophic. Consequently, after Kobe, LeMay returned to the precision bombing of the aircraft industry. This was exactly what General Hansell had done after the fire-raid on Nagoye. In other words, business as usual remained the order of the day. Obviously, fire-bombing was no panacea, no short-cut to the defeat of Japan.

This conclusion was more apparent than real. It was certainly not consistent with LeMay's line of thinking, nor with that of Norstad. In particular, LeMay remembered his experience with the Hankow fire-bombing when he commanded the XXth in India. It was not his idea to run this mission, and everything went wrong that could, including the failure of two-thirds of the bombers to reach their target, plus the fact that the bombing was in the wrong sequence, but the fact remained that the attack had burned out Hankow to the point where it was no longer useful as a major military base. Weren't all the Asian cities pretty much the same? If an attack as bad as this could burn out Hankow, didn't it stand to reason that fire was the optimum weapon for strategic air operations in the Pacific?

The more General LeMay thought about his problem as commander of strategic air operations in the Pacific, the more the idea of fire appealed to him. The solution might be to send the B-29s in low, too low for antiaircraft fire to bother them, and to send them in at night instead of the daytime, so they would not be menaced as much by enemy fighters. The available intelligence seemed to support this line of thinking. To the best of LeMay's knowledge, Japanese antiaircraft fire (contrary to the Germans) was ineffective at low levels, and also to the best of his knowledge there were only two units of night fighters in the whole of Japan. If this were true, it would mean that the vital targets could be attacked by night, at low level, using incendiaries instead of high explosive bombs. Also, lowering the altitude would save fuel, decrease the strain on the engines and greatly increase the accuracy of the bombing.

LeMay was well aware of the fact that this was a very radical way of thinking and that it entailed a complete reversal of everything he had been taught about Air Force doctrine, to say nothing of the "lessons" he had learned in Europe. In Europe, the Royal Air Force had staked its all on area bombing by night, including the full use of incendiaries, and the theory had been found wanting. On the other hand, the American doctrine of precision, daylight bombing had also been tried out in Europe and had not been found wanting. Was it prudent to fly in the face of both doctrine and experience? LeMay was not convinced that it wasn't. He was like the directors of the Bomb Survey who, when they were first presented with the problem of Japan, said "they were not at all sure that what held good in Germany would be equally valid in Japan."[37]

In discussing future bombing plans with General Arnold's representative, Lauris Norstad, General LeMay was very careful not to reveal all the details of the project he had in mind but used the occasion to sound out General Norstad as to General Arnold's probable reaction to something radically different in the way of bombing. LeMay said, "You know General Arnold. I don't know him. Does he ever go for a gamble? What do you think?"[38]

Norstad was just as careful as LeMay not to commit himself in their discussion of future bombing plans. Previously, he had assured LeMay that General Arnold wanted him to have "fullest latitude"[39] in mixing his blows any way he wanted to as long as he got results. On this occasion, he came back to this ploy and delivered himself of a little homily on the same subject. "You go ahead," Norstad said to LeMay,

> and get results with the B-39. If you don't get results, you'll be fired. If you don't get results, also, there'll never be any Strategic Air Forces of the Pacific--after the battle is finally won in Europe, and those ETO forces can be deployed here. If you don't get results it will mean eventually a mass amphibious invasion of Japan, to cost probably half a million more American lives.[40]

LeMay accepted this as a go-ahead for the low-level fire-bombing attack he had in mind. He also took full responsibility for the operation. "In any case," he said,

> it was my decision and my order which sent the B-29s to the task in the manner described. I was glad that I had not consulted General Arnold. I'd talked to

> subordinates, and some had gone along with me in the notion, and others had disapproved, sincerely and almost heart-brokenly. But again-- *My* decision and *my* order. There has to be a commander.[41]

LeMay staked his all on the success of the first mission in his fire-bombing plan, the massive strike against Tokyo on March 9-10. He stripped the B-29s down, taking off their bomb-bay tanks, taking out all the defensive armament, no guns and no ammunition, and taking out the gunners, too, leaving only the tail gunners for observation. Then the huge planes were loaded with incendiary bombs, an average of six tons per plane, and the pilots were told to bomb Tokyo by night at altitudes averaging about 7,000 feet, instead of the 30,000 feet to which they had become accustomed. The pilots were shocked into silence. It could be a suicide mission.

Field orders for the Tokyo raid were cut on March 8, 1945, and General Arnold was not even informed until the next day. LeMay was also true to the promise he had made to himself about keeping the secret. "If I do it," he said,

> I won't say a thing to General Arnold in advance. Why should I? He's on the hook, in order to get some results out of the B-29s. But if I set up this deal, and Arnold O.K.'s it beforehand, then he would have to assume some of the responsibility. And if I don't tell him, and it's all a failure, and I don't produce any results, then he can fire me. And he can put another commander in here and still have a chance to make something out of the B-29s. This is sound, this is practical, this is the way I'll do it: not say *one word* to General Arnold.[42]

The bombs began to fall on Tokyo a few minutes after midnight on March 9. The Japanese were caught completely by surprise, being totally unprepared for either the low-level attack, or its density. The bombing was good, and the losses were surprisingly low. In fact, and despite the low altitude, the loss rate for the B-29s was only a fraction above the average figure for all B-29 operations and well below the figure for the preceding month.

As for results, they were mind-boggling. Fifteen and eight-tenths square miles of Tokyo were destroyed, nearly 84,000 people were burned to death, not counting another 100,000 seriously injured and a million people made homeless. Tokyo radio said that the fire-bombing of March 10 was a good example of "slaughter bombing,"[43] and that it was reminiscent of the holocaust of Rome caused by the Emperor Nero. In other words, LeMay was pictured as being a modern Nero.

A good case can be made for the fact that the Tokyo fire-bombing was worse than the Roman disaster of 64 A.D. As the Air Force historians said,

> There are passages in Tacitus' famous account of the disaster of 64 A.D. that might have been applied to that of 10 March. But the physical destruction and loss of life at Tokyo exceeded that at Rome (where ten out of fourteen wards of a much smaller city were consumed) or that of any of the great conflagrations of the western world--London, 1666 (436 acres, 13,200 buildings); Moscow, 1812 (38,000 buildings); Chicago, 1871 (2,124 acres, 17,450 buildings); San Francisco, 1906 (four square miles, 21,188 buildings). Only Japan itself, with the earthquake and fire of 1923 at Tokyo and Yokohama, had suffered so terrible a disaster. No other air attack of the war, either in Japan or Europe, was so destructive of life and property.[44]

While this is true, the Hamburg raid of 1943 ran the Tokyo raid of 1945 a close second, and in some ways surpassed it. Again, fire was the principal agent of destruction. Albert Speer, in describing the effects of the Hamburg raid, said, "Huge conflagrations created cyclone-like firestorms; the asphalt of the streets began to blaze; people were suffocated in their cellars or burned to death in the streets."[45] The Bomb Survey reported that many of the German dead were found "in shelters where they were reached by carbon-monoxide poisoning."[46]

The life and property damage and destruction in Hamburg, as in Tokyo, reached heights that stagger the imagination. In Hamburg, over half the city was destroyed, 300,000 dwelling units were wiped out, and from 60,000 to 100,000 people were killed. This was greater in some respects than the damage in Tokyo, particularly in human fatalities. The difference between Hamburg and Tokyo was that whereas the latter was the result of a single raid, one night of terror, the former was the result of five major attacks. However, since the Hamburg attacks were on an around-the-clock basis and all took place between July 25 and August 2, they seemed like a single raid to the hapless Germans.

In any case, LeMay was pleased with his work in Tokyo, and his boss, General Arnold, was jubilant. He sent LeMay a telegram of "Congratulations" for a fine job, well done, and LeMay immediately set about getting his B-29s geared up for the next job. He wanted to get the B-29s started on the evening of their return from Tokyo for the next mission, but finding that this was impossible, settled for the next day, March 11, LeMay said, "it would be possible, I thought, to knock out all of Japan's major industrial cities during the next ten nights."[47]

LeMay's instinct to move ahead rapidly with the destruction of all the major industrial cities of Japan is reminiscent of the costly mistake that was made in Europe as an incident to the bombing of Hamburg in 1943. The Bomb Survey said that just as Hamburg was "the most damaging single city attack of the war," so it was that "documents show that German officials were thoroughly alarmed and there is some indication from interrogation of high officials that Hitler himself thought that further attacks of similar weight might force Germany out of the war." In the interrogation of Albert Speer, the Armaments Minister told the Survey, "I reported for the first time orally to the Fuehrer that if these aerial attacks continued, a rapid end of the war might be the consequence."[48] In his post-war Memoirs, Speer said, "fortunately for us, a series of Hamburg-type raids were not repeated on such a scale against other cities. Thus the enemy once again allowed us to adjust ourselves to his stragegy."[49]

It is clear that LeMay was in no mood to allow the Japanese to adjust themselves to his new strategy. Not only was Nagoya bombed on the night of the 11th, but Osaka on the 13th, and Kobe on the 16th. Then on the night of the 19th Nagoya was bombed again. This is when LeMay ran out of bombs, literally. For four weeks LeMay was unable to mount another incendiary attack because the stockpile of fire bombs was exhausted. The alternative was to go back to the non-incendiary work with high explosive bombs. This was distasteful to LeMay because he sincerely believed that in the fire-bombing he had hit upon a war-winning strategy. Nevertheless, there was nothing he could do about it until April 13, when the new stockpiles of incendiaries began arriving on the B-29 bases. To the end, LeMay believed that the war could have been shortened if it had not been for this delay in his work, and that if it had been shortened the issue of the atomic bomb would not have arisen.

However, it may be that LeMay was mistaken in thinking that the shortage of incendiary bombs lengthened the war. There was more to it than that. This was the invasion of Okinawa and the opportunity it gave Admiral Nimitz, under the Catch 22 of the formal charter for the Twentieth Air Force, to take over operational control of the entire B-29 forces. Admiral Nimitz proceeded to do this in early April and maintained his control for the next five weeks, or until May 11. Thus it would not have made any difference about the stockpiles of incendiaries. Okinawa was Nimitz's concern, not the fire-bombing of the Japanese cities.

When the control of the B-29s passed to Admiral Nimitz, much to the disgust of LeMay, they were turned into a tactical air force. "Our immediate mission," General LeMay said, "was to knock out the airdromes on Kyushu. We did this in a hurry."[50] But this was not enough. The Navy was understandably worried about the Kamikaze attacks on their ships and ordered LeMay to continue the bombing of the already cratered airfields. LeMay appealed and appealed, but got nowhere. "Day after day," he said, "we had to go out and bomb Target Nothing."[51]

It was the same story as the bombing of the sub-pens in Europe. The Germans were winning the Battle of the Atlantic, and even if the Flying Fortresses could not penetrate the sub-pens, bouncing bombs off of them did no harm. Here LeMay had a chance of winning the war with his B-29s, or at least thought he did, but all he was allowed to do was bomb Target Nothing. Finally, the Navy relented. "Admiral Nimitz released us from support of the Okinawa campaign on May 11th," LeMay said,

> and spoke kind words of thanks. Personally, I don't know why we should have received much in the way of thanks. The B-29 was not a tactical bomber and never pretended to be. No matter how we socked away at those airdromes, we could not reduce the *Kamikaze* threat to zero. In some proportion it was always there.[52]

From this point forward until he was relieved of his command on July 16 because of the major reorganization of the Twentieth Air Force following the victory in Europe, a reorganization designed to put the winning European team in the saddle, LeMay plugged away at the Japanese cities with his fire-bombs and continued the attack against industrial targets with visual and radar bombing techniques. He was also the major participant in a highly successful program of sowing minefields in channels and harbors. "In the aggregate," the Bomb Survey said,

> 104,000 tons of bombs were directed at 66 urban areas; 14,150 tons were directed at aircraft factories; 10,600 tons at oil refineries; 4,708 at arsenals; 3,500 tons at miscellaneous industrial targets; 8,115 tons at airfields and seaplane bases in support of the Okinawa operation; and 12,054 mines were sown.[53]

Even when his B-29s were in bondage to Admiral Nimitz, General LeMay was supremely confident about winning the war and about winning it without the atomic bomb, without help from the Russians and without a ground invasion. To him, it was clear that Air Power had come of age in the Pacific and could win the war alone. General LeMay expressed these views in a message to General Norstad on April 25, 1945. "I am influenced," General LeMay said,

> by the conviction that the present stage of development of the air war against Japan presents the AAF for the first time with the opportunity of proving the power of the strategic air arm. I consider that for the first time strategic

air bombardment faces a situtation in which *its strength is proportionate to the magnitude of its task.*"[54]

Very specifically, General LeMay continued,

I feel that the destruction of Japan's ability to wage war lies within the capability of this command, provided its maximum capacity is exerted unstintingly during the next six months, which is considered to be the critical period. Though naturally reluctant to drive my force at an exorbitant rate, I believe that the opportunity now at hand warrants extraordinary measures on the part of all sharing it.[55]

LeMay's vision was as clear as a bell at the time he informed Norstad that the power of the strategic air arm was such that it could bring about the defeat of Japan. It was obvious to him that the ten day fire-blitz of March 9-19 had been the turning point in the war, and now that Iwo Jima and Okinawa had been taken, the inner gateway to Japan, the last line of defense, had been breached, leaving the Japanese leaders only one choice, surrender or annihilation. The way the Potsdam Declaration expressed it was "unconditional surrender" or "complete and utter destruction."[56] The Potsdam Declaration also made it clear that these terms were not negotiable. "We shall not deviate from them," the document stated, "There are no alternatives. We shall brook no delay."[57]

This was not a very happy prospect for a proud people who had never lost a war, and who had their eyes glued on a world dominion, or at lease a condominuim with Germany. But it was the only choice, and this was clear, at least to LeMay, by April 25. There was not the shadow of a doubt in LeMay's mind that the destruction of the 66 urban areas of Japan would bring about the defeat of that country without help from any other quarter, not from the Russians, not from the atomic bomb, and not from the United States Army, Navy or Marines. The chameleonic nature of air force philosophy that this would involve did not bother Curtis E. LeMay at all. It merely proved that history was not always the best guide for the future and that there is and always will be something new under the sun, the old Biblical saying to the contrary notwithstanding.

FOOTNOTES--CHAPTER IX

[1]Roberta Wohlstetter, *Pearl Harbor Warning And Decision* (Stanford: Stanford University Press, 1962), 350.

[2]R.J. Overy, *The Air War 1939-1945* (New York: Stein and Day, 1981), 82.

[3]Alvin D. Coox, "Thc Rise and Fall of the Imperial Japanese Air Forces," *Air Power and Warfare* (Washington: Government Printing Office, 1978), 89.

[4]R.J. Overy, *The Air War 1939-1945* (New York: Stein and Day, 1980), 142.

[5]Alvin D. Coox, "The Rise and Fall of the Imperial Japanese Air Forces," 1978, 92.

[6]The United States Strategic Bombing Survey, *Summary Report: Pacific War* (Washington: Government Printing Office, 1946), 6.

[7]*Ibid.*, 7.

[8]B. H. Liddell Hart, *History of the Second World War* (New York: G.P. Putnam's Sons, 1970), 619.

[9]*Ibid.*, 628.

[10]Wesley Frank Craven & James Lea Cate, *The Army Air Forces in World War II* (Chicago: The University of Chicago Press, 1953), V, 7.

[11]*Ibid.*, V, 94.

[12]Curtis E. LeMay, *Mission with LeMay* (New York: Doubleday & Company, 1965), 322.

[13]*Ibid.*, 322.

[14]Winston S. Churchill, *Closing the Ring* (Boston: Houghton Mifflin Company, 1965), 328

[15]Wesley Frank Craven & James Lea Cate, *The Army Air Forces in World War II* (Chicago: University of Chicago Press, 1953), V, 21.

[16]*Ibid.*, V, 38.

[17]Henry H. Arnold, *Global Mission* (New York: Harper & Brothers, 1949), 348.

[18]Haywood S. Hansell, Jr., *Strategic Air War Against Japan* (Alabama: Airpower Research Institute, 1980), 26.

[19]*Ibid.*

[20]Wesley Frank Craven & James Lea Cate, *The Army Air Forces in World War II* (Chicago: University of Chicago Press, 1953), V, 170.

[21]Haywood S. Hansell, Jr., *Strategic Air War Against Japan* (Alabama: Airpower Research Institute, 1980), 27.

[22]The United States Strategic Bombing Survey, *Summary Report: Pacific War* (Washington: Government Printing Office, 1946), 8-9.

[23]Wesley Frank Craven & James Lea Cate, *The Army Air Forces in World War II* (Chicago: University of Chicago Press, 1953), 172.

[24]*Ibid.*, 112.

[25]*Ibid.*, 175.

[26]*Ibid.*, 98.

[27]Haywood S. Hansell, Jr., *Strategic Air War Against Japan* (Alabama: Airpower Research Institute, 1980), 70.

[28]The United States Strategic Bombing Survey, *Summary Report: European War* (Washington: Government Printing Office, 1945), 16.

[29]Haywood S. Hansell, Jr., *Strategic Air War Against Japan* (Alabama: Airpower Research Institute, 1980), 140.

[30]The United States Strategic Bombing Survey, *Overall Report: European War* (Washington: Government Printing Office, 1945), 10.

[31]Wesley Frank Craven & James Lea Cate, *The Army Air Forces in World II* (Chicago: The University of Chicago Press, 1953), V, 564.

[32]*Ibid.*

[33]Haywood S. Hansell, Jr., *Strategic Air War Against Japan* (Alabama: Airpower Research Institute, 1980), 58.

[34]*Ibid.*, 52.

[35]*Ibid.*

[36]*Ibid.*, 139.

[37]David MacIsaac, *Strategic Bombing In World War II* (New York: Garland Publishing, Inc., 1976), 100.

[38]Curtis E. LeMay, *Mission with LeMay* (New York: Doubleday & Company, 1975), 347.

[39]Wesley Frank Craven & James Lea Cate, *The Army Air Forces in World War II* (Chicago: University of Chicago Press, 1953), 568.

[40]Curtis E. LeMay, *Mission with LeMay* (New York: Doubleday & Company, 1965), 347.

[42]*Ibid.*, 355.

[41]*Ibid.*, 348.

[43]Wesley Frank Craven & James Lea Cate, *The Army Air Force in World War II* (Chicago: University of Chicago Press, 1953), V, 617.

[44]*Ibid.*, V, 617.

[45]Albert Speer, *Inside the Third Reich* (New York: The Macmillan Company, 1970), 337.

[46]The United States Strategic Bombing Survey, *Overall Report: European War* (Washington: Government Printing Office, 1945), 92.

[47]Curtis E. LeMay, *Mission with LeMay* (New York: Doubleday & Company, 1965), 353.

[48]The United States Strategic Bombing Survey, *Summary Report: European War* (Washington: Government Printing Office, 1945), 3.

[49]Albert Speer, *Inside the Third Reich* (New York: The Macmillan Company, 1970), 339.

[50]Curtis E. LeMay, *Mission with LeMay* (New York: Doubleday & Company, 1965), 370.

[52]*Ibid.*, 372.

[52]*Ibid.*, 372.

[53]The United States Strategic Bombing Survey, *Summary Report: Pacific War* (Washington: Government Printing Office, 1946), 17.

[54]Curtis E. LeMay, *Mission with LeMay* (New York: Doubleday & Company, 1965), 373.

[55]*Ibid.*

[56]Winston S. Churchill, *Triumph and Tragedy* (Boston: Houghton Mifflin Company, 1953), 644.

[57]*Ibid.*, 643.

CHAPTER X

WAS THERE AN ALTERNATIVE?

On April 25, 1949, when Curtis LeMay gave expression to the thought that Japan could be defeated by air power alone, and that this could be done by the fire-bombing of her cities by night, instead of by the selective bombing of her industries by day, Japan was already in a hopeless position. At this time, it could not count on any help from a defeated Germany, and Russia had already given notice (April 6) that a non-aggression pact of 1941 would not be renewed. This made it clear that an attack from that quarter could be expected in the near future. The Russians were still smarting from the defeat of 1904-05. Alone in the world, Japan had reached the point where she was almost defenseless. The U.S. Navy controlled the seas right up to the shores of Japan, and carrier planes ranged at will over the coastal areas of the home islands. The B-29s crisscrossed the whole of the Island Empire, and were burning out Japan's cities at will, as well as destroying her industrial centers one by one, that is those that the sea blockade had not closed down.

The only element of strength Japan had at this time was an undefeated Army. It comprised a force of some five million men plus a thousand suicide aircraft. Of these very large and still intact ground forces, some two million were stationed in the home islands, and another two million were in Korea, Manchuria, China proper and Formosa. The balance were widely scattered and dispersed in French Indochina, Thailand, Burma, the East Indies, the Philippines and the bypassed Pacific. The trouble was that these powerful forces were isolated, were cut off from each other and from the world at large by the sea and air forces of the enemy. In other words, the land power of the Japanese was only theoretical because they did not have the capacity to project their power to any quarter of the globe. This applied to the bypassed islands of the Pacific and to the home islands themselves, as long as the sea and air blockade prevailed. The outlook for the Japanese was nasty, brutish, and short, as Thomas Hobbes would say.

The sea-air blockade of the home islands was so effective that after April 1945, according to the Bomb Survey, Japanese shipping was "restricted to the Korean and Manchuria runs and to shallow inland waterways."[1] This meant that the outlying garrisons were left to fend for themselves. "Cut off from any supplies or reinforcements, except occasionally by submarines," the Bomb Survey said, "their food ran out. On certain of the islands, Japanese actually ate Japanese."[2] Survivors on the outlying islands were not the only ones to die from starvation. Food was getting short in Japan, and people were beginning to starve to death. This was also true of Japanese industry. Plants that were not bombed out of existence began to shut down because of a shortage of raw materials. In other words, Japan was dying a slow but certain death.

However, there was no lifting of the blockade, and the bombing not only continued, but increased in intensity. Then the bombers began to add insult to injury by dropping leaflets that accurately warned the people as to which cities would be next on the target list. Ostensibly, the warnings were given for humanitarian purposes, but in reality they served quite a different purpose. This was to demonstrate to the people that their government had ceased to exist from the point of view of furnishing elemental protection. A Japanese major general put it correctly when he told the Survey later that the Army had lost its purpose when it "could no longer protect the Japanese people from extermination."[3]

Japan was a defeated and defenseless nation. The home islands were no better off than the outposts in the Pacific, such as Truk and Rabaul, and Japan was also being subjected to the same kind of merciless pounding from the air. It is no small wonder that Curtis LeMay was convinced that Japan could be defeated by air power alone,[4] and that Fleet Admiral Ernest J. King reported later, "that the naval blockade alone would have 'starved the Japanese into submission'--through lack of oil, rice and other essential materials--'had we been willing to wait."[5] In other words, Japan was being killed twice over, once by air power and again by sea power. But both were slow processes, and the American people were not willing to wait. Action was what the people demanded, and the quicker the better. This is what motivated Curtis LeMay to stake his all on the low-level fire-bombing of the Japanese cities by night instead of the high-level precision-bombing of the Japanese industries by day. LeMay made this decision in early March 1945. The desire to get the war over quickly is also what motivated General Marshall to take the lead in supporting the Army's decision to launch a ground invasion of the home islands in November instead of the strategy favored by Arnold and King, which was to let the Japanese die on the vine. The White House gave its approval to the Marshall strategy on June 18.[6] Later, when President Truman decided to drop the atomic bomb in August, he did so on the basis of ending the war quickly, if possible. In other words, the political and military leaders of the United States completely ruled out the strategy of "waiting" as the way to end the war in the Pacific and instead opted for the strategy of action, any action that would end the war quickly, regardless of the cost, either to themselves or the enemy.

The cost was high. The "slaughter bombing"[8] alone, to use the expression coined by Tokyo radio after the fire-bombing of March 9, resulted in enormous casualties. The Bomb Survey said, "total civilian casualties in Japan, as a result of nine months of air attack, including those from the atomic bombs, were approximately 806,000. Of these, approximately 330,000 were fatalities. These casualties probably exceeded Japan's combat casualties which the Japanese estimate as having totaled approximately 780,000 during the entire war."[9] To put this another way, the Japanese battlefield casualties for the entire war were three percent lower than their civilian casualties. However, if the atomic victims are excluded from the calculations, the military casualties exceeded the civilians by a fraction of 25 percent.

Was there an alternative to the strategy of "slaughter bombing"? Of course, there was. It was the strategy of selective bombing, a strategy that had proved to be a huge success in the war against Germany, and one that fitted Japan like a glove. It was also a strategy that was theoretically sound since the results were cumulative rather than duplicative, as was the case with the area or "slaughter bombing," and it was a stragegy that guaranteed results. In other words, with selective bombing the economy in the target country is torn down progressively and scientifically, whether in steel, oil, chemicals, or armaments, whereas in area bombing the destructive progress is horizontal, non-selective, with a small segment of a number of essential industries and services lost with the destruction of each urban area, adding up to nothing in particular until all the pillars of the war economy are broken down to the critical level. As a strategy, it is wasteful, duplicative and non-scientific.

There were a number of good reasons why Japan was ideally suited to a program of reverse economic planning, which is another way of saying precision bombing in 1945. First off, the target country was very small in size, and the target complexes were few in number, which meant that a relatively small number of hits on vital installations would wreck enormous damage on the economy, no matter what segment of industry was brought under attack, provided of course that the attack was scientifically oriented to achieve a high degree of destruction in some particular segment of the economy, and was not aimed mindlessly at killing people and burning buildings. However, the country was small and so lacking in basic resources that even a non-plan, such as the one LeMay embarked upon, turned out to be adequate to get the desired results. It is hard to quarrel with "success." On the other hand, as General Hansell said, "but 66 cities?"[10] To make his point about strategy perfectly clear, Hansell went on to say, "I do think grand strategy was at fault, that the invasion should not have been considered the *sine qua non* of victory. Japan was headed for inevitable collapse, without any invasion. And I believe that Japan could have been rendered helpless without the extensive urban destruction."[11] He could also have said, without the atomic bomb.

One of the reasons the non-selective, urban area strategy was followed in Japan was because it was an easier product to "sell" than the precision bombing technique. It was easier to sell because the results were more dramatic than selective bombing. People cheered when they saw one Japanese city after another go up in flames, and the results were all visible, the number of dead and dying, the square miles gutted, and other material damage, but in the case of selective bombing there was nothing to show for the effort except some cold statistics such as production losses in aircraft engines, shortages of steel products, loss of oil production and so forth. In consequence, the political leaders and surface strategists, even some of the airmen, tended to favor the urban bombing, and it became the strategy of the day. So much so was this the case that for the war as a whole 83 percent of the bombing was on urban areas, leaving only 17 percent for industrial targets.[12]

The forces available for the bombing of Japan were small in the beginning, but from this small beginning they rose rapidly. The Bomb Survey said that on a monthly basis the tonnages available for use against Japan "increased from 13,800 tons in March to 42,700 tons in July, and, with the activation of the Eighth Air Force in Okinawa, would have continued to increase thereafter to a planned figure of 115,000 tons per month, had the war not come to an end."[13]

The 115,000 tons of bombs a month that could have been dropped on Japan if the war had not ended would have been equal to 1,380,000 tons of bombs per year. Only 1,400,000 tons of bombs were dropped on Germany during the entire course of the war.[14] And this was the result of the combined efforts of the Royal Air Force and the United States Air Force. In other words, the bombing effort that was available to bring about the defeat of Japan was many, many times greater than had been necessary to make a shambles of Germany.

If Japan was like Germany in that she could not wage war without a supporting war economy, and if she was also like Germany in that her factories and lines of communication were vulnerable to air attack, how could any reasonable person doubt that Japan could have been defeated by sea and air power without the necessity for a land invasion? Of course, the key word in this sentence is reasonable, and this was not the mood of the American public in 1945. However, as is the case with all statistics, the mind-boggling ones about American strategic air power in the Pacific in 1945 are open to question. The early bombing of Japan was beset by many difficulties, including the weather; and the bombing on the whole left much to be desired. In this respect, it did not differ from the experience in Europe. In other words, there were many lessons to be learned in the Pacific, but the commanders in the field were learning them fast, and after the highly successful bombing of the Akashi engine works on January 19, 1945, it was clear that all the major operational difficulties would be overcome in time. The two principal lessons, as General Hansell said, were that bombing altitudes had to be lowered to improve accuracy, and that there had to be increased reliance on the use of improved radar techniques.[15] Nevertheless, and even within the context of these lessons, the forces available in the Pacific in 1945 were so large that it would have been possible for the planners to have chosen any among many target systems, iron and steel, oil, transportation, chemicals, electric power, the destruction of any one of which would have produced the desired results. However, no serious effort was made to do this because the country as a whole, not LeMay alone, was so fascinated by the burning cities of Japan that they were indifferent to any and all other courses of action.

The Bomb Survey recapitulated the bombing effort against Japan by saying, "in the aggregate, 104,000 tons of bombs were directed at 66 urban areas; 14,150 tons were directed at aircraft factories; 10,600 tons at oil refineries; 4,708 at arsenals; 3,500 tons at miscellaneous industrial targets; 8,115 tons at airfields and sea-plane bases in support of the Okinawa operation; and 12,054 mines were sown."[16]

It is thus apparent that the whole effort in the Pacific was against the urban areas. No serious attempt was even made to destroy any of the "organic" target systems, and only two were really brought under attack of any kind. These were the aircraft and oil industries, and the effort against them was relatively small. The tonnage aimed at the aircraft factories was only nine percent of the total, and the oil refineries received even less, something like six percent. Nevertheless, the attacks against oil and aircraft were not disappointing in results.

The pattern of the attack on the aircraft industry in Japan was different from that followed in Germany. Learning something from the German experience, the engine plants in Japan were made the primary targets, not the final assembly plants, and the results were better. The attacks did enough physical damage to reduce production capacity considerably, but as in the case of Germany the indirect effects of the attacks were more serious than the direct effects. Indirectly, the effects of the attacks were to drive the Germans and the Japanese underground and to cause them to disperse their production facilities. Going underground and dispersing their production facilities proved to be costly and unproductive for both the Germans and the Japanese. In other words, the indirect effects of bombing the aircraft industry in Germany and Japan were more serious than the direct effects. Nevertheless, by the time the war ended, the physical damage to the aircraft engine plants in Japan was over 75 percent.[17]

Oil was another story. Both Germany and Japan went to war with very low stocks of oil. In Germany's case, the reserves were good for only three months,[18] and in Japan's case the situation was not very much better.[19] Before the war, Japan was almost wholly dependent on the United States and the Dutch East Indies for oil imports, having practically no indigenous sources. The American and Dutch oil embargo, after all, was what triggered the war in the first place, and after its outbreak oil imports began declining as the Japanese lost control of the sea approaches to the home islands. The Bomb Survey said, "oil imports from the south began declining in August 1943 and had been eliminated by April 1945. Crude oil stocks were virtually exhausted; refinery operations had to be curtailed; and stocks of aviation gasoline fell to less than 1,500,000 barrels, a point so low as to require a drastic cut in the pilot-training program and even in combat air missions."[20]

The cut in the training program for the Japanese Air Forces was one of the principal reasons for the defeat of Japan. Combat losses soared when inadequately trained Japanese pilots had to do battle against the skilled and highly trained American pilots. The same thing happened to the Germans, and for the same reason. They curtailed their flying training program as early as 1942 because of a shortage of gasoline, and as the Bomb Survey said, "this decision was reflected in a deterioration of quality of personnel, which was the principal cause of the defeat of the German Air Force."[21]

The Japanese shortage of oil not only resulted in the defeat of their Air Force, but played a major role in the final liquidation of the fleet. A case in point is the death of the *Yamato*--a suicide. The *Yamato* was the pride and joy of the Japanese Navy, and when she set sail on her last mission, to relieve the Japanese defenders on Okinawa, she took on just enough fuel for a one-way trip. There was also no air cover. The end came as expected. Steaming south from the Kyushu area, the *Yamato* was sighted by American search planes and promptly sunk. This marked the end of Japanese battleships in a combat zone for the rest of the war. Thus, the Imperial Navy died a suicide, a Kamikazi, Just like the Japanese Air Force.[22]

The oil shortage in Japan was so desperate that the government organized a program to convert acorns into fuel, and thousands of school children were put to the task of collecting bushels of acorns. In addition to this pathetic gesture, a work force of millions was also organized to grub out pine roots when it was discovered that they too could be forced to yield small amounts of crude oil.

The desultory and inconclusive bombing of Japan's oil and aircraft industries did not prove anything beyond the point that Japan was every bit as vulnerable to precision bombing as Germany. And the odd part of it is that the planners overlooked Japan's real weakness when they selected the targets for precision bombing. This was not in her aircraft and oil industries, vulnerable as they were, but in her land transportation system.

Just as in the case of England, Japan was an island empire that depended on overseas supplies for its very existence. This is why Churchill always insisted that the Battle of the Atlantic was the crucial conflict in World War II. He said that the loss of the Battle of the Atlantic would have meant "the starvation and ruin of the British Isles."[23] The war-time Prime Minister also said, "the Battle of the Atlantic was the dominating factor all through the war. Never for one moment could we forget that everything happening elsewhere, on land, at sea, or in the air, depended ultimately on its outcome."[24] This was twice as true with respect to Japan if for no other reason because Japan did not have the natural resources that are found in England. Thus when Japan lost control of the seas upon which she depended for the essentials of her existence her only hope for survival depended on her land transportation system.

Good examples of this are found in the movement of coal and bulk agricultural products within the home islands. These were formerly waterborne, and suddenly had to be shifted to the railroads. This opened up an opportunity for both carrier planes and B-29s to completely paralyze the Japanese war economy and pave the way for an invasion of the home islands, if that should prove necessary. The Bomb Survey said,

> A successful attack on the Hakkodate rail ferry, the Kanmon tunnels and 19 bridges and vulnerable sections of line so selected as to set up five separate zones of complete interdiction would have virtually eliminated further coal movements, would have immobilized the remainder of the rail system through lack of coal and would have completed the strangulation of Japan's economy.[25]

Finally, the Bomb Survey said,

> This strangulation would have more effectively and efficiently destroyed the economic structure of the country than individually destroying Japan's cities and factories. It would have reduced Japan to a series of isolated communities, incapable of any sustained industrial production, incapable of moving food from the agricultural areas to the cities and incapable of rapid large-scale movements of troops and munitions.[26]

There were almost no end to the opportunities open to the planners to bring about the defeat of Japan through the process of selective bombing rather than the blunderbuss method of urban area attacks, destroying 66 cities and resulting in 806,000 civilian casualties, exceeding Japan's combat casualties. But the opportunities were allowed to go by the board, so great was the fascination for burning up the cities of Japan, one by one. By the middle of June, LeMay boasted, "we have destroyed in Tokyo 56.3 square miles; in Nagoya, 12.4; Kobe, 8.8; Osaka, 15.6; Yokohama, 8.9; Kawasaki, 3.6; a total of 105.6 square miles."[27] The official Air Force historians said admiringly on the completion of Phase 1 of the urban area program, which was June 15, "the six most important industrial cities of Japan had been ruined. Great factories had been destroyed or damaged; thousands of household and feeder industrial units had gone up in smoke. Casualty lists ran into six figures. Millions of persons had lost their homes, and the evacuation of survivors had made it difficult to secure labor for those factories that remained."[28]

Major General J.F.C. Fuller may have been right when in commenting on man's intoxication with destruction for the sake of destruction, he said,

> When we think it out, it is the inventive genius of man which has obliterated his sense of moral values. From the javelin and the arrow to the superfortress and the rocket-bomb, the very power to destroy, first slowly and then at terrific speed, has intoxicated man. From out the first flint axe and bended bow has at length emerged a Frankenstein monster-- the inventiveness of today-- that is destroying man's own work, his own culture, his own civilization, his past, his present and his future.[29]

The pyrotechnics of Phase I made heroes out of the airmen and led many people to believe that the urban area program would lead to the defeat of the enemy. President Truman warned that if Japan did not surrender, all her cities, not only the six most important, but all of them would be destroyed, one by one. Japan replied defiantly,[30] but this was for public consumption, and privately the Japanese increased their efforts to find a way out of the war. They were convinced that the war was lost, and that the best that could be hoped for was a peace that would preserve the integrity of their home islands, and one that would be consistent with the dignity of their divine Emperor.

Japan's struggle to end the war was internal. The first break came in July 16, 1944, when General Tojo's government fell as a result of the loss of Saipan. This gave a great boost to the opponents of the war in Japan, but the succeeding government headed by General Kuniaki Koiso, was something of a disappointment. As Japan continued to fall back under the blows from the land, sea and air forces of the United States, and as the Okinawa landings were made on the first of April 1945, it was clear that the Koiso government could not endure. His successor was Admiral Kantoro Suzuki. Again, the hope was that here was a man who could stand up to the military clique and lead Japan out of the war. It was not an easy task, however, and there were many tribulations before it was accomplished.

The War Direction Council in Japan continued to be deadlocked on the issue of war and peace. The deadlock was more apparent than real, however, because all the cabinet members wanted peace, the only difference being that one group wanted a guarantee on the position of the Emperor before they would agree to stop fighting. The diehards or holdouts against unconditional surrender were the Army Minister and the Army and Navy Chiefs. On June 20 the Emperor himself tipped the scales against them by ordering that plans be made not only to defend the home islands, but also "to close the war at once."[31] It was the Emperor's intention "to send Prince Konoye to Moscow as a special emissary to negotiate for peace on terms less than unconditional surrender, but with private instructions from the Emperor to secure peace at any price."[32]

Meanwhile, all sorts of peace feelers were being put out by the Japanese. Russia was approached as a friendly neutral to act as an intermediary in arranging a peace between Japan and the Western Allies. This was revealed to Harry Hopkins when he was in Moscow at the end of May to hold personal discussions with Stalin as the special representative of President Truman. In other words, Japan was not only actively engaged in trying to get out of the war, but this was well known in Washington. "The Americans," Liddell Hart said, "were already aware of Japan's desire to end the war, for their intelligence services had intercepted the cipher messages from the Japanese Foreign Minister to the Japanese Ambassador in Moscow."[33]

In this atmosphere, an extraordinary meeting was held in the White House on June 18, 1945. The meeting was an extraordinary one because the main topic for discussion was a military one, and, contrary to the practice under Roosevelt, civilian advisers were present at the meeting. The question was the next step in the Pacific War: should it be the Army-backed invasion of the home islands, or was there some other more sensible way of winning the war?

Considering what was known of Japan's plight at this time, and of her frantic desire to end the war on almost any terms, the question borders on the absurd.

And none of this is hindsight. Not only was the White House reading the Japanese mail with respect to their frantic efforts to get out of the war, but it was common knowledge that Japan was being killed twice over, once by the strangulation effects of the blockade, and again by the rain of bombs on her cities and industries. Either pressure, the blockade or the bombs, would have been enough to defeat Japan, and the two together threatened annihilation. Was it really sensible to kill Japan a third time by an invasion of the home islands?

The plight of Japan was so bad by this time that on one occasion a flight of 1,000 planes from the Third Fleet joined with 500 Marianas-based B-29s and 300 Okinawa-based Army and Marine planes to blast at Kyushu and Honshu. It was also at about this time that the Secretary of Navy, James V. Forrestal, said, "we now control the sea right up to Japan."[34] In other words, the stage was now set for the direct bombardment of Japan by the fleet. Nevertheless, the meeting in the White House took up the question of the invasion of the home islands of Japan, whether it would be necessary to bring about the defeat of Japan, and if so when it should take place?

President Truman said that at the White House meeting of June 18th the Joint Chiefs of Staff laid before him their detailed plans for the invasion of Japan, which they considered necessary for her defeat, and that

> the Army plan envisaged an amphibious landing in the fall of 1945 on the island of Kyushu, the southernmost of the Japanese home islands. This would be accomplished by our Sixth Army, under the command of General Walter Krueger. The first landing would then be followed approximately four months later by a second great invasion, which would be carried out by our Eighth and Tenth Armies, followed by the First Army transferred from Europe, all of which would go ashore in the Kanto plains area near Tokyo. In all, it had been estimated that it would require until the late fall of 1946 to bring Japan to her knees.[35]

What kind of vision did America's political and military leaders have at this time? It could be called the tunnel vision of the old-line Army. The tunnel vision of the old-time Army was well protrayed or expressed by the Joint Army-Navy Board in 1941 when it reviewed the first tentative war plans drawn up by the three services. These plans were drawn up pursuant to a directive from the President to compile a statement of requirements to defeat the potential enemies of the United States. In reviewing the finished plans the Joint Army-Navy Board added some thoughts of its own, thoughts that bore the conservative imprint of the old-time Army. The Board warned,

> Naval and Air power may prevent wars from being lost, and by weakening enemy strength may greatly contribute to victory. By themselves, however, naval and air forces seldom, if ever, win important wars. It should be recognized as an almost invariable rule that only land armies can finally win wars.[36]

President Truman said that when General Marshall presented his plan for a two-phase invasion of Japan, Admiral King and General Arnold "supported the proposal heartily."[37] General Arnold, it was later revealed, did this not because he lacked confidence in air power to bring about the defeat of Japan, but for two other reasons. One was out of loyalty to General Marshall, who had always stood behind the Air Force, and the other was because he wanted Army support for an independent Air Force in the post-war world.[38] Admiral King's acquiescence in the Army plan is a little more difficult to understand in view of his conviction that naval blockade alone would have been enough to starve "the Japanese into submission.[39]

But this was the basic idea behind the decision of June 18 -- that only land armies can finally win wars. Otherwise, the decision made no sense. Invading the home islands of Japan had nothing to do with the security of the United States. Japan was not even a threat to Guam at this time in history. And the price that United States would have to pay for the folly of carrying out the Army plan and invading the home islands of Japan was well known. Secretary of War Henry L. Stimson said of his feelings at the time and of the whole question of invasion as he had seen it in 1945,

> Two great nations were approaching contact in a fight to the finish which would begin on November 1, 1945. Our enemy, Japan, commanded forces of somewhat over 5,000,000 armed men. Men of these armies had already inflicted upon us, in our break-through of the outer perimeter of their defenses, over 300,000 battle casualties. Enemy armies still unbeaten had the strength to cost us a million more. *As long as the Japanese Government refused to surrender,* we should be forced to take and hold the ground, and smash the Japanese ground armies, by close-in fighting of the same desperate and costly kind that we had faced in the Pacific islands for nearly four years.[40]

The two questions Stimson did not answer were, who was doing the forcing and why was it necessary to do battle on the grounds of the Japanese's own choosing? There was no power on earth that could force the United States to do anything it did not choose to do at this time, and it was almost insanity to even think of invading the home islands of Japan at a cost of perhaps a million American casualties. Was it worth a million American casualties to prove that only land armies can finally win wars? But if that was not the logic behind the proposed invasion of an already defeated Japan, then, pray, what was it?

One answer to the question of logic was suggested by Stimson himself. To him, war was like a boxing match. Clausewitz called it "a duel on a larger scale."[41] It's the same idea, of course. Carrying on, Stimson thought that just as in a boxing match the stronger fighter did not let up when his opponent showed signs of weakness, so it was or should be in war. Weakness should invite the harder blow with maximum speed.[42] Stimson did not believe it was the American responsibility to throw in the sponge for the Japanese. He felt that this was something the Japanese must do for themselves,[43] or suffer the consequences, which, in the language of the Potsdam Declaration, was "complete and utter destruction."[44]

The error in this kind of thinking is obvious. Was the lousy sponge worth a million American casualties?

worth twice as many as were suffered in the whole of the war in Europe? Apparently, in Stimson's mind this was the case, and the tragedy is that, for varying reasons, Stimson's views were accepted by the Joint Chiefs of Staff (unless they originated there), and were made law by the President on June 18. Much more than Decemver 7, 1941, June 18, 1945, was a day of infamy for the United States. On that day, the absolute nadir of reason and common sense was reached.

Less than two months after President Truman gave his approval (June 18) to the Army plan for the invasion of the home islands of Japan, an event took place at Almagordo Air Base in New Mexico that was to change the face and nature of things in the world. This was the first test explosion of an atomic bomb. The date was July 16, 1945. At the time, the new President of the United States, Harry S. Truman, was in Potsdam, Germany, meeting with Churchill and Stalin to consider, among other things, matters pertaining to a provisional settlement of the European War. The President said,

> The historic message of the first explosion of an atomic bomb was flashed to me in a message from Secretary of War Stimson on the morning of July 16. The most secret and the most daring enterprise of the war had succeeded. We were now in possession of a weapon that would not only revolutionize war but could alter the course of history and civilization. This news reached me at Potsdam the day after I had arrived for the conference of the Big Three."[46]

Secretary Stimson flew to Potsdam after the atomic test and brought the full details of the test with him. President Truman immediately convened a meeting of the military leaders and civilian secretaries to discuss the implications of the bomb from the point of view of military strategy and plans. Among the military, General Arnold's was a dissenting voice. He questioned the need to use the bomb "on the ground that the Japanese had already suffered defeat and would have to acknowledge it. Japan was mortally wounded and bleeding to death."[46]

The only justification Arnold could see for dropping the bomb was the straitjacket of the November invasion date. If that could be delayed or abandoned, he was sure that there would be no need to drop the bomb, but if nothing could be done about the invasion date, Arnold was in favor of dropping the bomb immediately. In other words, what Arnold was saying was that while the bomb was not necessary to bring about the defeat of Japan it might be necessary to save the Army "from the obsession with invasion."[47]

Also present at President Truman's meeting was chief of staff to the President, Admiral William David Leahy. He was vigorously opposed to the use of the bomb. Leahy agrued that the use of the bomb would "take us back in cruelty toward noncombatants to the days of Genghis Khan."[48] At a later date, Admiral Leahy said,

> The lethal possibilities of atomic warfare in the future are frightening. My own feeling was that in being the first to use it, we had adopted an ethical standard common to the barbarians of the Dark Ages. I was not taught to make war in that fashion, and wars cannot be won by destroying women and children.

Finally, Admiral Leahy was opposed to the use of the bomb because it was unnecessary. He said,

> It is my opinion that the use of this barbarous weapon at Hiroshima and Nagasaki was of no material assistance in our war against Japan. The Japanese were already defeated and ready to surrender because of the effective sea blockade and the successful bombing with conventional weapons.[50]

There were many others, of course, who were opposed to the use of the bomb, and among them was the noted Danish physicist, Neils Bohr. He was one of the leaders in the field of atomic research before the war, and like many other scientists had serious misgivings about the impact of the bomb on the future of mankind. But this was not what was uppermost in the minds of the men at Potsdam. Their concern was primarily with the strategy of the war and the relationship of the bomb to that strategy, not the long-range implications.

Of the men at Potsdam, probably the most influential was the Secretary of War, Henry L. Stimson. His view of the use of the atomic bomb was just as uncluttered as his concept of the land invasion, whether it should be launched on November 1, in the face of the expected casualties of a million men, or whether it would be better to bring about the defeat of Japan by the slow process of attrition, that is relying on the strangulation effects of the sea blockade, plus the tons of bombs that were being dropped on the defenseless cities and industrial facilities of the hapless country.

The Secretary said that the best course was to invade Japan on November 1 and accept the casualty toll of one million men, unless the Japanese had the good sense to throw in the sponge before that date. This was not an American responsibility, the Secretary said, but was something the Japanese would have to do for themselves. The Secretary had much the same thing to say about the use of the bomb. It was his feeling that if the bomb could be used to speed up the victory, it should be used, but if the question was the invasion on November 1 or the use of the bomb, the November date should stand, and America should simply accept the losses on the beaches and in the battles that followed. In other words, just as Stimson stood for the invasion no matter what, so he now favored the use of the bomb, but only if its use would not delay the November date for the invasion.[51]

To put this another way, Stimson agreed with MacArthur, that victory was the only objective in war, and disagreed with Clausewitz that the purpose of war was the attainment of a political objective, in this case the elimination of Japan as a threat to world peace. Stimson also disagreed with another tenet of Clausewitz, that the proper aim of all military action was "to disarm the enemy."[52] Except for the defense of the home islands, Japan was completely disarmed by this time and would become more so with the passage of each day. And without the invasion of the home islands. Time was not on the side of the Japanese. Cut off from the rest of the world by sea-power and facing absolute destruction by conventional airpower, Japan was doomed to a certain death by the process of inanition. Nevertheless, Stimson said that as far as he knew the view he had with respect to the use of the bomb was the one that was held by the President and all those who were in his confidence at Potsdam. In any case, President Truman said that his military advisers recommend "that we go ahead with the existing military plans for the invasion of the Japanese home islands."[53]

Later in life, when the President was interviewed by John Toland and asked to comment on the soul-searching that he must have gone through when he made the decision to drop the bomb, Truman denied that he had any such problem. Referring to the decision, Truman said, "Hell, no, I made it like"-- he snapped his fingers-- "that!"[54] President Truman probably made his decision on the use of the bomb exactly the way he said he did. According to Churchill's account, there was not even "a moment's discussion as to whether the atomic bomb should be used or not," adding, "the historic fact remains, and must be judged in the after-time, that the decision whether or not to use the atomic bomb to compel the surrender of Japan was never an issue." Finally, Churchill said, "There was unanimous, automatic, unquestioned agreement around our table; nor did I ever hear the slightest suggestion that we should do otherwise."[55]

There were at least two reasons for unanimity on this many-sided subject, the use of the atomic bomb at all and its hurry-up use in the war against Japan. One was the possibility, if not the probability, that the use of the bomb would bring the war to an end before the useless slaughter on the beaches and battlefields got started and the other was the fact that if the bomb succeeded in bringing about Japan's surrender the Russian demands for territorial concessions in the Far East would not have to be satisfied since, hopefully, the war would be over before Russia got a chance to be a belligerent.

On the first point Churchill said,

> To quell the Japanese resistance man by man and conquer the country yard by yard might well require the loss of a million American lives and half that number of British-- or more if we could get them there; for we were resolved to share the agony. Now all this nightmare picture

> had vanished. In its place was the vision--fair and bright indeed it seemed--of the end of the whole war in one or two violent shocks.[56]

As to the second point, the participation of Russia in the Pacific War, Churchill said,

> Moreover, we should not need the Russians. The end of the Japanese war no longer depended upon the pouring in of their armies for the final and perhaps protracted slaughter. We had no need to ask favours of them. A few days later I minuted to Mr. Eden: "It is quite clear that the United States do not at the present time desire Russian participation in the war against Japan." The array of European problems could therefore be faced on their merits and according to the broad principles of the United Nations. We seemed suddenly to have become possessed of a merciful abridgment of the slaughter in the East and of a far happier prospect in Europe. I have no doubt that these thoughts were present in the minds of my American friends.[57]

In other words, to Churchill at least, and to Truman to a lesser extent, the atomic bomb, as Churchill said was "a miracle of deliverance."[58] It was a deliverance for the Japanese because it would give them a face-saving opportunity to get out of the war without losing another million or so men in battle, a deliverance for the Americans because it might rescue them from the folly of their planned invasion, and a deliverance to the world at large because it would not give the Russians what they wanted, an opportunity to play the jackal in the Pacific War at the expense of China and Japan.

Churchill had another reason for pushing the atomic bomb at this stage of the war. This was his concern about the power position of England in the post-war world. It was clear to Churchill that Russia would emerge from the war as the dominant power in Europe. In fact, in 1943 he said, "I think it inevitable that Russia will be the greatest land power in the World after this war, which will have rid her of the two military powers, Japan and Germany, who in our lifetime have inflicted upon her such heavy defeats."[59] It was also clear to Churchill that the good old days of the balance of power for England were gone forever unless a new equalizer could be found. The atomic bomb, in Churchill's way of thinking, held out some hope in this connection. Otherwise, it seemed clear to Churchill the way "would be open to the Russians in a very short time to advance if they chose to the waters of the North Sea and the Atlantic."[60]

Nearly thirty years before this, in 1917, General Jan C. Smuts in South Africa had given expression to much the same thought in a report he was asked to prepare on the subject of air power. Even at this early date, Smuts saw very clearly what it took others much longer to see, that air power would become the dominant and decisive factor in war, and that England's future hung in the balance. He said,

> It is important for the winning of the war that we should not only secure air predominance, but secure it on a very large scale; and having secured it in this war we should make every effort and sacrifice to maintain it for the future. Air supremacy may in the long run become as important a factor in the defense of the Empire as sea supremacy.[61]

Churchill thought that the time had come, and that the atomic bomb could be England's salvation in the post-war world.

Churchill was undoubtedly right about the atomic bomb being a miracle of deliverance from the very heavy casualties that would have followed upon an invasion of the home islands of Japan. He may also have been right about the bomb being a miracle of deliverance from an immediate push to the Channel on the part of the Soviet forces. However, he was dead wrong from the point of view of the bomb being a miracle of deliverance from the spread of Communism in the post-war world. As professors Norman J. Padelford and George A. Lincoln pointed out during the American monopoly period, "some 750 million people have gained independence from the Western Powers since 1945; in the same period roughly the same number have come under Communist domination."[62]

Churchill was not the only one with problems of a more or less personal nature that bore on the atomic bomb. The problem that puzzled Truman the most was Stalin. What should he tell him about the bomb? How should be impart the news to him? Was it necessary to tell him anything in advance of the weapon's use? These were tricky questions because up to this time it was thought that the implementation of the Army plan for the invasion of the home islands could not be undertaken without the help of the Russian armies to tie down the million and a half Japanese soldiers in Manchuria. Now with the bomb, was Stalin's help necessary or even desirable? Churchill said, "the President and I no longer felt we needed his aid to conquer Japan."[63] If this was so, a further question was, could the Western Allies act swiftly enough to actually prevent the Russians from getting into the Pacific War and thereby earning for themselves a place at the peace settlement?

Naturally, there were sharp differences of opinion among those who advised the President as to how to proceed in this delicate matter. The hard-liners did not believe the Russians should be told anything about the bomb in advance, nor indeed even after the event, except in return for an important *quid pro quo*. They were in favor of using the bomb immediately and surprising both the Russians and the Japanese. On the other hand, there were the liberals and some of the scientists, particularly Neils Bohr, who believed that the best course was to reveal all the secrets of the bomb to Russia at once to obtain international control of the great weapon, and to prevent a ruinous arms race. "A unique danger," Bohr said, "presented a unique opportunity."[64]

President Truman consulted Churchill as to what to do in this perplexing dilemma. They decided to straddle the question, that Stalin must be "informed" in advance about the bomb, but that he whould not be given any of the "particulars." Churchill said the President decided, "I had best just tell him after one of our meetings that we have an entirely novel form of bomb, something quite out of the ordinary, which we think will have a decisive effect upon the Japanese will to continue the war."[65]

This is just what Truman did. After one of the formal sessions of the Potsdam Conference, Truman strolled over to Stalin, and carefully not mentioning the words "atomic" or "nuclear," informed the Soviet dictator that the United States had just perfected a new weapon that it could use against the Japanese to help them make up their minds about the continuation of the war. In his *Memoirs*, the President recalled this incident by saying, "On July 24 I casually mentioned to Stalin that we had a new weapon of unusual destructive force. The Russiam Premier showed no special interest. All he said was that he was glad to hear it and hoped we would make 'good use of it against the Japanese.'"[66]

Truman was astonished at Stalin's apparent indifference to the fact that the United States was now in possession of an atomic bomb, and later attributed it to the fact that Stalin did not understand the significance of what he was being told. This was one of the biggest mistakes Truman ever made. It was not Stalin who was a stranger to the bomb, but Truman.

The Soviet nuclear research program had been initiated in June 1942,[67] and by this time Stalin was quite well informed as to the physics and chemistry of the bomb. On the other hand, Truman was a complete stranger to the bomb, and prior to being sworn into office in 1945, did not even know that his country was striving to produce an atomic bomb. Just the opposite was the case with Stalin. Not only did he have complete knowledge of the work on the bomb going on in the USSR, but was also up to date on what the United States was doing because the Soviets had penetrated the U.S. Manhattan Project almost from its inception.

Once July 24 was out of the way and the Soviet ally had been "informed" of the new weapon, the work could go ahead on actually delivering the dread bomb. The fulfillment came on August 6 and 9, when two bombs were dropped, one on Hiroshima and the other on Nagasaki. The Bomb Survey went to great lengths to record for posterity mankind's first experience with atomic bombing. In its special report on "The Effects of Atomic Bombs on Hiroshima and Nagasaki" the Survey said,

> Teams of experts were selected to study the scenes of the bombings from the special points of emphasis of physical damage, civilian defense, morale, casualties, community life, utilities and transportation, various industries, and the

> general economic and political reprecussions. In all, more than 110 men--engineers, architects, fire experts, economists, doctors, photographers, draftsmen-- participated in the field study of each city, over a period of 10 weeks from October to December 1945.[68]

The Bomb Survey said that, according to eye witness accounts of the explosion over Hiroshima on August 6, the bomb,

> exploded with a tremendous flash of blue-white light, like a giant magnesium flare. The flash was of short duration and accompanied by intense glare and heat. It was followed by a tremendous pressure wave and the rumbling sound of the explosion. This sound is not clearly recollected by those who survived near the center of the explosion, although it was clearly heard by others as much as fifteen miles away. A huge snow-white cloud shot rapidly into the sky and the scene on the ground was obscured first by a bluish haze and then by a purple-brown cloud of dust and smoke.[69]

The casualties in the first atomic attack were 57 percent of the total population of Hiroshima. Out of this total population of approximately 245,000, some 70,000 to 80,000 people were killed and another 70,000 injured. Sixty-nine percent of all the building structures in the city were totally destroyed, including 42 of the 45 hospitals in the urban area. The casualty rate among the doctors was over 90 percent and among the nurses 93 percent. In addition, 80 percent of all the people in the city were rendered homeless.

Three days after the attack on Hiroshima, that is August 9, a second atomic bomb was dropped in anger. This time the target was Nagasaki. Thirty-five percent of the total population were casualties in Nagasaki and about 40 percent of the residential buildings were destroyed or badly damaged. After the bombing, the Nagasaki Prefecture Report said the city looked like "a graveyard with not a tombstone standing."[70]

It would be fair to say that Hiroshima and Nagasaka were gutted by the two atomic explosions, and since only one bomb was involved in each case, it is not surprising that reputable political and military analysts jumped to the conclusion that this kind of carnage and physical destruction was unique in the history of the world. In fact, it was this kind of thinking that gave rise to all the loose talk at the end of World War II about one world or none, the twilight of geopolitics and the choice between the quick and the dead. While this may be the case today, it was certainly not the case in 1945.

The Bomb Survey, operating with the data at hand, put the atomic bombs in 1945 in proper perspective when it published a table on conventional and atomic bombing entitled "Effort and Results."

	Hiroshima	Nagasaki	Tokyo	Average of 93 Urban Attacks
Planes	1	1		173
Bomb load	1 atomic	1 atomic	1,667 tons	1,129 tons
Population Density per square mile	35,000	65,000	130,000	unknown
Square miles destroyed	4.7	1.8	15.8	1.8
Killed and missing	70/80,000	35/40,000	83,600	1,850
Injured	70,000	40,000	102,000	1.830
Mortality rate per sq. mile destroyed	15,000	20,000	5,300	1.000
Casualty rate per sq. mile	32,000	43,000	11,800	2,000[71]

What the table of "Effort and Results" shows is that the atomic bombs of 1945 were not as revolutionary as many at the time imagined. Tokyo is a good example of this. Two hundred seventy-nine B-29s dropped 1,667 tons of conventional bomba on Tokyo, and the results were far more devastating than they were for the atomic bombs in Hiroshima and Nagasaki. The casualties in the Tokyo fire-raid were close to 200,000, whereas they were only 150,000 in Hiroshima and 80,000 in Nagasaki. Also, the number of square miles destroyed in Tokyo was over 15 whereas in Hiroshima and Nagasaki the numbers were under five and two respectively. All in all, therefore, there was nothing unique in history about the physical destruction and human carnage wrought by the atomic bombing. The fire-bombings of Cologne, Hamburg and Tokyo were every bit as devastating, and in some cases more so.

How were the atomic bombs different from ordinary bombs? The atomic bombs were not supernatural or incomprehensible in their operation, the Bomb Survey said, and differed from ordinary bombs in only one respect, radiation. Continuing, the Bomb Survey said, "no other more mysterious or immeasurable forces acted; these were all."[72] In other words, the atomic bombs worked by explosion as do ordinary bombs, and they released energy in the form of heat and blast or pressure as do other bombs. Except for radiation, therefore, the atomic bombs can be equated with ordinary high explosives, and calculations can be made to determine how many ordinary bombs would be required to do the same amount of damage to life and property as the atomic bomb.

Before going into that, however, the Bomb Survey looked at the atomic bombs of 1945 from the point of view of radiation and came up with the finding that though the radiation effects were lethal, they proved to be so only "for an average radius of 3,000 feet from ground zero." In addition, the Bomb Survey said,

> The radiation apparently had no lasting effects on the soil or vegetation: seeds later planted within a few hundred feet of ground zero grew normally. Examination of subsurface soil in the immediate area showed presence of earthworms and other life only a few inches below the surface.

Finally, the Bomb Survey noted,

> Stories of harmful effects on people who came into the area after the explosion have been disproved by investigation.[73]

Continuing on the theme of the non-effects of radiation in the Hiroshima and Nagasaki bombings, the Survey said that in Hiroshima,

> a mass flight from the city took place as persons sought safety from the conflagration and a place for shelter and food. Within 24 hours, however, people were streaming back by the thousands in search of relatives and friends and to determine the extent of their property loss. Road blocks had to be set up along all routes leading into the city to keep curious and unauthorized people out.[74]

The only caveat the Bomb Survey made with respect to the non-effects of radiation was the finding, "in tests of the ground and bones of victims of radiation disease, certain substances--phosphorus, barium, strontium, rare earths--have shown radioactivity. Though evidence of lingering radioactivity is slight, it is strong enough to leave open the ominous possibility of a different situation had the bomb exploded at ground level."[75]

Another factor the Bomb Survey could have taken into consideration in investigating the difference between atomic and non-atomic bombing was the shock effect of the atomic bombs. Nothing is more important in weakening resistance or turning defeat into rout than a powerful shock. General Marshall in his testimony on the use of the bomb, according to Secretary of War Stimson, "was emphatic in his insistence on the shock value of the new weapon."[76] Accordingly, everything that could be done to *maximize* the shock effect of the bomb was done. In both Hiroshima and Nagasaki the bombs were delivered without warning to a surprise target and delivered in such a way that they would catch the people out in the open, not in their bomb shelters. The idea was to hit the Japanese with an all-at-once shock, the likes of which they had never experienced before. Hopefully, the blow would be the *coup de grace*, if not the first bomb, then certainly the second.

The bomb lived up to the expectations of the planners and exceeded them in fact, but the shock effects did not, at least as far as the people of Hiroshima and Nagasaki were concerned. Why was this? It all goes back to Coventry (1940), Cologne and Hamburg (1942), and the great fire-bombing of Tokyo (March 9-10, 1945). Each was a shock in its own way, and none, with the exception of Hamburg, took more than a few minutes to pull it off. In Coventry it was one night of terror, in Cologne one hour, and in Tokyo (where the casualties were far greater, and the physical destruction three times the total of Hiroshima and Nagasaki), the bombing was completed in three hours.[77] In other words, it should have been obvious to the planners that from the point of view of the shock effects the atomic bomb was nothing special. After all, and way back in 1942, the Royal Air Force had put a thousand bombers over Cologne in a matter of ninety minutes. Could anyone really expect a big shock difference between a one-minute attack (atomic bomb) and a ninety-minute attack?

If there was no "shock" effect to amount to anything as a result of the atomic bombing of Hiroshima and Nagasaki, and if only a small percentage of the deaths was due to radiation effects, not less than seven percent nor more that 20,[78] the Bomb Survey was probably on safe ground when it sought to determine what the relationship was between atomic and non-atomic bombing. In other words, how many B-29s carrying what tonnages of bombs would have been required to bring about the loss of life and physical damage suffered in Hiroshima and Nagasaki?

Using objective statistical evidence, the Bomb Survey found out that 220 B-29s, carrying 2,100 tons of conventional bombs, would have been enough to do the job in Hiroshima, and that 125 B-29s carrying 1,200 tons of "conventional" bombs would have been able to do as much damage to Nagasaki as one atomic bomb. Therefore, roughly the calculation was, one atomic bomb was the equivalent of 2,000 tons of "conventional" bombs.[79]

Given these statistics, and they seem to have a conservative bias, the 104,000 tons of conventional tons that General LeMay dropped on the 66 urban areas of Japan were equivalent to over 50 atomic bombs. Making a forward projection, as soon as the planned figure of 115,000 tons of bombs a month was reached, which would have been when the Eighth Air Force was activated on Okinawa, Japan could have been put on the receiving end of 57 atomic bombs a month for the indefinite future. This underscores the folly of the planners when they decided on June 18 that a million American casualties would be put on the line to force the Japanese to throw in the sponge, and thereby prove that the Joint Army-Navy Board was right in 1941 when it laid down the dictum, "that only land armies can finally win wars."[80]

As it was, Japan surrendered long before she was subjected to this kind of air attack. In fact, the surrender came right on the heels of the second atomic bomb. The terms of the surrender were a great victory for the drafting team. The sovereignty of the Emperor was fully recognized, but at the same time the Emperor himself was made subject to the Supreme Commander of the Allied Forces. However, this had nothing to do with the basic question of the atomic bomb, what did it contribute, if anything, to the surrender of Japan?

In doing research on this question, the Survey, as a starting point, began with the survivors of the bombings in Hiroshima and Nagasaki, and presented them with a standard set of questions designed to determine what effect the bomb had on their willingness to continue fighting. The results were surprising. The Survey said that a tabulation of the results from the many hundreds interviewed in all parts of Japan clearly showed "that the will to resist had indeed been higher in the 'atomic bomb cities' than in Japan as a whole."[81] This, of course, refers to the period before the atomic bombing. After the bombing, the results were even more surprising. The Survey said,

> Even in the target cities, it must be emphasized the atomic bombs did not uniformly destroy the Japanese fighting spirit. Hiroshima and Nagasaki, when compared with other Japanese cities, were not more defeatist than the average. The bombs were tremendous personal catastrophes to the survivors, but neither time nor understanding of the revolutionary

threat of the atomic bomb permitted them to see in these personal catastrophes a final blow to Japan's prospects for victory or negotiated peace.[82]

This suggests very strongly that as far as the people of Hiroshima and Nagasaki were concerned there was no straight-line relationship between the atomic bombings (August 6 and 9) and the offer to surrender (August 10). It also suggests that the "shock" effect of the atomic bombings was considerably less than Secretary Stimson and General Marshall thought it would be, at least insofar as the people of Hiroshima and Nagasaki were concerned.

Another non-effect of the atomic bombs, as the Survey pointed out, is the fact that the atomic bombs "did not convince the military that defense of the home islands was impossible"[83] In fact, on the evening of August 9, after the second bomb had been dropped and Russia had declared war, the War Minister General Korechika Anami, supported by the Army and Navy Chiefs of Staff, opposed accepting the Potsdam terms of unconditional surrender and argued that Japan should continue to fight with the means at her disposal. General Anami was convinced that they Navy could sink at least a third of the invasion fleet and that American casualties in the savage fighting on the beaches and in the subsequent battles would be so high that the country would eventually come to its senses and offer Japan an honorable peace. In short, the War Minister's plea was for a last decisive battle on the mainland, atomic bombs or no atomic bombs.

The Emperor felt otherwise. He had not been keen on the war in the first place, but the constitutional limitations on his power were such that he was unable to contain the headstrong and power-hungry military fanatics. However, as the tide of battle turned, and the string of military defeats grew longer, the influence of the peace party on the Emperor grew stronger. Eventually, the peace party became strong enough to enable the Emperor to take a personal stand on Japan's struggle to end the war. It was at this time that the Emperor decided to dispatch Prince Fumimaro Konoye to Moscow to negotiate the best peace terms he could get even if he had to settle for unconditional surrender.[84] Of course, this effort came to naught because of bad timing, plus the fact that Russia was of no mind to let the defeated Japanese get off the hook without reversing the outcome of the war of 1904-05. Consequently, the Russians simply kept stalling until the time was ripe for the stab in the back.

How did the atomic bomb fit into these complicated maneuverings? It didn't. After the loss of Saipan on July 9, 1944, and the invasion of Okinawa in April 1945, it was clear that the war was lost, and "early in May 1945, the Supreme War Direction Council began active discussion of ways and means to end the war, and talks were initiated with Soviet Russia seeking her intercession as mediator."[85] Thus it can be accepted that "early in May 1945" the strategic defeat of Japan was a *fait accompli,* and all that needed to be worked out was the details of the surrender. It was not until two months later, on July 16 to be exact, that the atomic bomb was ever field tested. Thus it is difficult, if not impossible, to say that there was a straight-line connection between the bomb and the defeat of Japan. On the other hand, the evidence is overwhelming that the atomic bomb did play an important part in working out the details of the final surrender.

It was clear to the Emperor after Hiroshima and Nagasaki that, given an adequate supply of atomic weapons, the Americans had enough B-29s in the Marianas to destroy in a single day all the cities of Japan with a population of thirty thousand or more. It was also clear to the Emperor "that no army without the weapon could possibly resist an enemy who had it."[86] Given these "facts," the problem that the Emperor faced was greatly simplified. This was that he had to convince the Army leaders that it was not dishonorable to lay down their weapons in circumstances such as these and that even unconditional surrender would not reflect adversely "on the competence of Japanese industrialists or the valor of the Japanese soldier."[87] In other words, the Emperor made the bomb an argument that he used against his own military forces, and he was successful in doing this because the bomb was a hard argument to refute.

When the Emperor made his decision to accept the terms of the Potsdam Declaration there was little, if anything, that the Army could do about it. The Emperor's arguments, supported by the logic of the bomb, enabled the Army to save "face," and give way to the inevitable. Also, there was really no alternative since obedience to a divine Emperor was the highest law of the Army. That is why a palace revolution to overturn the Emperor's decision was such a complete failure. On balance, then, what part did the atomic bomb play in bringing about the end of the war in the Pacific? The Bomb Survey was undoubtedly right when it said that "the atomic bombings considerably speeded up these political maneuverings within the government"[88] on the subject of surrender. But did it do more than that? The chances are it did. In giving its answer to the question, the Bomb Survey reported that there was a quip in high government circles at the time, "that the atomic bomb was the real Kamikaze, since it saved Japan from further useless slaughter and destruction. It is apparent that in the atomic bomb the Japanese found the opportunity which they had been seeking, to break the existing deadlock within the Government over acceptance of the Potsdam terms."[89]

More seriously, the Bomb Survey also said,

> It seems clear that, even without the atomic bombing attacks, air supremacy over Japan could have exerted sufficient pressure to bring about unconditional surrender and obviate the need for invasion. Based on a detailed investigation of all the facts, and supported by the testimony of the surviving Japanese leaders involved, it is the Survey's opinion that certainly prior to 31 December 1945, and in all probability prior to 1 November 1945, Japan would have surrendered even if the atomic bombs had not been dropped, even if Russia had not entered the ear, and even if no invasion had been planned or contemplated.[90]

On the other hand, when the Emperor, in an unprededen-ted radio address, announced the surrender and the government's action, he put the entire blame for the defeat on the atomic bomb. In the Imperial Rescript, the Emperor said,

> Now the war has lasted for nearly four years. Despite the best that has been done by everyone -- the gallant fighting

> of military and naval forces, the diligence and assiduity of Our Servants of the State and the devoted service of Our one hundred million people--the war situation has developed not necessarily to Japan's advantage, while the general trends of the world have all turned against her interest. Moreover, the enemy had begun to employ a new and most cruel bomb, the power of which to do damage in indeed incalculable, taking the toll of many innocent lives. Should we continue to fight, it would not only result in an ultimate collapse and obliteration of the Japanese nation, but also it would lead to the total extinction of human civilization.[91]

The Emperor continued by saying, "such being the case, how are We to save the millions of Our subjects; or to atone Ourselves before the hallowed spirits of Our Imperial Ancestors? This is the reason why We have ordered the acceptance of the provision of the Joint Declaration of the Powers."[92] In closing, His Majesty said, "it is according to the dictate of time and fate that We have resolved to pave the way for a grand peace for all the generations to come by enduring the unendurable and suffering what is unsufferable." And finally, "unite your total strength to be devoted to the construction for the future. Cultivate the ways of rectitude; foster nobility of spirit; and work with resolution so as ye may enhance the innate glory of the Imperial State and keep pace with the progress of the world."[93]

From the point of view of the progress of the world, the verdict of history has not been handed down on the atomic bomb. It was both a curse and a deliverance. Churchill called the bomb "a miracle of deliverance."[94] And to a certain extent it was. Certainly from the point of view of the millions who would have died on the beaches and in the pitched battles on the mainland of Japan, the bomb was a deliverance. Likewise, if it was the atomic umbrella that saved Western Europe from going behind the Iron Curtain, the bomb was also a deliverance. On the other hand, to the quarter of a million casualties in Hiroshima and Nagasaki, the bomb was an unmitigated disaster. It may also be so for the world population of today.

The atomic bomb of 1945 set in motion a series of breakthroughs in the field of atomic energy that led to the development of the hydrogen bomb, a bomb that could lead to the one big bang that could bring about the end of the human experience. This could hardly be called any kind of a deliverance. On balance, however, it seems that the "lessons" of the Pacific War are clear. There was an alternative strategy, selective bombing, and it was far better than the one pursued. Not only was the planned invasion of the home islands not necessary, but it was also not necessary to kill over half a million Japanese civilians in the incendiary raids of the sixty-six cities in Japan. Nor was it necessary for this country to stigmatize its proud history by being the first to employ the atomic bombs in warfare. It is said that a country usually gets the kind of leadership it deserves, but surely no country ever deserved the kind of leadership the United States got in the closing days of the Pacific War. It is fortunate that the old Biblical proverb did not prevail in this instance, "where there is no vision, the people perish." Perhaps Bismark was right when he said that God ordained a "special providence" for drunkards, fools and the United States of America. That "special providence" is being put to a severe test by the balance of terror that exists in the world today. The "miracle of deliverance" has become a sword of Danocles.

FOOTNOTES--CHAPTER IX

[1]The United States Bombing Survey, *Summary Report: Pacific War* (Washington: Government Printing Office, 1946), 11.

[2]*Ibid.*, 13.

[3]*Ibid.*, 22.

[4]Curtis E. LeMay, *Mission with LeMay* (New York: Doubleday & Company, 1965), 373.

[5]B.H. Liddell Hart, *History of the Second World War* (New York: G.P. Putnam's Sons, 1970), 697.

[6]Harry S. Truman, *Year of Decisions* (New York: Doubleday & Company, 1955), 416.

[7]*Ibid.*, 420.

[8]Wesley Frank Craven and James Le Cate, *The Army Air Forces in World War II* (Chicago: University of Chicago Press, 1963), V, 617.

[9]The United States Strategic Bombing Survey, *Summary Report: Pacific War* (Washington: Government Printing Office, 1946), 20.

[10]Haywood S. Hansell, Jr., *Strategic Air War Against Japan* (Alabama: Airpower Research Institute, Air War College, 1980), 75.

[11]*Ibid.*, 91.

[12]*Ibid.*, 66.

[13]The United States Strategic Bombing Survey, *Summary Report: Pacific War* (Washington: Government Printing Office, 1946), 17.

[14]The United States Strategic Bombing Survey, *Statistical Appendix to Overall Report European War* (Washington: Government Printing Office, 1947), 13.

[15]Haywood S. Hansell, Jr., *Strategic Air War Against Japan* (Alabama: Airpower Research Institute, 1980), 52.

[16]The United States Strategic Bombing Survey, *Summary Report: Pacific War* (Washington: Government Printing Office, 1946), 17.

[17]*Ibid.*, 18.

[18]Haywood S. Hansell, Jr., *The Air Plan That Defeated Hitler* (Atlanta: Higgins-McArthur/Longino & Porter, Inc., 1972), 219.

[19]John Toland, *The Rising Sun* (New York: Random House, 1970), 108.

[20]The United States Strategic Bombing Survey, *Summary Report: Pacific War* (Washington: Government Printing Office, 1945), 15.

[21]The United States Strategic Bombing Survey, *Overall Report: European War* (Washington: Government Printing Office, 1945) 25.

[22]Robert W. Shugg and H.A. DeWeerd, *World War II* (Washington: The Infantry Journal, 1946), 379.

[23]Winston S. Churchill, *The Grand Alliance* (Boston: Houghton Mifflin Company, 1950), 461.

[24]Winston S. Churchill, *Closing the Ring* (Boston: Houghton Mifflin Company, 1951), 6.

[25]The United States Strategic Bombing Survey, *Summary Report: Pacific War* (Washington: Government Printing Office, 1946), 19.

[26]*Ibid.*

[27]Curtis E. LeMay, *Mission With LeMay* (New York: Doubleday & Company, 1965), 375.

[28]Wesley Frank Craven & James Lea Cate, *The Army Air Forces in World War II* (Chicago: University of Chicago Press, 1953), V, 643.

[29]Major General J.F.C. Fuller, *Armament and History* (New York: Charles Scribner's Sons, 1945), xiv.

[30]Robert W. Shugg and H.A. DeWeerd, *World War II* (Washington: The Infantry Journal, 1946), 396.

[31]The United States Strategic Bombing Survey, *Summary Report: Pacific War* (Washington: Government Printing Office, 1946), 26.

[32]*Ibid.*

[33]B.H. Liddell Hart, *History of the Second World War* (New York: G.P. Putnam's Sons, 1970), 695.

[34]Roger W. Shugg and H.A. DeWeerd, *World War II* (Washington: The Infantry Journal, 1946), 388.

[35]Harry S. Truman, *Year of Decisions* (New York: Doubleday & Company, 1955), 416.

[36]Robert Frank Futrell, *Ideas, Concepts, Doctrine: A History of Basic Thinking in the United States Air Force 1907-1964* (Alabama: Maxwell Air Force Base, Air University, 1974), 61.

[37]Harry S. Truman, *Year of Decisions* (New York: Doubleday & Company, 1955), 416.

[38]Haywood S. Hansell, Jr., *Strategic Air War Against Japan* (Alabama: Airpoer Research Institute, 1980), 69.

[39]B.H. Liddell Hart, *History of the Second World War* (New York: G.P. Putnam's Sons, 1970), 697.

[40]Henry L. Stimson and McGeorge Bundy, *On Active Service in Peace and War* (New York: Harper and Brothers, 1947), 631.

[41]Karl von Clausewitz, *On War* (New York: The Modern library, 1943, tr. O.J. Matthijs Jolles), 3 (Bk. I, ch. 1).

[42]Henry L. Stimson and McGeorge Bundy, *On Active Service in Peace and War* (New York: Harper & Brothers, 1947), 629.

[43]*Ibid.*

[44]Winston S. Churchill, *Triumph and Tragedy* (Boston: Houghton Mifflin Company, 1953), 644.

[45]Harry S. Truman, *Year of Decisions* (New York: Doubleday & Company, Inc., 1955), 415.

[46]Haywood S. Hansell, Jr., *Strategic Air War Against Japan* (Alabama: Airpower Research Institute, 1980), 92.

[47]*Ibid.*

[48]William D. Leahy, *I Was There* (New York: McGraw-Hill Book Company, 1950), 442.

[49]*Ibid.*, 441.

[50]*Ibid.*

[51]Henry L. Stimson and McGeorge Bundy, *On Active Service in Peace and War* (New York: Harper & Brothers, 1947), 629.

[52]Karl von Clausewitz, *On War* (New York: The Modern Library, 1943), tr. O.J. Matthijs Jolles, 3 (Bk. I, ch. 1).

[53]Harry S. Truman, *Year of Decisions* (New York: Doubleday & Company, 1955), 415.

[54]John Toland, *The Rising Sun* (New York: Random House, 1970), 946.

[55]Winston S. Churchill, *Triumph and Tragedy* (Boston: Houghton Mifflin Company, 1953), 639.

[56]*Ibid.*, 638

[57]*Ibid.*, 639.

[58]*Ibid.*

[59]Winston S. Churchill, *Closing the Ring* (Boston: Houghton Mifflin Company, 1951), 129.

[60]Winston S. Churchall, *Triumph and Tragedy* (Boston: Houghton Mifflin Company, 1953), 573.

[61]Group Captain R.S. Mason, RAF, "The British Dimension," in *Air Power and Warfare* (Washington: Government Printing Office, 1979), 27.

[62]Norman J. Padelford and George S. Lincoln, *The Dynamics of International Politics* (New York: The Macmillan Company, 1962), 184.

[63]Winston S. Churchill, *Triumph and Tragedy* (Boston: Houghton Mifflin Company, 1953), 640.

[64]Herbert Feis, *The Atomic Bomb and the End of World War II* (Princeton: Princeton University Press, 1971), 31.

[65]Winston S. Churchill, *Triumph and Tragedy* (Boston: Houghton Mifflin Company, 1953), 640.

[66]Harry S. Truman, *Year of Decisions* (New York: Doubleday & Company, 1955), 416

[67]Adam B. Ulam, *Expansion and Coexistence* (New York: Praeger Publishers, 1971), 387.

[68]The United States Strategic Bombing Survey, *The Effects of Atomic Bombs on Hiroshima and Nagasaki* (Washington: Government Printing Office, 1946), 1.

[69]The United States Strategic Bombing Survey, *Summary Report: Pacific War* (Washington: Government Printing ing Office, 1946), 22.

[70]The United States Strategic Bombing Survey, *The Effects of Atomic Bombs on Hiroshima and Nagasaki* (Washington: Government Printing Office, 1946), 4.

[71]*Ibid.*, 33.

[72]*Ibid.*, 24.

[73]*Ibid.*, 28.

[74]*Ibid.*, 6.

[75]*Ibid.*, 28.

[76]Henry L. Stimson and McGorege Bundy, *On Active Service in Peace and War* (New York: Harper & Brothers, 1947), 617

[77]Wesley Frank Craven and James Lea Cate, *The Army Air Forces in World War II* (Chicago: University of Chicago Press, 1953), V, 616.

[78]The United States Strategic Bombing Survey, *The Effects of Atomic Bombs on Hiroshima and Nagasaki* (Washington: Government Printing Office, 1946), 15.

[79]The United States Strategic Bombing Survey, *Summary Report: Pacific War* (Washington: Government Printing Office, 1946), 24-25.

[80]Robert Frank Futrell, *Ideas, Concepts, Doctrine; A History of Basic Thinking in the United States Air Force 1907-1964* (Alabama: Maxwell Air Force Base, Air Uniersity, 1974), 61.

[81]The United States Strategic Bombing Survey, *The Effects of Atomic Bombs on Hiroshima and Nagasaki* (Washington: Government Printing Office, 1946), 21.

[82]*Ibid.*, 22.

[83]*Ibid.*, 23.

[84]The United States Strategic Bombing Survey, *Summary Report: Pacific War* (Washington: Government Printing Office, 1946), 26.

[85]*Ibid.*, 26.

[86]The United States Strategic Bombing Survey, *The Effects of Atomic Bombs on Hiroshima and Nagasaki* (Washington: Government Printing Office, 1946), 23.

[87]*Ibid.*

[88]*Ibid.*

[89]*Ibid.*

[90]The United States Strategic Bombing Survey, *Summary Report: Pacific War* (Washington: Government Printing Office, 1946), 26.

[91]John Toland, *The Rising Sun* (New York: Random House, 1970), 1037

[92]*Ibid.*, 1037-38.

[93]*Ibid.*, 1038.

[94]Windson S. Churchill, *Triumph and Tragedy* (Boston: Houghton Mifflin Company, 1953), 639.

CHAPTER XI

A PREVENTIVE WAR

In his famous essay on history, Lord Macaulay said, "facts are the mere dross of history. It is from the abstract truth which interpenetrates them like gold in the ore, that the mass derives its whole value."[1] Additionally, he said that separating the gold from the ore was a task of the greatest difficulty, even though this was almost the sole justification for the writing of history. Emphasizing this point, Macaulay went on to say, "what do we mean when we say that one past event is important and another insignificant? No past event has any intrinsic importance. The knowledge of it is valuable only as it leads us to form just calculations with respect to the future."[2] In this sense, Macaulay was a prophet looking backward but pointing forward. The way Thucydides expressed it was that "history is philosophy teaching by examples."[3]

Taking the cue from Macaulay's point that Clio was a faithful Muse only to the extent that she enabled the present generation to form just calculations with respect to the future, how should the events of World War II and the immediate postwar years be interpreted? What past events stand out as being important from the point of view of the future? One stands out as being head and shoulders above the rest, and this is the emergence of air power as the dominant and decisive force in war.

Up to this time it had been possible to think in terms of land power, sea power and air power, but after the events of 1945 (the atomic bomb), 1952, the hydrogen bomb) and 1957 (the inter-continental ballistic missile), all the forms of national power merged into one--airpower, or if the expression is preferred "aerospace power." Bernard Brodie, a skeptic about the importance of airpower from its inception, reversed his thinking after the war, and shocked his large following by admitting that "the effectiveness of strategic bombing as a way of war could no longer be questioned. It at once became, incontrovertibly, the dominant form of war."[4] However, this did not mean that armies and navies had become obsolete or obsolescent, and in no sense detracts from the indispensableness of ground forces and sea forces. A country that put its dependence on air power alone, to the exclusion of land power and sea power, would be like the pugilist who, because he had a lethal right hand, refused to lead with his left or to use his legs to position himself for delivering the decisive blow. Military air power can and should be used in the future not as an exclusive and monopolistic force but as an integral element --possibly the long-range decisive striking element-- of a well-balanced and perfectly coordinated team of land, sea and air forces.

Korea and Vietnam are shining examples of the fact that even a first-class military power can be defeated by a weaker opponent if the will to use the great force at its disposal is lacking. This would have meant the full use of air power against Hanoi in the Vietnam War and against the Chinese "volunteers" in the Korean War. In both instances air power could have been decisive, but in both cases it might have widened the war, leading to the necessity of using the atomic bomb against targets in the interior of Red China and Communist Russia. Since the national will was not up to this in either Vietnam or Korea, stalemate and defeat resulted. So in these days of weak resolve about the use of air power, strong armies and navies are absolutely essential for national security.

The transition from the non-atomic past to the atomic present, symbolized by the emergence of air-atomic power, is one of the great revolutions of mankind comparable to the agricultural revolution in the days of pre-history, the gunpowder revolution of the 15th Century and the industrial revolution of the 18th Century. However, it is by no means the most important or the most radical revolution in the human experience. George Sarton argued that the earliest period in history was more important than the latest. What he was talking about was the period of pre-history. He said that the most critical period in man's experience was when he was "gratifying his urgent needs and slowly emerging out of the darkness, when his instinctive craving for power and for knowledge was beginning to appear. Who first thought of kindling a fire? Who invented the earliest stone implements? Who domesticated the animals which have shared our lives ever since?

How did language develop? And later, much later, writing? Who conceived the wheel?" He concluded by saying:

> Just think of these discoveries and of their infinite implications. Without articulate language man remained an animal. Without writing, the safe transmission and preservation of knowledge were impossible. Progress implies safekeeping of what we already have. Without writing, the accumulation of knowledge was precarious and limited, progress small and uncertain. Can any one of our modern discoveries, however startling, begin to compare with those which made possible all the others?

Viewed in the light of George Sarton's historical observations, it is clear that however startling the discoveries of today may seem, looking at them from the long perspective of the human experience they are not so startling after all. In other words, just as there were kings before Agamemnon, so there were inventions and discoveries that affected the human race as profoundly, if not more so than fission, fusion and ballistic missiles. However, the past never seems as important as the present, and to every generation their own achievements make those in history pale by comparison, so the present generation is inclined to look at their achievements as being unparalleled in history.

Taking a short look backward at the power position of the United States at the end of World War II, before the demobilization of its armed forces took place, it is clear that this country did not shoulder its responsibilities wisely or well when it was the acknowledged master of the destinies of the world. The United States was supreme in all the elements at this time, land, sea and air, and in addition had its atomic monopoly. Even before the period of the atomic monopoly, however, in fact before the bomb was actually tested, Winston Churchill said he had no doubt whatsoever that from a military point of view "the United States stood on the scene of victory, master of world's fortunes, but without a true and coherent design."[6] Stalin agreed with him. He showed his true feelings in this respect in 1948 after the power of the bomb had been demonstrated at Hiroshima and Nagasaki, and at a meeting in the Kremlin with some party functionaries from Yugoslavia. Stalin demanded that a Communist uprising in Greece be stopped because of U.S. opposition. Specifically, Stalin said,

> What do you think, that Great Britain and the United States--the United States, the most powerful state in the world --will permit you to break their line of communication in the Mediterranean Sea: Nonsense. And we have no navy. The uprising in Greece must be stopped, and as quickly as possible.[7]

Khrushchev was even more emphatic than Churchill or Stalin in describing the power position of the United States at this time. Looking back, he said,

> America's air force at that time was the best in the world both in the quality and quantity of its planes. The American "flying fortress" and "super-fortresses" had played a big part in winning the war against Germany and Japan, and they were still unmatched by any other planes in the world. I would even say that America was invincible.[8]

As to the power position of the Soviet Union at this time, Khrushchev said it was very weak from the ravages of the war and was so weak in fact that "in the Soviet Union there was considerable alarm that the United States might send its troops into Czechoslovakia and try to restore the capitalist government which had been overthrown in 1948 by the working classes under the leadership of the Communist Party."[9] More than that, Krushchev said, "America was conducting its foreign policy from a position of strength. The Americans had the atomic bomb, and they knew we didn't."[10] Finally, Khrushchev said that the relative power position of Russia and the United States did not change until the Soviet Union developed its own atomic bomb and space rockets. These developments, he said,

> Had a sobering effect on the aggressive forces in the United States, England, France and, of course, inside the Bonn government. They knew that they had lost their chance to strike at us with impunity.[11]

To say that the United States had the power to strike at the Soviet Union with "impunity" until the Russians had their own atomic bomb and space rockets is to say what Churchill did and Stalin corroborated, that the United States was "master of world's fortunes." The time period is also important. Churchill spoke his piece before the atomic bomb was even on the scene; Stalin's reference point was the year 1948; and Khrushchev said that the relative power position of Russia and the United States did not change until the Soviet Union developed its own atomic bomb (1949) and space rockets (1957).

During this period of more than a decade when, on the testimony of Churchill, Stalin and Khrushchev, the United States had the power to dictate the course of human events, what did it do? What actually happened, of course, is that the United States did nothing. During the whole period of the air-atomic monopoly the United States reacted instead of acting positively. It was not that the flesh was weak and the spirit willing. Quite the contrary. The flesh was strong but the spirit was weak to the point where national interests and the interests of the whole free world were gravely imperiled. In other words, there was no will in America at this time to use the great power it possessed.

The red herring that was drawn across the path of serious thinking in the United States in 1945 and 1946 was the false hopes and promises of the new World Organization. What happened in Iran and to the Baruch proposal for the control of atomic energy should have been enough to dispel all illusions. The Russians did not give up the idea of partitioning Iran until President Harry S. Truman used the baseball bat of atomic diplomacy to force the Russian hand. However, in the case of the Baruch Plan, Truman let the Russians have their way. Their way was to sabotage an honest effort on the part of the United States to put the atomic genie back in the bottle and thereby to rescue the world from the folly of having created the agent of its own destruction.

Baruch proposed to do this through the creation of an International Atomic Development Authority under the United Nations. Once the Authority was established and functioning satisfactorily, he proposed to give it full authority over the existing supply of atomic bombs as well as the manufacturing facilities. The only things he asked in return were that the Authority be given the right to police all atomic activities anywhere and everywhere in the world, including on-site inspections of facilities, and in addition he wanted the United Nations to impose sanctions on any country violating the agreement. These sanctions, according to the plan, could be invoked by a majority of the Security Council, thereby depriving the Permanent Members, including Russia, of their veto power. Bernard Baruch introduced his plan in the United Nations in an address delivered on June 14, 1946, in which he said;

> We are here to make a choice between the quick and the dead. This is our business. Behind the black portent of the new atomic age lies a hope which, seized upon with faith, can work our salvation. If we fail, then we have damned every man to be the slave of Fear. Let us not deceive ourselves: We must elect World Peace or World Destruction.[12]

True, Baruch's proposal could be twisted to mean almost anything, and the Soviet Union lost no time in doing so. The first step was Andrei Gromyko's counterproposal. Gromyko was the Soviet Union's delegate to the United Nations Atomic Energy Commission, and he rejected the Baruch proposals lock, stock and barrel. Instead, Gromyko proposed that the United States show its good faith by destroying all its atomic bombs, plus the facilities for manufacturing them and then enter into an international convention banning the production or use of atomic bombs. As for the guarantees in the Baruch Plan, inspecting, sanctions and the abolition of the veto, Gromyko simply rejected them all, saying that they were "not reconcilable with the principles of national soverignty."[13]

In all fairness to the Soviet Union, it probably should be said, as Chalmers M. Roberts did, that at the time of the Baruch Plan the Soviet view was so negative because they could see that the atomic bomb "had upset the balance of power achieved by the end of World War II between the Red Army and the forces of the Western Allies," and they honestly believed that "the Baruch Plan was nothing more than an American attempt to impose on the world a nuclear Pax Americana, a device to relegate the Soviet Union forever to seond-class status."[14] The reasonsing was that if the

U.N. controlled the bomb, and the Americans controlled the U.N., the United States would control the world, since the U.S. would then have the power and authority to perpetuate the American monopoly of atomic weapons.

Given this stand on the part of the Soviet Union, would the United States have been justified in resubmitting the Baruch Plan not as a plan but as an ultimatum to the Soviet Union? Is the use of force ever morally justified in a democracy? A consensus had developed over the ages that the use of force is always justified in self-defense, in accordance with the first law of Nature, which is the law of survival. In this case, the alternative to the use of force was to sit back and let the world drift into what now exists, which is truly a choice between the quick and the dead, and the democracies no longer have the choice. The Kremlin can choose to destroy the world, and there is nothing that can be done to stop them, at least not up to now. In other words, the question of whether it would have been better to have confronted the Russians with an ultimatum in 1946 or not, cannot be answered without taking into consideration the alternatives. Moreover, when these are taken into consideration, it seems obvious that if ever there was a time when the nation would have been justified in using its air-atomic power in the service of peace and humanity, the was it. Not only would the Free World have been better off, but the Communist world and their satellites as well. The alternative is the specter that confronts the world today, the specter that J. Robert Oppenheimer referred to when he said that the Superpowers today are like "two scorpions in a bottle, each capable of killing the other, but only at the risk of his own life."[15]

In the light of history, particularly American history, it is not just a foolish idea to think that the better part of wisdom would have been to have confronted the Soviet Union with a real choice between the quick and the dead in 1946, when it was within the power of the United States to do so with its air-atomic monopoly. Many responsible people had this idea in the 1950s, and in 1970 a study was produced under the auspices of Harvard University's Center for International Affairs, in which Professor George H. Quester surprisingly suggested just such a course of action. He said that the idea had been dismissed too lightly when it was first broached, and suggested further that the terms of the ultimatum could have been "three cities a month."[16] He also pointed out that the United States should have learned from the German and Japanese experience that "the occupation and political remolding of a foreign country to be less strenuous than had been supposed."[17] In other words, at the time of the Baruch Plan it would certainly have been possible, and might have been highly desirable, for the United States to have used its air-atomic power to impose a freeze on the nuclear arms race before it ever got started.

If the atomic diplomacy of Harry Truman failed in the dispute over the Baruch Plan and succeeded in the case of the Iranian crisis, it succeeded again in the watershed crisis that led to the promulgation of the famous Truman Doctrine. This crisis really began on February 24, 1947, when the British Ambassador informed Washington that British aid to Greece and Turkey would come to an end on March 31. This was very alarming because it meant that the whole of the Eastern Mediterranean, including Italy and possibly France as well, might go behind the Iron Curtain. The questions were what could be done about it, and if something could be worked out, how could Congress and the American people be persuaded that their vital interests were at stake in that faraway corner of the Eastern Mediterranean? These were not idle questions because the American people were returning to an isolationist frame of mind at this time and wanted reductions in the armed forces plus a return to economy in government, not foreign programs that would involve large sums of money and possibly military commitments as well. More and more the American people were leaning in the direction of the advice of the founding fathers to keep a distance between themselves and foreign affairs.

One of the State Department's most alert young officers at this time, Joseph M. Jones, recognized the problem for what it was, a crisis in world affairs, and so on February 26, 1947, he recommended to Assistant Secretary of State William Benton that the Administration embark upon an immediate program on a scale hitherto unimagined "to inform the people and convince the Congress adequately with respect to today's crisis.[18] Jones thought that Secretary of State George C. Marshall should present the program to Congress as he, even more than the President, had the necessary "prestige" to impress Congress. The idea caught on, and the result was a spate of activity on the part of the government that began at once with the immediate crisis over Greece and Turkey, went on to the Truman Doctrine, and did not subside until the United States and Europe were fully joined in their post-war position and structure. But for the Truman Doctrine there would not have been a Marshall Plan; but for the Marshall Plan there would not have been an Atlantic Pact, and but for the Atlantic Pact there would not today be a NATO army in Europe.

A year before Jones began his campaign to get some action on the crisis threatening the country, in fact almost a year to the day, February 22, 1946, the American charge d'affaires in Moscow, George F. Kennan, sent what came to be known as the long telegram to his superiors in Washington. This was in response to a Departmental request for his opinion as to why the Russians behaved the way they did. Kennan's "long telegra" ascribed to the Communists a fanatical hatred of the West, a hatred made necessary because of their own feelings of insecurity, and predicted that there would be no diminution of hostility as long as the Communists were in power. The only course of action open to the United States, Kennan said, was to resist Soviet expansion with every means possible, including miliary force and at the same time to wait for some change in the Soviet leadership, a change that would make it possible for the rest of the world to coexist with Russia.[19]

The argument Kennan advanced in the "long telegram" became the basis for American policy and attitudes toward the Soviet Union. Kennan extended and expanded on the subject in the "By X" article in *Foreign Affairs* entitled "The Sources of Soviet Conduct." Walter Lipman, in his book, *The Cold War*, identified Kennan as the author of the "By X" article in the July 1947 issue of *Foreign Affairs*.

It was against the background of the Kennan argument in the "long telegram" that President Truman approached the problem of Greece and Turkey in 1947. The problem was how best to turn the tide of Soviet aggression in the Eastern Mediterranean and also how to secure the backing of a Republican Congress and the people on what amounted to a fundamental break with tradition and history going back to the time of James Monroe and George Washington.

Washington called the new Republic a "political fortress" and said that isolationism (not a dirty word to the founding fathers) should be "the great rule of conduct for us in regard to foreign nations."[20] It is significant that Washington said the great rule, not a great rule. James Monroe's message was not a rule of conduct for future generations of Americans as much as it was a warning to the European powers not to interfere with the affairs of the New World. It was also a pledge that the United States would not interfere in the affairs of the Old World. In other words, Monroe promised that if the European powers refrained from meddling in the affairs of the Western Hemisphere the United States would stay out of the affairs of the Old World. This included Greece and Turkey, of course. Now President Truman was about to tell the American people that their security would be threatened if they did *not* decisively interfere in the affairs of Greece and Turkey. Oddly enough, the Monroe Doctrine was a warning to Russia against encroachments on the West Coast down from Alaska.

To lay the groundwork for the proposed basic change in the foreign policy of the United States, a change made possible by the American air-atomic monopoly, the President, on February 27, called a bipartisan group of congressional leaders to the White House. As Joseph Jones had recommended, Secretary of State George C. Marshall made the principal address to these leaders, and as everyone could see it fell with a dull thud. Amid mutterings about pulling British chestnuts out of the fire, the Undersecretary of State Dean Acheson asked for permission to speak.

"In the past eighteen months," Acheson said,

> Soviet pressure on Iran and on northern Greece had brought the Balkans to the point where a highly possible Soviet breakthrough might open three continents to Soviet penetration. Like apples in a barrel infected by one rotten one, the corruption of Greece would infect Iran and all to the east. It would also carry infection to Africa through Asia Minor and Egypt, and to Europe through Italy

> and France, already threatened by the strongest domestic Communist parties in Western Europe. The Soviet Union was playing one of the greatest gambles in history at minimal cost. It did not need to win all the possibilities. Even one or two offered immense gains. We and we alone were in a position to break up the play. These were the stakes that British withdrawal from the eastern Mediterranean offered to an eager and ruthless opponent.[21]

In the group Acheson was addressing by far the most important person was Senator Arthur Vandenburg, the key to the Republican Congress. At the end of Acheson's presentation, the Senator was obviously moved, and solemnly said, "Mr. President, if you will say that to the Congress and the country, I will support you and I believe that most of its members will do the same."[22] They did. President Truman appeared before Congress on March 13, 1947, and made his case for coming to the aid of Greece and Turkey. But he did not limit it to Greece and Turkey. The President made precisely the arguments Senator Vandenburg said would be convincing. The President said:

> At the present moment in world history nearly every nation must choose between alternative ways of life. The choice is too often not a free one.
>
> One way of life is based upon the will of the majority and is distinguished by free elections, guarantees of individual liberty, freedom of speech and religion and freedom from political oppression.
>
> The second way of life is based upon the will of a minority forcibly imposed upon the majority. It relies upon terror and oppression, a controlled press and radio, fixed elections and the suppression of personal freedom.
>
> I believe that it must be the policy of the United States to support free people who are resisting attempted subjugation

> by armed minorities or by outside presures.[23]

President Truman rationalized the strong stand he took on Greece and Turkey by saying that it was not only Greece and Turkey

> but the situation had even wider implications. Poland, Rumania, and the other satellite nations of eastern Europe had been turned into Communist camps because, in the course of the war, they had been occupied by the Russian Army. We had tried, vainly, to persaude the Soviets to permit political freedom in these countries, but we had no means to compel them to relinquish their control, unless we were prepared to wage war. Greece and Turkey were still free countries being challenged by Communist threats both from within and without. These free peoples were now engaged in a valiant struggle to preserve their liberties and their independence.[24]

President Truman said he came to the conclusion,

> America could not, and should not, let these free countries stand unaided. To do so would carry the clearest implications in the Middle East and in Italy, Germany and Greece. The ideals and the traditions of our nation demanded that we come to the aid of Greece and Turkey and that we put the world on notice that it would be our policy to support the cause of freedom wherever it was threatened. The risks which such a course might entail were risks which a great nation had to take if it cherished freedom at all.[25]

Finally, Truman accepted the full responsibility for his famous doctrine, and was just as immodest in appraising its importance as Washington had been with his. Whereas the First President referred to his doctrine as "*the* great rule of conduct for us in regard to foreign nations," the thirty-third President said that the Truman Doctrine "was, I believe, *the* turning point in America's foreign policy, which now declared that wherever aggression, direct or indirect, threatened the peace, the security of the United States was involved."[26]

Without the support of Senator Vandenburg in the Republican Congress, and the help of Joseph Jones, Dean Acheson and George Kennan, there would not have been a Truman Doctrine, and the Middle East would probably have gone to the Soviet Union by default. The chances are also great, as Dean Acheson said, that the loss of the Middle East would have led to the loss of North Africa and possibly Western Europe as well, certainly Italy and France. This would have been the worst-case scenario depicted by Halford J. Mackinder in 1904, 1919 and 1943. As Mackinder said:

> Who rules East Europe commands the Heartland:
> Who rules the Heartland commands the World-Island;
> Who rules the World-Island commands the World.[27]

George Kennan's article in *Foreign Affairs* appeared in a matter of weeks after President Truman's presentation to Congress on March 12. Repeating his conviction that the Russians were leading from weakness not strength in all their aggressions, Kennan described the Kremlin's foreign policy as being "a fluid stream which moves constantly, wherever it is permitted to move, toward a given goal. Its main concern is to make sure that it has filled every nook and cranny available to it in the basin of world power."[28] As to how to cope with Soviet pressure against the free institutions of the Western World, it was perfectly apparent to President Truman that, considering the Soviet internal weakness and the absolute power of the atomic weapon, the United States could block Soviet aggression at times and places of its own choosing. What Kennan said in his article was that the United States would be warranted in "entering with reasonable confidence upon a policy of firm containment, designed to confront the Russians with unalterable counter-force at every point where they show signs of encroaching upon the interests of a peaceful and stable world."[29]

The Berlin Blockade of April 1, 1948, was another example of the fluid stream of Soviet diplomacy, only this time it was not moving against a nook and cranny in the world, but against the heart of Western Europe, the German people as a whole. The purpose of the Berlin Blockade was to deliver a shattering blow to Western prestige by sending the democracies running from Berlin, and thus demonstrating the fact that the Western powers did not have the backbone to stand up against the Soviet Union in a crunch. If this worked, the Kremlin hoped that it would forestall the creation of a viable state in West Germany and nurture the feelings among the Germans that if they aligned themselves with the West they would be on the losing side in the global conflict. In other words, what Stalin was relying on was that the democracies would repeat their performance at Munich, when the love of peace and the fear of war was so great that they gave in to Hitler without a struggle. Khrushchev later explained, "I always operated on the principle that I should be clearly against war but never frightened of it."[30] In the Russian view, fear was for the democracies, not for the Soviets. That is why Khrushchev was so contemptuous of Stalin's fear of the Germans. Khrushchev said that when Germany invaded the Soviet Union in 1941, Stalin's nerve cracked, and he stood "paralyzed by the fear of Hitler, like a rabbit in front of a boa constrictor." It was Stalin's hope that this was what would happen to the democracies when they were confronted with the Berlin Blockade.

The Western Allies did not react to the Berlin Blockade the way Stalin had hoped for, and probably not the way they should have reacted. President Truman moved three squadrons of atomic bombers to British bases, and at the same time did nothing to augment the size of American ground forces in Europe. This made it clear that if it came to war over Berlin, the American response would be with its nuclear option rather than ground forces.[32] In sharp contrast to this apparent resolve in connection with Berlin, the United States only a few months before had stood idly by while Czechoslovakia had been gathered into the Communist fold by being forced to yield to a Soviet ultimatum. The Kremlin reasoned that if the Berlin Blockade proved to be as much of a success as the Czech coup the whole of Western Europe might very easily be brought under the sway of the Soviet Union, and in that event the end of the democracies might well be in sight, as Mackinder had gloomily predicted, unless the democracies showed greater foresight in connection with their peril.

The air-atomic option was not the only one President Truman had in connection with Berlin. "What was at stake in Berlin," Truman said,

> was not a contest over legal rights, although our position was entirely sound in international law, but a struggle over Germany and, in a larger sense, over Europe. In the fact of our launching of the Marshall Plan, the Kremlin tried to mislead the people of Europe into believing that our interest and support would not extend beyond economic matters and that we would back away from any military risks.[33]

There were several choices open to President Truman in connection with the Berlin Blockade. One was to recognize some stubborn facts, and get out of the city, lock, stock and barrel. The facts were that Berlin was a hundred miles inside the Soviet zone and was untenable from a military point of view. Getting out would simply be a recognition of the fact that it had been a mistake to agree to the joint occupation in the first place. The second option was to stand on the American rights of access to Berlin by sending a military convoy through the blockade. If this involved shooting, it would mean war, if the Soviets were not bluffing. Was Berlin worth the risk of war? Two such diverse persons as General Lucius Clay, the American military commander in Berlin, and the left-wing British labor leader, Aneurin Bevan, thought it was. This is the course Churchill is said to have advocated. The third option was the one the President took. This was the airlift. Instead of using the "absolute weapon," the atomic bomb, to subject Moscow and Leningrad to the fate of Hiroshima and Nagasaki, the United States flew into the beleaguered capital two and a half million tons of food and fuel. This saved Berlin and forced Stalin to eat his words. Previously, the Soviet dictator had boasted to Milovan Djilas and others of the inner circle, "that all of Germany must be ours, that is, Soviet, Communist."[34]

Worse was to follow from the Soviet point of view. The airlift, not the land blockade, was what inspired the Germans and captured the imagination of the other peoples of Western Europe. In other words, the West Germans, instead of being cowed by the power of the

Soviet Union, closed ranks against Russia and put Konrad Adenauer in power by an overwhelming vote. The Communists won only fifteen seats out of a total of three hundred fifty-four in the election. The other Western European powers also closed ranks against Russia as a result of the Berlin Blockade. The North Atlantic Treaty was signed in Washington pledging a united front of the ten Western European nations plus the United States and Canada. The treaty did not exclude the use of military force to safeguard the freedom of the Atlantic community, and specifically provided that an armed attack against one or more would be construed as an attack against all. The North Atlantic Treaty Organization was brought into existence in the summer of 1948 to put teeth into the pact. Adam B. Ulam's considered opinion is that had this force been in existence at the time of the Berlin Blockade it is altogether possible that the challenge would have been met by something more drastic than an airlift.[35]

As it was, the airlift, while it was a complete success from a narrow tactical point of view, was at the same time less than a success from the larger strategical point of view. By resorting to an airlift the United States backed down from the strong stand it had taken in other crises (Iran, Greece, Turkey) and demonstrated that it would not stand up for its rights in Europe under certain circumstances. President Truman put the emphasis on the wrong point when he said the main question in Berlin was, "how could we remain in Berlin without risking all-out war?" The right point was the one he made when he said that certain risks "were risks which a great nation had to take if it cherished freedom at all."[36] Professor Ulam said Stalin was bluffing at the time of the Berlin Blockade, and if Truman had called his bluff, that not only would there not have been a war, but the whole "subsequent history of Germany would have been different."[37]

Having succeeded in bluffing the first atomic President, Hary Truman, but also having failed to achieve his primary objective in Europe, Stalin turned his gaze eastward, perhaps mindful of a saying that has often been attributed to Lenin, that the shortest road to Paris is through Peking.[38] The assault on South Korea began in June 25, 1950, when sixty thousand North Korean troops spearheaded by over one hundred Soviet tanks invaded South Korea. This started what could be called the War of the 38th Parallel. The invasion of South Korea was a movement that was led and directed by the Russians, not the Chinese. The priorities of the Chinese Communists at this time were the consolidation of their gains on the mainland following the defeat of Chiang Kai-shek on December 7, 1949, and after that the conquest of Formosa and Tibet. The Chinese did not even have an Ambassador in North Korea until after the invasion began. Not only did the Chinese have no part in planning the invasion of South Korea, but it was not in their best interest to stir up trouble in Korea at this time.

On the other hand, Stalin had every reason to believe that the conquest of South Korea would be a cheap and easy victory, particularly since American forces had been withdrawn, and also one of immense value to the Soviet Union. A Communist China was already in the fold, and with a unified Korea in the Communist camp, Stalin was sure that Japan would jump on the bandwagon. This would mean that the whole of the Far East would be his, and his without the loss of a single Russian soldier. A risk-free gamble, it was almost too good to be true.

Harry Truman's violent reaction to the invasion of South Korea caught the Kremlin by surprise. After all, only six months before, on January 12, 1950, Dean Acheson had made it clear that Korea was outside the sphere of America's vital interest in the Pacific. Why should American power, land, sea and air, suddenly be committed to an area of no strategic importance and to one ruled over by a tinhorn dictator Syngman Rhee? It did not make any sense.

Sense or no sense, this is what happened, and the only thing that saved the Russians from total disaster was the folly of General Douglas MacArthur. The God-like general, after winning a brilliant victory at Inchon, apparently forgot he was human and divided his forces for the dash to the Yalu River. The results were what could have been expected-- a total disaster, one of the worst defeats in the history of American arms.

The Chinese had issued repeated warnings that if the "foreign devils" in South Korea crossed the 38th parallel and approached the borders of China proper at the Yalu all hell would break loose. It did on November 26 when two hundred thousand Chinese Communist "volunteers" crossed the Yalu River and put the Americans to rout. The forces of General MacArthur did not stop running until they were seventy miles deep into South Korea. However, when the rout came to an end the American troops closed ranks and mounted a counter-attack that drove the Chinese back across the 38th parallel. It was at this point that the famous dispute between President Truman and General MacArthur broke out. When the Chinese poured across the Yalu with their two hundred thousand "volunteers" against the divided forces of MacArthur, the God-general quite rightly said, "we face an entirely new war,"[39] and insisted that the country recognize the fact that it was a war with Communist China.

President Truman refused to widen the unpopular war largely because of domestic politics, but also because of the opposition of the western allies and the fear that Soviet Russia would come to the aid of the Chinese Communists. Instead, the President fired MacArthur not because he was insubordinate, which he certainly was, but because he had failed to clear all policy statements through the Defense Department. In substance, of course, General MacArthur was entirely correct about the fact that the country faced an entirely new war when the Chinese Communists joined the fray, and the President was wrong. The President was probably wrong again when he did not suport an American capability to drive the Chinese Communists completely out of Korea. Instead, he opted for the cease fire and armistice talks that were started in July 1951. These dragged on until July 1953, when the armistice was finally signed at the line of the 38th Parallel. This, then the 38th Parallel was the final victor in the Korean War. It was an inglorious war, badly led on the field and miserably directed from the White House. But at least it was not a defeat. Only a stalemate.

The atomic bomb did not play a large part in the Korean War. It did not deter Stalin from initiating the war, and the United States did not use it during the war. One of the reasons the United States refrained from using the bomb may have been the actual shortage of nuclear weapons in the American stockpile. Also, all during the war there was a persistent fear that the conflict was a clever trick or feint on the part of the Russians the object of which was to get the United States mired down in Asia so the Russians would have a free hand in Europe. Using the atomic bombs in Korea, according to this line of thinking, would be playing right into the hand of the Kremlin. Perhaps this was what Omar Bradley had in mind when he characterized the Korean conflict as "the wrong war, at the wrong place and with the wrong enemy."[40]

There was also another factor that may have mitigated against the use of the atomic bomb in Korea. This was the fact that the Russians probably had a few atomic bombs of their own by this time. If the Americans took the lead in using their weapons in Korea the Russians might retaliate by using a bomb or so against American targets, if not in the zone of the interior at least against American bases in Japan or Korea. Thus Korea may have been the beginning of the "balance of terror" concept.

Before Korea, and with the possible exception of Churchill, there were few in the West who gave expression to the thought that the only effective way to deal with the Russians was to go beyond the defensive concept of containment and act directly against the Soviet Union by pre-emptive strikes or preventive war, thus getting at the source of the problem. This feeling was benaath the surface in the United States at the time of the Czech crisis and the land blockade of Berlin. It took the Korean war, however, to bring it out into the open. One of the most clear-headed and imaginative thinkers in the Air Force, Major General Orvil Anderson, advocated just such a policy in 1950 and offered to lead the American bombers in a pre-emptive strike against the Soviet targets, particularly in the atomic energy field. General Anderson said, "we're at war, damn it." He also said, "give me the order to do it and I can break up Russia's five A-bomb nests in a week."[41] At roughly the same time, the Secretary of Navy Francis Matthews advocated "instituting a war to compel cooperation for peace," and further said, "we would become the first aggressors for peace."[42]

President Truman and the rest of America were not in any mood to become the first aggressors for peace. In fact, at this time they had all the war on their hands they wanted. During the late August and early September period of 1950, which was when Matthews and Anderson made their proposals, the American and South

Korean forces were fighting to defend their last toehold in Korea at Pusan. The last thing they wanted to do was widen the war to include Russia and possibly Communist China. Therefore, an infuriated President Truman moved quickly to silence the talk about a preventive war. General Anderson was suspended, and Secretary Matthews was pressured to resign, even though he "explained to Truman that he had heard preventive war talked so much that he had used the phrase in his speech without realizing its full implications to the administration's policy."[43]

In addition to Matthews and Anderson, there were other important Americans who were at least giving some thought to the idea of getting at the source of the trouble in Asia by taking a swipe at the Soviet Union. The rumor was that Secretary of Defense Louis A. Johnson was sympathetic to the idea, and Douglas MacArthur was making noises about being able to "dominate with air power every Asiatic port from Vladivostok to Singapore." Secretary of the Air Force Stuart Symington suggested that stronger action be taken against the Soviets "on the basis of America's preponderance in atomic air power."[44]

Machiavelli gave his Prince sound advice when he stressed the value of "foresight" in planning the affairs of state, with particular reference to the virtues of pre-emptive strikes and preventive war. He said, "hence the Romans, seeing their troubles far ahead, always provided against them and never let them continue in order to avoid a war because they knew that such a war is not averted but is deferred to the other side's advantage. Hence they chose to urge war against Philip and Antiochus in Greece in order not to have to wage it against them in Italy."[45] Clausewitz had much the same thing to say about preventive war and the factor of time, "if it is to the interest of the one commander to act, it must be to the interest of the other to wait."[46] Machiavelli had another way of expressing himself with respect to pre-emptive strikes and preventive war. He said,

> In these instances the Romans did what all wise princes do: these take thought not merely for present discords but also for future ones, and the latter they forestall with every sort of ingenuity; when foreseen far ahead, discords easily

> can be remedied, but when you wait until they are upon you, the medicine is not in time; they have grown incurable. It is the same as with the hectic fever; the physicians say that when the disease begins it is easy to cure but hard to recognize, but in the course of time, when not recognized and treated at the beginning, it becomes easy to recognize and hard to cure. So it is in things of state.[47]

In all fairness to President Truman, it probably should be said that a Pax Americana was not his idea of a well-ordered world. It could also be said that he did his best according to his lights, even though his best was not always good, and his lights in some instances were few and dim. This was particularly true with respect to his dealings with the Russians on their rejection of the Baruch Plan. The future of the world, that is of planet earth, may have been what was at stake in those negotiations. As Clausewitz said, "war is nothing but a continuation of political intercourse with an admixture of other means."[48]

Actually, being the first atomic president and having a global responsibility, may have been more of a handicap to President Truman than an asset. It encouraged a false sense of security and distracted attention from some basic problems of the world, such as the urgent need to make the world safe from the danger of being blown up. In fact, it can be said that the American atomic monopoly was like the French Maginot Line in its influence on the government. The French felt safe behind their Maginot Line, no matter what the Germans did, and the Americans felt safe behind their atomic monopoly, no matter what the Russians did. In other words, the French were not seriously alarmed when the Germans marched into the Rhineland because they had the protection of their Maginot Line, and the Americans were not seriously alarmed by whatever the Russians did because they felt safe behind their atomic monopoly. Professor Adam B. Ulam put it in a nutshell when he said,

> Had the French been without this wretched line of fortifications in 1936, they might have felt the necessity of making Hitler pull back his troops from the Rhineland.

> Had the mirage of a long monopoly on atomic weapons not exercised its baleful fascination over the minds of American policy-makers, it is possible they would have tackled in a more urgent spirit the need to set up that balance of power without which in this sinful world of ours any world order is but a snare and delusion. As it was, the North Atlantic Treaty Organization was set up when the American monopoly was a thing of the past.[49]

On balance again, what can be said about the presidency of Harry Truman? He certainly struck out with the bases loaded when he failed to make the world safe for democracy by forcing the Russians to enter upon a workable plan for the control of atomic energy, even if this meant using the bomb against them in the process of establishing a Pax Americana. Truman struck out a second time when he failed to join with Churchill in the "standstill" order, which would have prevented the whole of East Europe from going behind the Iron Curtain. And the man from Missouri also struck out for a third time when he failed to send a land convoy to the rescue of Berlin in 1948. Finally, the thirty-third President struck out for the fourth time when he settled for a stalemate in Korea. On the other hand, Harry Truman belted a home run when he faced Stalin down in Iran, Greece and Turkey, thus setting the United States on the proper course for future dealings with the Soviet Union. This was the course of containment. In other words, Harry Truman's record was good enough to get him into the Big League of presidents, even if it did not justify a place for him in the Presidential Hall of Fame along with George Washington, Abraham Lincoln and Franklin D. Roosevelt. It cannot be said that Harry Truman left the world better than he found it. That would have required decisive action at the time of the rejection of the Baruch Plan.

FOOTNOTES--CHAPTER XI

[1]Lord Macauley, *Critical and Historical Essays* (Boston: Houghton, Mifflin and Company, 1900), 247.

[2]*Ibid.*, 277.

[3]Edmund A. Walsh, *Total Power* (New York: Doubleday and Company, 1948), 98.

[4]Bernard Brodie, *Strategy in the Missile Age* (Princeton: Princeton University Press, 1956), 152.

[5]George Sarton, *The History of Science and the New Humanism* (Bloomington: Indiana University Press, 1962), 64-65.

[6]Winston S. Churchill, *Triumph and Tragedy* (Boston: Houghton Mifflin Company, 1953), 456.

[7]Milovan Djilas, *Conversations with Stalin* (New York: Harcourt Brace & World, Inc., 1962), 182.

[8]Nikita Khrushchev, *Khrushchev Remembers* (Boston: Little, Brown and Company, 1970), 362.

[9]*Ibid.*

[10]*Ibid.*, 261.

[11]*Ibid.*, 571.

[12]Chalmers M. Roberts, *The Nuclear Years* (New York: Praeger Publishers, 1970), 123.

[13]*Ibid.*, 10.

[14]*Ibid.*, 17.

[15]*Ibid.*, 7.

[16]George H. Quester, *Nuclear Diplomacy* (New York: Praeger Publishers, 1970), 123.

[17]*Ibid.*

[18]John Lewis Gaddis, *The United States and the Origins of the Cold War* (New York: Columbia University Press, 1972), 346

[19]*Ibid.*, 302-03.

[20]Washington's Farewell Address, *Documents of American History* (New York: Appleton-Century-Crofts, Inc., 1958, ed. Henry Steel Commager), 174.

[21]Dean Acheson, *Present at the Creation* (New York: W.W. Norton & Company, 1969), 219.

[22]*Ibid.*

[23]John Lewis Gaddis, *The United States and the Origins of the Cold War* (New York: Columbia University Press, 1972), 351.

[24]Harry S. Truman, *Years of Trial and Hope* (New: York: Doubleday & Company, 1956), 101.

[25]*Ibid.*

[26]*Ibid.*, 106.

[27]Halford J. Mackinder, *Democratic Ideals and Reality* (New York: W.W. Norton & Company, 1962), 160.

[28]George F. Kennan, "The Sources of Soviet Conduct," *Foreign Affairs* (July 1947), 575.

[29]*Ibid.*, 581.

[30]Nikita Khrushchev, *Khrushchev Remembers* (Boston: Little, Brown and Company, 1970), 518.

[31]*Ibid.*, 169.

[32]George H. Quester, *Nuclear Diplomacy* (New York: The Dunellen Company, Inc., 1970), 49-52.

[33]Harry S. Truman, *Years of Trial and Hope* (New York: Doubleday & Company, 1956), 123.

[34]Milovan Djilas, *Conversations with Stalin* (New York: Harcourt, Brace and World, Inc., 1963). 153.

[35] Adam B. Ulam, *Expansion and Coexistence* (New York: Praeger Publishers, 1971), 497.

[36] Harry S. Truman, *Years of Trial and Hope* (New York: Doubleday & Company, 1956), 101.

[37] Adam B. Ulam, *The Rivals* (New York: The Viking Press, 1972), 148.

[38] Norman J. Padelford and George A. Lincoln, *The Dynamics of International Politics* (New York: The Macmillan Company, 1962), 80.

[39] James McGovern, *To the Yalu* (New York: William Morrow & Company, 1972), 109.

[40] Bernard Brodie, *Strategy in the Missile Age* (Princeton: Princeton University Press, 1965), 317.

[41] Robert Frank Futrell, *Ideas, Concepts, Doctrines: A History of Basic Thinking in the United States Air Force 1907-1964* (Alabama: Air University, Maxwell Air Force Base, 1974), 148.

[42] *Ibid.*

[43] *Ibid.*

[44] *Ibid.*

[45] Niccolo Machiavelli, *The Prince* (Durham: Duke University Press, 1965, tr. Allan Gilbert), I, 17.

[46] Karl von Clausewitz, *On War* (New York: The Modern Library, 1943, tr. O.J. Matthijs Jolles), ii (Bk. 1, ch. 1).

[47] Niccolo Machiavelli, *The Prince* (Durham: Duke University Press, 1965, tr. Allen Gilbert), I, 16-17.

[48] Karl von Clausewitz, *On War* (New York: The Modern Library, 1943, tr. O.J. Matthijs Jolles), 596 (Bk. VIII, ch. 6).

[49] Adam B. Ulam, *Expansion and Coexistence* (New York: Praeger Publishers, 1971), 413.

CHAPTER XII

"THE BALANCE OF TERROR"

One of the most baffling problems to challenge the wisdom of man from earliest times has been the relationship between the physical environment and the governance of man. To a certain extent, Aristotle, Julius Caesar and Herodotus, were all exponents of the idea that there was a direct connection between geography and politics. In the modern period, Halford J. Mackinder developed a theory that the pivot of world power is in the "heartland" of the greater continent of Euro-Asia, and that if it came under the sway of a single great power the empire of the world would be in sight. However, Mackinder was not a "determinist" in the sense that some of the other geopoliticians were, and always qualified his findings with the caveat that the physical features of the world were not static but variable forces, and therefore that each century had its own geographical perspectives and its own outlook on the future. Accordingly, he said that "as we live through each new epoch, we see all the past and all the present through new eyes and from new standpoints."[1]

This is certainly true of the present generation. Its relationship to nature is different from what it was in the past and different because of two revolutions, one in firepower and the other in mobility. The first was the product of developments in the field of fission and fusion giving man the atomic and hydrogen bombs, and the other was in the ability of man to penetrate outer space with his giant rockets. Together, these two revolutions have transformed the whole world of power, and they may have given man Jupiter's ability to flash the lightning and hurl the thunderbolt. Never before in history has mortal man had such awesome power, which may raise some serious questions about the old Biblical saying: "The thing that hath been, it is that which shall be; and that which is done is that which shall be done: and there is no new thing under the sun."[2]

The impact of these two revolutions on the art of war is best understood against the background of the aphorism coined by Nathan Bedford Forrest. The dashing Confederate general defined war as the ability to "git thar fustest with the mostest."[3] Though there is some question as to whether Forrest ever used these exact words there is no question that getting the right force to the right places at the right time is the secret of success in war. The "fustest" is defined in war as "mobility," and the "mostest" as firepower. That is why the two revolutions, coming as they did at the same time, are so important to the future of mankind. Ordinarily, firepower and mobility are opposites in the sense that mobility is usually achieved by the sacrifice of firepower. In this case, however, the two were joined, and that is why the event was almost without precedent in the experience of mankind. The same result would have been achieved if the Gunpowder Revolution and the Industrial Revolution had come at the same moment. The whole future of mankind is under the dark cloud of the great revolution in firepower and mobility.

One of America's outstanding airmen, General Thomas D. White, looked at the revolution in firepower and mobility and predicted that "in the future, the people who control space will control the world."[4] At the same time, he reminded his listeners that in the course of history the people who had controlled the land had controlled the world; then the people who had controlled the seas had controlled the world; and then the people who had controlled the air had controlled the world. By the time of General White's retirement exercises in 1961 the race for control of outer space was well underway. It began in the fall of 1957 when the Soviets put the first artificial satellite into orbit. Prior to this, voyages beyond the atmosphere were largely in the realm of fantasy because they were considered useless for any serious military purpose. The German V-2 rockets were a spectacular achievement of science and industry, but their payload was so small that the military effects were almost nil. However, with the creation of the atomic and hydrogen bombs, rockets began to take on an awesome importance. Thus it was the convergence of the two revolutions that put an end to the geography of the past, and opened up a new epoch in the history of man.

General White's prediction that the people who control space would control the world was made in 1961.

Twenty years later, in 1982 to be exact, a distinguished historian, Walter A. McDougall, reaffirmed what General White had said about the importance of space and went so far as to call the present the space age. He also took cognizance of the fact that in the strategic literature of the time, Soviet as well as American, the importance of space power was fully recognized. Expressing full agreement with General White's analogy, Professor McDougall said that "as the Roman empire dominated the land by its road system, the British empire the seas, and the American empire the air, so now was outer space the decisive medium, whoever ruled it could dictate events on earth as well."[5]

The thinking in the Kremlin was in keeping with or possibly in advance of military thought in the United States at this time. Stalin was tremendously impressed by American air power in World War II and particularly by American atomic power as it was demonstrated at Hiroshima and Nagasaki. He thought that America's air power made her "invincible";[6] that her big bombers were "unmatched"[7] by any in the world; and that as long as she maintained her air-atomic monopoly, the United States could strike at the Soviet Union with "impunity."[8] Therefore, as early as 1945 Stalin began to force-feed the Soviet atomic bomb program as well as the program for the development and production of short- as well as medium- and long-range rockets and missiles.[9] To Stalin, it was clear that there was an urgent need for a defense against the threat posed by the American bomber bases encircling the Soviet Union, and that if it could not be brought into existence quickly, all the gains of World War II might be lost. It was also clear that whereas the revolution in the "mostest" element of warfare was in the hydrogen not the atomic bomb, that is in fusion not fission, so it was evident that the revolution in the "fustest" element was not in the field of the manned bomber but in the field of missiles. Thus it was that the Soviets hedged their bets from the beginning on the air-atomic age, and put their major efforts into developing the hydrogen bomb and the inter-continental ballistic missile. It is not surprising, therefore, given these priorities and the long-term interest of the Russians in atomic energy and rocketry,[10] that they won the race to be first in space, and that they also finished ahead of the United States in the race to produce a hydrogen bomb, not merely a hydrogen device.

What it amounted to was that the Russians planned to more or less skip the air-atomic age, and to put their major effort into being first in the hydrogen-space age. Thus they were satisfied with a token investment in conventional air power and had only a passing interest in the atomic bomb, while their major concentration was on missiles and the production of the hydrogen bomb. Michael Mandelbaum commented on the success of their efforts in the missile field when he said:

> The Soviet Union did not bother to build a substantial fleet of inter-continental range jet aircraft in the 1950s, as did the United States. Instead, they concentrated their efforts on perfecting ballistic missiles, and after launching the first earth orbiting vehicle in 1957, they seemed to lead the United States in this weapon for a few years.[11]

As to the hydrogen bomb, it should have been apparent to the United States from the beginning what the Soviets were doing in the field of nuclear weapons. In fact, when the dust particles were scooped from the air in 1949 to see if there had been an unfriendly atomic explosion, the analysis quickly revealed that there had not only been one, but that the Soviet explosion "had exceeded the force of the Hiroshima bomb by a factor of six."[12] From this, it should have been evident that the Soviets were already heading for the hydrogen bomb, and that their first atomic bomb (1949) was only a step in this direction. In fact, this is the way it worked out. When the Russians made their first test of hydrogen explosive in August 1953, it was not of a device, but of a plausibly deliverable hydrogen bomb. They successfully tested an air-dropped bomb in November 1955.

The Americans set off a hydrogen explosive in late 1952, but it was clearly a device, not an operational bomb. In fact, it was not until May 1956 that the first American air-drop took place. Thus it was that the Americans did not win the race for the hydrogen bomb. They were about six months behind the Russians in getting the first deliverable bomb.[13] This was not a very good beginning for the subsequent arms race. It was not as bad, however, as the great leap forward the Soviets made when they launched Sputnik in 1957. This really shook up the bewildered Americans.

While the race to be the first in the space-hydrogen age was underway, the United States was in the throes of a race to cut down on government spending and at the same time to preserve some semblance of national defense. These were Eisenhower's objectives when he took office on January 20, 1953, and in addition he understood that he had a mandate from the people to go to Korea and personally put an end to the fighting-talking deadlock that had been in effect ever since July 1951. He made his three-day trip to Korea in December 1952, but it was not until after Stalin's death on March 5, 1953, that the Korean peace talks began to show signs of going anywhere, and it was not until July 27 that the armistice was signed which ended the three years of fighting, two of which were truce-talking. With the signing of the armistice, Communist pressure shifted to Indo-China, the scene of a violent clash between the Communist forces of Ho Chi Minh and the non-Communist forces supported by the French.

The Geneva Conference of April 1954 was an attempt to resolve the Indo-China conflict, as well as the dispute over the unification of Korea. En route to this conference, Ho Chi Minh and the Chinese Prime Minister Chou En-lai met with Khrushchev to settle on a united front for the forthcoming Geneva Conference. The meetings were held in the Catherine Hall of the Kremlin. At one of their sessions a dialogue took place between Chou En-lai and Khrushchev that has an important bearing on history. The Russian leader said that Chou engaged him in a very intimate tete-a-tete at the end of one of the meetings, and said to him.

> Comrade Ho Chi Minh has told me that the situation in Vietnam is hopeless and that if we don't attain a cease-fire soon, the Vietnamese won't be able to hold out against the French. Therefore, they've decided to retreat to the Chinese border if necessary, and they want China to be ready to move troops into Vietnam as we did in North Korea. In other words, the Vietnamese want us to help them drive out the French.[14]

Then Chou made a surprising statement. He said, "we

simply can't grant Ho chi Minh's request. We've already lost too many men in Korea -- that war cost us dearly. We're in no condition to get involved in another war at this time."

Chou's testimony about the weakness of China makes it perfectly clear that General Van Fleet was probably right when he said that if he had been given a free hand in Korea he could have driven the Chinese back across the Yalu from whence they came and unified Korea. By the same token, Ho's testimony also makes it clear that there was a time in 1954 when the Vietnam War could have been won, and won decisively, if American aid had been forthcoming. In other words, the meeting in the Kremlin suggests that America's two unwinnable wars were winnable after all.

Khrushchev said that the "miracle"[15] of Dien Bien Phu was the only thing that saved the cause of Ho Chi Minh and rescued his movement from certain defeat at the hands of the French. This is not exactly the case. Dien Bien Phu was not a miracle of any kind, but a blunder of the first magnitude on the part of the French, and a similar blunder on the part of the Americans. It all began when the French made the mistake of committing the bulk of their first-line troops to an isolated fortress on the border of Laos directly west of Hanoi, Dien Bien Phu. The object of this concentration was to have a base in the north from which they could control the insurgent rebellion. It was a good idea except for the fact that Dien Bien Phu could not be supplied or reinforced except by air, and this made it highly vulnerable to enemy siege and frontal attack. This was not lost on the resourceful enemy commander, General Vo Nguyen Giap, and he proceeded to close the few mountain roads leading to Dien Bein Phu and soon had the French outnumbered by a margin of four to one. Not only that. Giap had the single airfield under the guns of his artillery and was prepared to move in for the kill when the French belatedly realized the predicament they were in and frantically began calling on the United States for help.

The United States was not unprepared to give the French the kind of help they needed at Dien Bien Phu, which was air support. Two American carriers equipped with nuclear weapons were standing by in a position to relieve the besieged garrison, and all they needed was the order to go.[16] However, this was not forthcoming. On April 3, Eisenhower convened a meeting of congressional leaders, as Harry Truman did in the case of aid for Greece and Turkey, and the result was that no aid was forthcoming, not even if it meant the loss of Vietnam. Only a year before this, Eisenhower had rescued the United States from the "folly" of "Mr. Truman's War," and he was in no mood to get himself involved in what could easily become "Mr. Eisenhower's War," not at this time in his political career. Thus it was that Eisenhower turned a deaf ear to the frantic calls for help that came from the oldest ally, France. And they were frantic indeed. On April 5, two days after Eisenhower made his wrong-headed decision about not coming to the aid of the French, and made it because of domestic politics, Douglas Dillon, the American ambassador in Paris, cabled Washington that the French Government had officially requested "immediate armed intervention of U.S. carrier aircraft at Dien Bien Phu," and further that without American help by the middle or the end of the week the "fate of Dien Bien Phu will probably be sealed."[18] Ambassador Dillon also informed Washington that the French military commander in the field, General Henri Navarre, made it clear "that a relatively minor U.S. effort could turn the tide but naturally hopes for as much help as possible."[19] Finally, Dillon reported that the French Foreign Minister Georges Bidault said, "that for good or evil the fate of Southeast Asia now rested on Dien Bein Phu. He said that Geneva would be won or lost depending on outcome of Dien Bien Phu."[20]

The French Foreign Minister turned out to be right about Dien Bein Phu. For failure of the Americans to help even in a "minor" way, Dien Bein Phu fell on May 7, and as a reault Geneva was lost. The French also lost. Leslie H. Gelb put it correctly when he said, "Dien Bein Phu was the shock that ended the first Indo-China war. While Dien Bien Phu did not spell military defeat for French forces, it did decisively crack what was left of the shell of French public endurance and tolerance for the war."[21] With the withdrawal of France from Vietnam, it became an American responsibility to be the guardian of Southeast Asia, and in this sense the Americans also lost at Dien Bien Phu.

In 1954, at the time of Dien Bein Phu, the would-be architect of American foreign policy was Eisenhower's Secretary of State John Foster Dulles. He had come on board at the beginning of the Eisenhower administration, and his stock in trade was a very vigorous new foreign policy for the United States. As early as February 1952, Dulles had talked of "massive retaliation" as a correct foreign policy for the United States. This was before the H-bomb had been tested anywhere but was at a time when the American test was just about to be made. It was also at a time when the United States had begun to have large stockpiles of atomic bombs, plus a growing number of tactical nuclear weapons. In other words, when Dulles came on the scene America was at the peak of its atomic power (other than in the atomic-monopoly period) vis-a-vis the Soviet Union and was threatening to surge even further ahead depending on the outcome of the hydrogen test. Therefore, as far as the balance of power went, this was a time when the United States had the "big stick" and could afford to indulge in big talk about "massive retaliation," "rollback," and the "liberation" of oppressed people everywhere. The only pity is that it was all sound and fury signifying nothing.

When Eisenhower entered the White House it ended twenty years of Democratic rule in Washington, and the Republicans were determined that it would end a lot of other things, including the negative, futile and immoral policy of "containment." Instead, Dulles promised to seize the initiative and to roll back the tide of Communist aggression, not only in Europe, but in Asia as well. However, he did not propose using the iron fist of crude military power to accomplish these lofty objectives (with the possible exception of unleashing Chiang Kai-shek to liberate the Chinese on the mainland), but instead to use the powers of moral suasion backed up by the threat of atomic and hydrogen bombs. In other words, the diplomacy of the new Republican administration sought to turn Teddy Roosevelt's famous policy around entirely, and instead of speaking softly and carrying a big stick, Dulles apparently thought that the stick was not necessary if the voice was loud enough. Dulles said that this deception was the art of diplomacy. The name he gave tbis kind of artistry was "brinkmanship."[22] The art of "brinkmanship" is like the adolescent game of "chicken" played by juvenile delinquents in their souped-up jalopies, who drive toward each other on a collision course with their throttles opened wide. As soon as one loses his nerve the game is over, unless of course there is an actual head-on-collision. For a Secretary of State to play this game with the fortunes of the nation, is only to demonstrate that the wrong man is in office.

The doctrine of massive retaliation was aired in a speech Dulles made before the Council of Foreign Relations on January 12, 1954. Shortly after the January address, the climactic battle of Dien Bien Phu got underway, and while the assault was in progress Dulles gave another address in which he made it clear that Southeast Asia was an area of vital concern to the United States, saying that its fall "would be a grave threat to the whole free community."[23] President Eisenhower obviously agreed with him because in a news conference on April 7 the President said that the fall of Vietnam would have a "domino" effect on the other states in the area, endangering "the American defense system from the Philippines to Japan."[24] At the same time, "two American aircraft carriers, bearing nuclear weapons, sped toward Indo-China."[25]

However, Dien Bien Phu fell on May 7, with no aid from the Americans. Dulles later boasted that Dien Bien Phu was a good example of the art of "brinkmanship." However, it is difficult to follow Dulles' line of reasoning about Dien Bien Phu being a fine example of anything, unless, of course, the reasoning was that as long as the Philippines and Japan did not also fall, Dien Bien Phu was a success.

In the address before the Council of Foreign Relations, Secretary Dulles presented an overall view of the new national security policy for the Eisenhower administration. He said that the essence of the New Look was the doctrine of "instant, massive retaliation," and further, that this meant that a potential aggressor must know that his aggression would be met by "massive retaliatory power," not necessarily at the point of the aggression, but possibly at the source.[26]

The clear implication of this was that the Eisenhower administration was not going to put up with proxy wars in the future and that punishment would be meted out, not necessarily against the local pawns only, but also against the principals. "Local defense," Dulles said, "will always be important. But there is no local defense which alone will contain the mighty land power of the Communist world. Local defense must be reinforced by the further deterrent of massive retaliatory power. A potential aggressor must know that he cannot always prescribe battle conditions that suit him." Dulles concluded by saying, "the way to deter aggression is for the free community to be willing and able to respond vigorously at places and with means of its own choosing."[27] These brave words were no sooner spoken than they caused a storm of protest. Adlai Stevenson said that the new policy put the United States in the position of having to choose between "inaction or a thermonuclear holocaust," and predicted that Moscow would "nibble us to death." Henry Kissinger said the new policy showed signs of "Maginot mentality," or worse still a tendency to slide back to a way of thinking that reminded him of "Fortress America."[28] On the other hand, Dean Acheson found fault with the idea of having the United States threaten to take the lead in plunging the world into a nuclear holocaust under any circumstances, much less one that merely involved some minor infraction of the status quo.[29]

As a result of the essential absurdity of the massive retaliation doctrine and of the sharp criticism of it from nearly every quarter, foreign as well as domestic, the President took it upon himself to issue a denial that there had been any new basic decision on foreign policy. And it was not long before Dulles himself got into the act of denying his own words. In fact, this was done in the April 1954 issue of *Foreign Affairs*, which contained an article by the Secretary that explained away most of what he had said in the massive retaliation speech. "To deter aggression," Dulles said, "it is important to have the flexibility and the facilities which make various responses available." To emphasize the point, he added, "the free world must have the means for responding effectively on a selective basis when it chooses. It must not put itself in the position where the only response open to it is general war."[30] He continued by saying, "that does not mean turning every local war into a world war. It does not mean that, if there is a Communist attack somewhere in Asia, atom or hydrogen bombs will necessarily be dropped on the great industrial centers of China or Russia. It does mean that the free world must maintain the collective means and be willing to use them in the way which most effectively makes aggression too risky and expensive to be tempting."[31]

The article in *Foreign Affairs* clarifying the massive retaliation doctrine put the country back at aquare one from the point of view of foreign policy. Containment was still the basic policy, and the nuclear weapons were the last-resort guarantee of the policy. However, Dulles did put to rest any idea that the United States might not be the first to use its atomic weapons, everything else failing, and in this sense the massive retaliation speech was a reaffirmation of the policy of Hiroshima and Nagasaki.

That massive retaliation was not effective as a foreign policy for the United States was demonstrated every time it was put to a test. However, there may have been one or two exceptions to this. China may have been dissuaded from invading Formosa in 1954 and again in 1958 by the atomic threat. In the latter instance, Khrushchev took cognizance of the threat by warning President Eisenhower that

> those who carry out plans of atomic attack on the Chinese People's Republic should not forget that not only the U.S. but the other side possesses not only atomic but hydrogen weapons and also the corresponding means of delivery, and should such an attack be delivered on the Chinese People's Republic, then the aggressor will receive a fitting rebuff by the same means.[32]

Another occasion when massive retaliation may have had a bearing on a crisis was in the wrangling over the armistice terms in the peace talks over Korea. According to an undocumented story, Dulles and Eisenhower used the threat of the atomic bomb to get the armistice signed, thus allowing the United States to back out of Korea with the face-saving formula of a stalemate.[33]

In any case, at the very time Dulles was delivering his address to the Council of Foreign Relations on massive retaliation, January 12, 1954, the Soviets were making the transition from the atomic to the hydrogen bomb and were in the process of beating the United States to the first deliverable H-bomb.[34] In other words, Dulles was way behind the times with the logic of massive retaliation. The doctrine would have made a lot of sense during the period of the American atomic monopoly, which was roughly from 1945 to 1949, but it made absolutely no sense in 1954, when the Soviet Union had its own nuclear weapons and was fast moving to a position where it would be ahead of the United States in the field of ballistic missiles. Thus, it is probably fair to say, as Louis J. Halle did, that "massive retaliation" was "a phrase rather than a policy."[35]

Massive retaliation was a phrase or an idea that Khrushchev doted on, only it was the Soviet variety, not the American. Missiles were uppermost in Khrushchev's thinking. In fact, even before the days of Sputnik and the world encircling missiles, Khrushchev got out his short- and medium-range missiles at every possible opportunity and rattled them with obvious glee. Missiles were a passion with Khruchchev.

Khrushchev almost ruined a state visit to England in the spring of 1956 by his rocket diplomacy, to use the word diplomacy very lightly. For example, at a gathering sponsored by the First Lord of the Admiralty, Khrushchev was asked to make some remarks, and he did so with a vengeance. To this naval group, Khrushchev said, "gentlemen, here you are, the representatives of Great Britain, and the whole world knows there was a time when Britannia ruled the waves. But all that era is now past, and we must look at things realistically. Everything has changed."[36] Having thus set the stage, Khrushchev went on to say,

> Your specialists have told me how much they admire the battle cruiser which brought us here. Well, I'll tell you something. We will be happy to sell you this cruiser if you really want it because it's already obsolete. Its weapons have been outdated by new weapons. Besides, cruisers like ours no longer play a decisive role. Nor do bombers. Now it's submarines that rule the seas, and missiles that rule the air -- missiles that can strike their targets from great distances.[37]

On another occasion, Khrushchev was invited to a dinner at which Anthony Eden's wife, Churchill's niece, was the hostess. According to Khrushchev's account, "during dinner Mrs. Eden asked us, tell me, what sort of missiles do you have? Will they fly a long way? Yes, I said, they have a very long range. They could easily reach your island and quite a bit farther."[38] Referring to his hostess, Khrushchev said, "She bit her tongue. It was a little rude of me to have answered her as I had. Perhaps she took it as some sort of a threat. We certainly didn't mean to threaten anyone. We were simply trying to remind other countries that we were powerful and deserved respect, and that we wouldn't tolerate being talked to in the language of ultimatums."[39]

Khrushchev was deadly serious about his missiles, and the effect they would have on the character of war. In fact, he went overboard on the subject to such an extent that under him the size of the Soviet Army was sharply reduced. In explaining this reduction, Khrushchev said, "now, I know you can find people--especially in the military-- who will tell you that our reduction of the Soviet Union's armed forces was a mistake. They'll tell you that the imperialist camp has been dreaming for years of a chance to annihilate the Soviet Union and that the only thing holding back the aggressors has been our armed might. Well, people who say this are wrong."[40] Khrushchev went on to say,

> Once it was important how many troops, how many rifles, how many bayonets a country had. But we live in a new and different age. The number of troops and rifles and bayonets is no longer decisive. Now the important thing is the quality and quantity of our nuclear missile arsenal. The defense of our country and our ability to deter imperialist aggression depends on our nuclear and thermonuclear fire power.[41]

Later the same year, during the Suez crisis, Khrushchev got his rockets out again and threatened to rain them down on England if she did not desist from her predatory attack on Egypt. "What would be the position of Britain itself," he asked, "if she were attacked by stronger states possessing destructive weapons of all kinds?" Khrushchev went on to say that the days were long gone when England's insular position was a guarantee of her safety. He boasted that in this day and age "other means, for instance rocketry," could be used to bring England down and to accomplish this without even "sending a navy or an air fleet to British shores."[42] As for the United States, Khrushchev was all smiles. He vigorously applauded the fact that Eisenhower had come out against England, France and Israel in the Suez crisis and wondered aloud whether it would not be a good idea for the United States and Russia to "take joint action against the aggressors who had attacked Egypt?" He then put his thoughts on paper in a formal note to the mixed-up American President, proposing that "we should join forces under the flag of the United Nations" against the outlaw nations of England, France and Israel.[43]

The bewildered Eisenhower was so horrified at where his wrong-headed policy had led him that he ordered a world-wide alert of American forces. In other words, "the Soviet threat to bring medium-range missiles to bear on Britain or France in fact served to pull the United States into more of a confrontation with the U.S.S.R., with American bombers alerted and deployed to forward bases."[44] A footnote to this crisis is the fact that General Curtis E. LeMay (of all people) was the only one to take exception to this war-alert order on the part of President Eisenhower. He persuaded his mixed-up President that it would be unwise to tip his hand on war mobilization in a crisis as contrived as this, and thus paradoxically the Strategic Air Command was, as Professor George H. Quester said, "the only major American military command not fully alerted during the crisis."[45] Nevertheless, Khrushchev was satisfied with his ploy. So with the right amount of sarcasm and comedy, Khrushchev concluded, "so our note had done exactly what it was supposed to do, it had put the lie to the Americans' claim of being fighters for peace and justice and nonaggression. They may have been fighters in words, but not in deeds, and we have unmasked them."[46]

The crazy incident of the Suez crisis was not spent with the public humiliation of England, France and Israel. As Professor Adam B. Ulam so rightfully said, "one outcome of the Suez crisis was to be a conviction on Khrushchev's part that one could freely and profitably indulge in this kind of atomic blackmail, and this conviction stayed with him till one fall day in 1962."[47]

Long before the fall day in 1962 arrived, the day that marked the beginning of the Cuban missile crisis,

General Eisenhower's administration had been hard pressed to find a solution to a whole series of problems relating to national defense. All these problems touched upon the issue of massive retaliation in one way or another. The basic problem was the inability of the President to reconcile the conflicting goals of his campaign pledges to balance the budget and strengthen national defense. Apparently the only way this could be done was to go the atomic route and get "more bang for the buck." To this way of thinking, any defense that was not atomic was not considered good defense. On one occasion, Secretary of Defense Charles E. Wilson explained to a group of Congressmen that his budget contained very little "for the procurement of so-called conventional weapons" because "we are depending on atomic weapons for the defense of the Nation."[48]

General Eisenhower's concern about balancing the budget was so great that it took on the proportions of a religious crusade. He honestly believed that the story that was making the rounds at the time:

> That national security depended, in the first instance, on a sound economy, which in turn depended on fiscal policies that limited expenditures to a level substantially lower than those that the Democratic Administrations had accepted. The purpose of the Communists, it was maintained, was to induce the United States to bankrupt itself by its military expenditures, whereupon they would be free to take over the world. The conclusion was that one had to decide in advance how large a defense-budget the country could afford and then design its defense accordingly. The Democrats, on the other hand, insisted that one had first to decide what the requirements of defense were and then meet those requirements at whatever cost. There was, here, a clear issue in which the prevailing opinion in the Republican Party differed from the prevailing opinion in the Democratic Party.[49]

Since the Soviet Union had already become an atomic power and had successfully tested a hydrogen bomb, it was clearly necessary to go back to basics and rethink the problem of national defense. Obviously, it was a different problem from the days of the American atomic monopoly, and whereas massive retaliation might have been a practical policy in the good old days, it did not make any sense in the greatly altered circumstances. The same thing could be said about the problem of target selection in the event of a war. Now that American cities and industries were under the atomic gun, were the lessons of World War II still applicable to strategy in the event of a World War III? Bernard Brodie felt very strongly that this was not the case. Speaking of the atomic capabilities of the Russians, Dr. Brodie said, "one immediate result of the new development was the realization that questions inherited from World War II concerning appropriate selection among industrial target-systems were now irrelevant."[50] In a sense, Dr. Brodie is right, of course, but in another sense, he is wrong. Missiles still have to be aimed at targets, and targets have to be selected on the basis of doctrine. This is not too different from what it was in World War II.

For example, targets in World War II were selected on the basis of there being three broad objectives, two of which were strategic in nature and one tactical. The tactical targets were those designed to affect the outcome of a specific battle, and the strategic ones focussed on the war as a whole, rather than any one battle. Examples of tactical targets in World War II were enemy troop concentrations, command and control centers, and enemy lines of communications. As opposed to these, there were the two broad categories of strategic targets, one designed to cut down production in some vital area, such as the munitions of war, and the other aimed at enemy morale or general war-making capabilities. The same is true today to a very large extent. The battlefield targets are still selected for the same purpose, to kill enemy soldiers, to prevent reinforcements from being moved up and to turn an orderly retreat into a disorderly rout. In other words, the tactical targets of a World War III would be identical with the tactical targets in World War II, and it would not make any difference whether conventional or nuclear forces were used. As long as the objective is to win a particular battle the targets would be the same.

Would the doctrinal basis for the selection of strategic targets in a World War III be the same as it was in World War II? Here again, there is no real difference. The targets in the zone of the interior that were attacked in World War II were those that supported some existing military capability such as the ability of the enemy to produce the end-product of war, such as planes, tanks, or naval vessels, plus the airfields, naval bases and army staging areas. Today, these targets are called "counter-force" objectives, and the main targets in this category are the missile silos and launching pads for the large nuclear missiles of the enemy, again plus the bomber and naval forces, with particular reference to submarines. The target category for such attacks in World War II could very easily have been called "counter-force" objectives. Finally, there is the objective of "assured-destruction," which is exactly the same as the basic objective underlying the mission of the Combined Bomber Offensive in World War II. There in the famous Casablanca Directive of February 4, 1943, it was spelled out that the objective consisted of "the progressive destruction and dislocation of the German military, industrial and economic system, and the undermining of the morale of the German people to a point where their capacity for armed resistance is fatally weakened."[51] This is exactly what it means today when the expression "assured-destruction" is used.

In other words, the target doctrine of World War II recognized three distinct categories of targets: (1) battlefield targets; (2) counter-force targets; and (3) counter-value targets. It is the same today. However, another concept of the airmen that had its ups and downs in World War II was "command of the air," a concept the airmen borrowed from Alfred Thayer Mahan's "command of the sea." Giulio Douhet took the phrase and popularized it when he published his book *The Command of the Air*[52] in 1921. Oddly enough, his book was an argument against "command of the air" as a doctrinal basis for air power. Douhet's philosophy was that no time should be wasted trying to police the air, that is denying it to an enemy air force, but that a great blow should be struck at the counter-value target system, which, if destroyed, would make the people "rise up and demand an end to the war -- this before their army and navy had time to mobilize at all."[53]

The airmen in the United States entered World War II convinced that Douhet was right, and that they had the Flying Fortresses to prove it. After Schweinfurt, however, they were willing to admit that Douhet was wrong about "command of the air" and that their Flying Fortresses were not really flying fortresses at all. Then the American airmen turned their backs on Douhet, and in the Big Week proceeded to win "command of the air" with their long-range fighters. However, after Hiroshima and Nagasaki, the airmen reversed themselves again and reinstated Douhet as the prophet of air power. Even a non-believer such as Bernard Brodie was willing to admit that for all his errors in World War II Douhet was the designer of a "framework of stragegic thought which is considered by many responsible airmen to fit the atomic age astonishing well."[54] One of these forward-thinking airmen to whom Brodie referred was General Orvil A. Anderson, who astonished a group of fellow airmen in 1949 by telling them to forget about "command of the air" in order to concentrate on the "counter-force" targets, because "you'll fight them at the launching site or you won't fight them."[55]

There was another point that was raised at about this time, and that was whether the emergence of the Soviet Union as an atomic power created such a balance of terror in the world that it forced mankind to forget about war, at least nuclear war, as an instrument of foreign policy. This was the idea of the "nuclear stalemate," and it raised a series of interesting questions, some of which were very old questions. Was balanced power neutralized power? Was J. Robert Oppenheimer right with his simile about the two scorpions in a bottle? Had the nuclear weapons succeeded in accomplishing an objective that had eluded the statesmen of 1928, when they had sought to outlaw war with their famous (and forgotten) Pact of Paris? One analyst, Oskar Morgensterm, was so enamored of this doctrine that he seriously proposed "that it would be beneficial to maintain a nuclear stalemate, even by the expedient of strengthening Soviet forces by weakening U.S. power."[56]

In any case, the ambiguity and uncertainty about the doctrine of nuclear stalemate put General Eisenhower on the spot with respect to the problem of organizing the national defense and at the same time staying within the limits of what he conceived of as national solvency. Which had the priority, the atomic forces that made the nuclear stalemate credible, or the non-atomic forces that were required in local conflicts? The resolution of the problem of budget sharing for the armed forces touched off a very lively debate at the national level, among the services, as well as within the services. The globalists were the general war thinkers, and massive retaliation was their doctrine. On the other hand, there were many thoughtful analysts that challenged the doctrine of massive retaliation and advocated a mix of the armed forces that would enable the United States to be effective in fighting limited wars. It was argued that this was a more practical view, and indeed it was if it could be assumed that the era of nuclear stalemate had arrived. In addition to the debate at the national level on general versus limited wars, there was also a debate within the air force on the doctrine to be followed upon the outbreak of general war.

If the primary mission was Douhetan, the targets that would be struck first would be those in the "counter-value" system with the object of putting an end to the war in one fell swoop. On the other hand, if the primary mission was to protect the United States from the nuclear blows of the enemy, the "counter-force" targets would be the ones struck first. This would be in accordance with Orvil Anderson's concept. The dilemma of the choice was reflected in the views of some of the airmen. General Nathan F. Twining, for example, reflected on some of the operational problems when he said, "in the past it has been difficult enough to impose a new strategy on top of an old strategy. To impose now the old strategy on top of the new is out of the question."[57] On the other hand, General James H. Walsh was more flexible in his thinking and more resigned to the changing requirements of the times. His comment was "that Air Force thinking had turned full circle away from the Mitchell-Douhet doctrines to waging strategic air war against industrial capabilities and had returned to the older doctrines of Clausewitz and Schlieffen that considered enemy forces in being as the primary objectives of war effort."[58]

General Thomas D. White found the middle ground between these two extremes when he said that, given enough warning of the Soviet attack, his target priorities would be, "one, to destroy the enemy's capability to destroy us-- that would be the first priority; next would be to blunt the enemy attack against our deployed military forces in Europe and in Asia; and, third, systematically destroy the Soviet Union's ability to wage war."[59] On the other hand, White said that in the event of a surprise first-strike, the mission of the surviving forces in the United States would have to be improvised "to do the greatest possible damage to the Soviet Union as a whole with attention to applying that destruction in such a way as to do as much damage as possible to their residual military striking force."[60]

While the debate was going on over strategic doctrine within the Air Force, and at the national level, a new figure appeared on the scene, and he had some very definite and highly unpopular things to say about strategy and the mix of the armed forces. The new arrival was General Maxwell D. Taylor. He became Army Chief of Staff on July 1, 1955, and immediately made his presence felt by presenting to his staff a position paper called "A National Military Program." This was the first coherent statement of the new strategy of Flexible Response, a strategy that was designed to replace the old strategy of Massive Retaliation. General Taylor called the latter a Great Fallacy, and dedicated his book, *The Uncertain Trumpet*, to demonstrating that it was indeed a great fallacy.

The logic of General Taylor's National Military Program was

> that the deterrence of war is the primary objective of the armed forces. It recognizes deliberate general war as a grave threat, but, in view of the implications of a situation of mutual deterrence, it rates deliberate general war as less likely than lesser forms of aggression. Under such conditions it is extremely important to be able to cope quickly with limited aggressions since, if unopposed, they will lead to the attrition of our world position or, if resisted with inadequate means, they may expand into that general war which we are most anxious to avoid.[61]

General Taylor's fundamental point was that in an era of atomic plenty there would be a nuclear stalemaet, which neither side would risk upsetting, and that under these circumstances the ability to fight and win limited war would be the decisive favtor. He said that the alternatives were to let the world be burned up in an all-out exchange of nuclear weapons, or to watch creeping aggression take over the world the way the cat ate the grindstone, one grain at a time. Thus General Taylor thought that of the two kinds of wars in prospect it was more important to make preparations to fight limited rather than general wars, if for no other reason because they were more likely to happen.

At first, General Taylor's line of reasoning did not make much of an impression on anyone outside of Army circles. At a very special meeting of the Joint Chiefs of Staff in Puerto Rico to consider the subject of doctrine, General Taylor put his paper on "A National Military Program" on the table for discussion, and as he said, "My colleagues read this Army study politely and then quietly put it to one side."[62] There was no question about it, "Massive Retaliation" was still too firmly in the saddle to be shaken loose by a staff study. However, he was not deterred. He insisted that the "nuclear stalemate" was real, and that it even put a new face on the old doctrine of sword and shield. Taylor's argument was that

> Initially, the concept had been that the ground forces in Europe and the Far East were the shield behind which the U.S. could deliver the devastating blows of its atomic sword. Now the role was being reversed. The Atomic retaliatroy forces had become the shield of protection warding off the threat of hostile atomic attack, while the forces of limited war provided the flexible sword.[63]

It followed from this that the limited-war forces were the ones that had to play an active role in future military operations, and the atomic retaliatory forces were the ones that were relegated to a passive role. Thus it was clear (at least to General Taylor) that once the retaliatory forces had reached an optimum size and quality, the bulk of the defense dollar should go to the limited-war forces, the new and "flexible sword of parry, reposte and attack."[64]

Failing to make any headway with the Joint Chiefs on the idea of a Flexible Response, Taylor decided to take his case to the public, and with this in mind arranged with the editor of *Foreign Affairs* for the publication of an article on "Security Through Deterrence." The article was a reasonably low-key restatement of the Army's position as outlined in the "National Military Program." Taylor did not expect any trouble with security because the Army paper on the "National Military Program" had already been declassified. However, when his article came up for review, State and Defense disallowed it for publication on the grounds that the views expressed therein were "in conflict with approved policy."[65] General Taylor said of his suppressed article, "faced with this opposition to my article and knowing the futility of an appeal to the Secretary of Defense, I dropped the matter."[66] However, the general had the last word. Upon his retirement in 1959, and the publication of his book, *The Uncertain Trumpet*, General Taylor included the article in its entirety as Appendix A.

All of these conflicting views that split the Joint Chiefs up and down the middle were part and parcel of the fight for the almighty dollar. The fight was bitter and intense because it almost boiled down to a fight to the finish. The budget of funds for new equipment, for example, controlled the rate of modernization in the three services, and in the years of General Taylor's tenure from 1955 to 1959, the Air Force, as he said, "had consistently received some 60 percent of the available resources, the Navy and Marine Corps about 30 percent and the Army about 10 percent."[67]

General Taylor felt that this inbalance in the modernization funds was a direct result of the doctrine of Massive Retaliation and the theory of "more bang for a buck." Nevertheless, it rankled him, and the general refused to accept the fact that the defense world could be neatly divided into two spheres, one atomic and the other largely non-atomic, with the former getting all the dollars. For this reason, among others, General Taylor felt that at the heart of the defense problem was the practice of budget-making by the three services acting independently of each other instead of going at the problem from the point of view of the job that had to be done, that is from the point of view of the functional categories of the forces. By functional categories General Taylor meant the operational functions, such as "the atomic retaliatary force, continental air defense, overseas deployments, limited-war forces and the like." In each case, he said, the fundamental question is "how much of these forces is enough?"[68] Put another way, General Taylor said that it would never be possible for the Joint Chiefs "to produce an agreed tabulation of the forces needed for our security without first settling the basic question of how much is enough in the various operational categories. These yardsticks of sufficiency are the building blocks necessary to provide a solid foundation for defense planning."[69] "In other words," General Taylor said,

> The three services develop their forces more or less in isolation from each other, so that a force category such as the strategic retaliatory force, which consists of contributions of both the Navy and Air Force, is never viewed in the aggregate. Similarly, it is impossible to tell exactly how much continental air defense is being obtained from the defense budget since this is another category to which several services contribute. In other words, we look at our forces horizontally when we think of combat functions, but we view them vertically in developing the defense budget.[70]

General Taylor took it upon himself to calculate how much was enough for the atomic retaliatory forces, and found that the calculation was simplicity itself. He said that the force requirements could be

> determined fairly accurately, figuring back from the enemy targets which, if destroyed, will represent the loss of his war-making capacity. These targets amount only to a few score, at the most a few hundred. After the allowance has been made for aborts, enemy action and failures both human and mechanical, the number of atomic vehicles needed to destroy this target system can be determined by simple arithmetic. Even after adding a heavy factor of safety to cover imponderables, the size of the required atomic retaliatory force will be found to be much smaller than the bombers and missiles of our present force.[71]

By 1960, when General Taylor called attention to the "overkill" factor in current war planning, the Navy had switched sides in the debate and had come round to accepting Taylor's point of view on Flexible Response. Arleigh A Burke, even though the Navy had a part of the retaliatory force, said he agreed with Taylor that such forces were out of all proportion to the forces required for the mission. In fact, Burke said that the Navy could accomplish the "assured destruction" mission with its Polaris submarines. When asked how many such submarines would be needed to accomplish the mission, Burke replied,

> You can take from the number of Russian cities the number of megatons it takes to destroy a Russian city, the reliability of the missiles, the accuracy of the missile, and you can compute it pretty accurately yourself. And then you double it just to make sure and you come out some place in the neighborhood of perhaps 30.[72]

While this debate was going on over the issue of limited and general wars and over the role of air power in keeping the peace, Khrushchev was moving to take advantage of a changing of the guard at the White House. He got the idea by carefully studying the reaction of John F. Kennedy to the crisis in Cuba set off by the Bay of Pigs invasion of April 17, 1961. The trouble in Cuba began on January 1, 1959, before Kennedy came to office, when the corrupt regime of Fulgencio Batista suddenly collapsed, and power literally fell into the hands of the insurgent leader, Fidel Castro. He did not win a military victory in order to gain power; Castro simply moved into the vacuum that was created when the Batista government collapsed of its own weight. First, the Cuban people hailed Castro as a liberator, and his supporters in the United States succeeded in getting his government recognized in a week. In fact, the euphoria of his coming was so great that the simple-minded in the State Department and all the so-called liberals in the country called Castro the Robin Hood of the Caribbean. When his true colors began to show, however, and they were all Red, responsible Cubans began to turn against him in large numbers. They felt that Castro had betrayed the Cuban Revolution. Having a card-carrying Communist as their absolute dictator was not what the freedom-loving Cuvans had bargained for when they forced the corrupt Batista to flee. And out of these feelings the mass exodus of Cuban revolutionaries to the United States began. For some of the Cuban refugees the United States was a sanctuary and a staging base for the counter-revolution. Under the auspices of the Central Intelligence Agency, these Cubans were organized, trained and equipped not only in the United States, but in Guatemala and Nicaragua as well. The result was the abortive Bay of Pigs invasion on April 17, 1961. The object was to overthrow the regime of Premier Fidel Castro and bring the Cuban Revolution back to its starting place.

The invasion was a failure. The initial air strikes did not succeed in taking out all the Cuban planes, and President Kennedy refused to come to the aid of the men on the beach with planes from the carriers standing by. The result was that the mother ships had to flee or be sunk, and the men on the beach were left stranded without ammunition or supplies. The further result was the disaster of the Bay of Pigs, a disaster because of a lack of air support. Letting the Bay of Pigs invasion fail sent a message to the Kremlin about the character of John F. Kennedy and confirmed Khrushchev in his first impression that the new American President was a weakling and a fool. This is the same impression Hitler had formed of the Western leaders at Munich. "Our enemies," he said, "are little worms. I saw then at Munich."[73] On the basis of this impression of Kennedy's character, Khrushchev conceived of a bold plan to reverse the balance of power between the United States and Russia by secretly putting missile bases in Cuba. This would accomplish that purpose because it would bring the United States within range of the short- and medium-range missiles that were so abundant in the Soviet arsenals, and not make it necessary for the Soviet Union to rely solely on its long-range missiles to get at the United States. If successful, Khrushchev said, "our missiles would have equalized what the West likes to call 'the balance of power.'"[74]

The deployment of a substantial number of short- and medium-range missiles to Cuba would have made it possible for the Russians to have taken out a large portion of the U.S. strategic bomber force in a first-strike operation and to have done this without warning because from such a deployment the Soviets could have outflanked the U.S. radar warning net.[75] It was almost too good to be true.

Khrushchev's original plan was to deploy eighty missiles in Cuba on forty launching pads.[76] Accordingly, in the late summer and early fall of 1962, and with the full support of Castro, the Russians began to build the launching pads in Cuba and to secretly bring in their short- and medium-range missiles armed with atomic warheads.

In addition to the construction crews, the Russians brought in a force of twenty thousand regular troops to protect the work force at the missile sites. The troops were equipped with short-range tactical nuclear weapons. Even with the tactical nuclear weapons, however, security at the sites was not absolute, simply because the undertaking was too large to prevent word of it from leaking out of Cuba and eventually finding its way to the United States. When that happened, the rumor mill began to churn in the United States, and finally on October 14, it was discovered that the Soviet Union was constructing twenty-four launching pads for medium-range missiles and sixteen for intermediate ones. Not only that. It was clear from the aerial photography that the work was well along on these launching sites. It was also evident that the work was being rushed to completion. In fact, the work was so far along that on the Sunday of the Soviet capitulation from twelve to eighteen of the Soviet medium-range missiles "apparently had been readied for action,"[77] complete with their atomic warheads.

The question that confronted President Kennedy at this time was what should be done about the presence of the Russian missiles in Cuba? It was probably the most important question any American President ever faced. President Kennedy finally came to the conclusion that the missiles would have to be withdrawn from Cuba, and that in the event of a refusal, force would have to be used against Cuba, or Russia, possibly both. In short, there seemed to be no alternative to a World War III--unless, of course, the Russians backed down.

President Kennedy's first move in the crisis was like a token of earnest money in a business transaction. He imposed a naval blockade (calling it a quarantine), and notified the Soviets (October 22) that the response of the United States to the launching of any missiles from Cuba would be to initiate a full-scale nuclear war. The war-alert message to American missile commanders was sent out in the clear. To the Soviets, Kennedy said quite bluntly, "It shall be the policy of this nation to regard any nuclear missile launched from Cuba against any nation in the Western Hemisphere as an attack by the Soviet Union on the United States, requiring a full retaliatory response upon the Soviet Union."[78]

President Kennedy's courageous reaction to the presence of the Russian atomic missiles in Cuba came as quite a surprise and a shock to Nikita Khrushchev. In the first place, he had not expected his little secret to be discovered until all was ready, and in the second place, even if it were discovered, Khrushchev, recalling the Bay of Pigs, was confident he would be dealing with a weakling and a fool in the White House. Now he found that he had been wrong on both counts. Later, recalling this "fall day in 1962," as Professor Ulam called it, Khrushchev put the best possible face on it and said,

> The Americans began to make a belligerent show of their strength. They concentrated their forces against Cuba, completely surrounding the island with their navy. Things started churning. In our estimation the Americans were trying to frighten us, but they were no less scared than we were of atomic war. We hadn't had time to deliver all our shipments to Cuba, but we had installed enough missiles already to destroy New York, Chicago and the other huge industrial cities, not to mention a little village like Washington. I don't think America had ever faced such a real threat of destruction as at that moment.[79]

Khrushchev was right about the threat of destruction that the United States faced at the time of the Cuban missile crisis, but what he failed to mention was his own over-extended position. What sort of threat did America's might, with all the missiles on the ready, pose for the Soviet Union? While Khrushchev was speaking, the United States had one hundred fifty-six inter-continental missiles "ready to fire," and in addition "144 missiles aboard nine Polaris submarines."[80] Khrushchev described the situation accurately when he said that the two most powerful nations in the world were "squared off against each other, each with its finger on the button. You'd have thought that war was inevitable."[81]

Khrushchev was also quite correct as to what it would have meant if he or Kennedy had pushed the button. "I remember," Khrushchev said, "that President Kennedy once stated in a speech at a press conference that the United States had the nuclear missile capacity to wipe out the Soviet Union two times over, while the Soviet Union had enough atomic missiles to wipe out the United States only once."[82] The Soviet Premier continued by saying, "when journalists asked me to reply to Kennedy's statement, I said jokingly, 'yes, I know what Kennedy claims, and he's quite right. But I'm not complaining as long as the President understands that even though he may be able to destroy us two times over, we're still capable of wiping out the United States, even it it's only once. I'm grateful to the President for recognizing that much."[88]

One of the interesting things about the Cuban missile crisis is the role that Defense Secretary Robert S. McNamara played in its resolution. The reconnaissance flight of Sunday, October 14, 1962, revealed that the Soviets were beginning their missile build-up in Cuba. At first, McNamara disbelieved the evidence and felt that the case was not good enough to warrant alerting the President. By Tuesday, October 16, however, the President had been alerted, and a special task force had been organized to deal with the problem. Naturally, the most important question up for discussion was the military effect of Khrushchev's move. If allowed to go forward, how would it affect the balance of power between the United States and Russia?

The one person at the Tuesday meeting who should have known the most about the subject and who should have been more concerned than anyone else, was Defense Secretary Robert S. McNamara. Exactly the opposite turned out to be the case. McNamara all but denied that there was any relationship between the Soviet build-up in Cuba and the balance of power between Russia and the United States. He argued that the only military effect of having the missiles in Cuba "would be to reduce America's warning time in the event of war by just a few minutes."[84] Secretary McNamara did not seem to be at all cognizant of the fact that there were important differences between the various types of missiles and that the short- and medium-range missiles in Cuba, only ninety miles offshore the mainland, would give the Soviets such an enormous advantage over the United States that it would upset the balance of power between the two countries. The Secretary did not seem to appreciate the fact that by the move Khrushchev had put the United States under the gun of all the Russian missiles, not the relatively small number of her inter-continental nuclear weapons. On the basis of this misunderstanding of the strategic threat, McNamara came up with his famous *non sequitur*, "a missile is a missile. It makes no great difference whether you are killed by a missile fired from the Soviet Union or from Cuba."[85]

Another interesting thing about the Cuban missile crisis is the effect it had on the formulation of basic defense policy in and for the United States. The New Frontiersmen had run against the Eisenhower-Dulles policy of massive retaliation with the same vigor the Republicans had displayed in running against Harry Truman's policy of containment. Yet now that the chips were down the New Frontiersmen were coming back to the policy of massive retaliation just as Dulles and Eisenhower had come back to the policy of containment. The reversal of form on the part of the New Frontiersmen was not only on the policy of massive retaliation, but also on another point that was dear to Dulles. This was the point that the "full retaliatory response" would be against the source of the trouble, in this case the Soviet Union and would not be limited to their pawns, in this case the Cubans. For the Cubans at this stage of the crisis (October 22), the United States had assembled a force of a quarter million men in Florida and neighboring states for another Bay of Pigs operation, only this time it would be one where the presidential hand would not falter, and it was also one that would be backed up by the United States Navy, plus the full weight of Air Defense Command.

The continuity of American foreign policy is striking. All the administrations that had followed in the wake of Harry Truman's have had the same policy. The different slogans have been meaningless. "New look" policies have not been new, and there has never been any substance in the catchwords of diplomacy, "massive retaliation," "flexible response," "unleashing Chiang," "agonizing reapparisal," and the "liberation" of captive nations, or "rollback." It all goes back to a sincle sentence in President Truman's address to the Congress on March 12, 1947, "it must be the policy of the United States to support free peoples who are resisting attempted subjugation by armed minorities or by outside pressures."[86]

In the case of Cuba, the policy of the United States was a reaction to a Soviet attempt to upset the balance of power by putting the missiles in Cuba. The Russian attempt failed largely because President Kennedy did not lose his nerve again and had the courage to back Khrushchev down. Before capitulating fully, however, Khrushchev played a trick on the new President by attaching a proviso to the Russian withdrawal, making it conditional upon a promise by the United States that it would not invade Cuba. This was tantamount, of course, to making Cuba a Russian satellite in perpetuity. In agreeing to the face-saving condition, Kennedy did himself and his country a grave disservice. Earlier, Khrushchev had said that the Monroe Doctrine was dead and that the only thing to do was "to bury it, just as you bury anything dead, so it will not poison the air."[87] What Khrushchev was doing at this time was inviting Kennedy to be one of the pallbearers at the funeral of the Monroe Doctrine. From that time onward, not only did Cuba belong to Russia, but the whole of the Caribbean, Central America, and even South America, became fair game for the forces of Communism. The sword of the United States remained, but the shield of the Monroe Doctrine lay "a mould'ring in the grave."

Therefore, it is not surprising that Khrushchev, in his Memoirs, should be lavish in his praise of the Kennedys. "In our negotiations with the Americans during the crisis," Khrushchev said, "they had, on the whole, been open and candid with us, especially Robert Kennedy." Khrushchev also said, "the episode ended in a triumph of common sense. I'll always remember the late President with deep respect because, in the final analysis, he showed himself to be sober-minded and determined to avoid war."[88] Khrushchev continued by saying that President Kennedy

> didn't overestimate America's might, and

> he left himself a way out of the crisis. He showed real wisdom and statesmanship when he turned his back on right-wing forces in the United States who were trying to goad him into taking military action against Cuba. It was a great victory for us, though, that we had been able to extract from Kennedy a promise that neither America nor any of her allies would invade Cuba.[89]

The Cuban missile crisis in 1962 was a watershed in the affairs of man. For the first time, the nuclear Superpowers came right to the brink and backed away without plunging the world into a nuclear holocaust. Up to this time, the old rules of warfare were still in effect, and it was possible to think in terms of World War II; land power, sea power and air power. By 1962, however, a new dimension of warfare had been added, and a decisive one as far as the major powers were concerned. This was the dimension of aerospace power, or would it suffice to call it merely an extension of air power? General Thomas D. White had no doubts on this score and made his position clear when he said, "there are no barriers between air and space. Air and space are an indivisible field of operations."[90]

There was another thing that was made clear by 1962, and this was that the New Frontiersmen had put the United States on a new course in its relations with the rest of the world. The seeds that General Taylor had sown were beginning to sprout, and Robert S. McNamara took cognizance of this when he said "that the doctrine of massive retaliation is therefore useless as a guarantee of our security and must continue to give way to both the theory and the practice of flexible response."[91] In this connection, it was also clear by 1962 that the nuclear stalemate was probably an illusion and that nuclear war would not be ruled out simply because it was too horrible to contemplate. There is no assurance that all the missile crises of the future will be resolved in the same happy way the Cuban crisis was. In other words, all the Cuban crisis proved was that nuclear war cannot be ruled out as a possibility. In fact, it could be argued that the Cuban missile crisis strongly suggests that nuclear war is a high probability in the future.

This is not to say that there can be any winners in a nuclear war of the future. The only way this could be brought about is for one side to be strong enough to destroy all the "counter-force" targets (including those on the battlefields) in one fell swoop, and then be able to proceed with the destruction of the "counter-value" targets to the point where the enemy's capability and will to continue fighting are fatally weakened or broken beyond repair. This is why Defense Secretary McNamara put so much emphasis on the "assured-destruction" mission. He said,

> It is important to understand that assured destruction is the very essence of the whole deterrence concept. We must possess an assured-destruction capability and the capability also must be credible. The point is that a potential aggressor must believe that our assured-destruction capability is in fact actual, and that our will to use it in retaliation to an attack is in fact unwavering. The conclusion, then, is clear, if the United States is to deter a nuclear attack on itself or its allies, it must possess an actual and a credible assured-destruction capability.[92]

What does "assured-destruction" mean? An extreme example of it took place in the Punic Wars when the Romans took Carthage, killed or enslaved the inhabitants and ploughed the site with salt. A lesser example was Hiroshima in 1945, when 61 percent of the people were killed or injured in the atomic attack, and 60 percent of the industry in the urban area was destroyed. A third example was Nagasaki where the damage was so widespread that an official Prefectural Report described the city as looking "like a graveyard with not a tombstone standing."[93] When the expression "assured-destruction" is used in the strategic literature it means just this kind of destruction.

How vulnerable are the United States and Russia to the kind of "assured-destruction" that befell Carthage, Hiroshima and Nagasaki? Exceedingly vulnerable as it turns out. In the United States, two-thirds of the industrial capital is concentrated in 54 metropolitan areas, and the population and industry in the U.S.S.R.

is roughly equivalent to that of the United States, though much less urbanized. For example, the Soviet Union has only four cities with a population of a million or more, and the United States has 14. What this amounts to, as Professor Brodie said, is that "the number of cities that account for the so-called economic war potential of either the U.S. or the U.S.S.R. is small, possibly fifty or less, and certainly not over two hunred."[94] Thus it should be possible to knock out the United States or the Soviet Union with a finite number of bombs, 200 at the maximum, or 50 at the minimum. These are bombs on targets, of course, and do not make allowance for aborts, enemy action and failures both human and mechanical, such as navigational and guidance errors, plus all the other hazards of bombing. Arleigh Burke thought that it might be necessary to double the number of bombs required to make sure that there were enough bombs in the stockpile. However, even if they were tripled, the number of bombs would be only 600 for the 200 target cities, or 150 for the 50 cities.

These rough calculations did not satisfy Defense Secretary McNamara because of his passion for exact figures, and also because of his desire to be "cost-effective" in all his programs for national defense. Thus he had his "quiz kids" apply themselves to the problem of "assured-destruction," and the conclusion they came to was that the mission could be accomplished if the bomb damage added up to a loss of "50 percent of the Soviet Union's industrial capacity and 25 percent of its population."[95] It was thought that these figures were accurate because a generous allowance had been made for error, and the figures were greater than any of the losses suffered by the belligerents in World War II. The figures also corresponded roughly with the actual losses at Hiroshima, when 69 percent of the industry in the urban area was lost and 61 percent of the population was killed or injured.

The calculations the "quiz kids" made were probably accurate from the point of view of what it takes to collapse a state in modern society. However, they were unrealistic from the point of view of the military situation facing the planners today. The fact is that the Hiroshima bomb had a yield of only 13 kilotons, about 13,000 tons of TNT,[96] while the bomb that is in common use today is a one-megaton bomb, with a yield of 1,000,000 tons of TNT.[97] This is equivalent to 77 Hiroshima-type bombs. In other words, if each of the 200 target-cities in the United States and Russia were hit with a one-magaton bomb today, it would be the same as if each were hit by 77 Hiroshima bombs. This would be enough to positively guarantee reaching the goal of "assured-destruction" in any sense of the term. And for the country as a whole, exploding 200 one-megaton bombs over 200 cities, with each bomb being the equivalent of 77 Hiroshimas, would be the same as having 154,000 Hiroshima bombs going off all at once. This would certainly be adequate to achieve the goal of "assured-destruction" for the country as a whole.

Looked at from this point of view, the minimum requirement to carry out the "assured-destruction" mission in the United States or for Russia would be 200 warheads of one-megaton each. If there is a bias in this calculation, it is on the conservative side. Professor Brodie made it quite clear that taking out only 50 target-cities in Russia and the United States, rather than 200, would bring about a social and economic collapse in both countries,[98] and take each out of the war. If this is so, only 50 bombs of the one-megaton type would be required, not the 200. This is the case for both the U.S. and the U.S.S.R.

This is about as close as it is possible to get to the core question of the number of weapons required to achieve the goal of "assured-destruction" in the United States or in Russia. It is also the answer to the question of how much is enough or when is enough enough? Obviously, when a country has been destroyed to the point of no return from a military point of view, there is no need to go on piling on additional destruction ad infinitum and ad nauseam. It would not be "cost-effective."

The point that enough is enough is the one General Taylor and Admiral Burke tried so hard to make in the late 1950s and early 1960s. Since that time, there has been no increase in the number of target cities, but there has been a decrease in the number of weapons required because of the vast improvements that have been made in the weapons themselves and the accuracy of delivery. However, the stockpiles have grown much larger, not smaller. Why? If the point of no return as far as sufficiency goes had been reached and surpassed in the early 1960s, as Burke and Taylor both pointed out, why should it be considered necessary to double or triple the already bloated stockpiles?

The basic problem the United States has had (and still has) with atomic and hydrogen bombs is that the weapons were ready before the mind was prepared to cope with them. Logically, doctrines would be formulated that would establish military requirements for certain types of weapons, and then scientists and engineers would bestir themselves to produce the weapons. When this process is reversed, as it was in the case of the nuclear weapons, it creates a problem. Military doctrine then has to adjust itself to technological development. This is a case of the tail wagging the dog. A recent report by the Secretary-General of the United Nations commented on this development in weaponry by saying,

> The ongoing technological development of nuclear-weapon systems is sometimes described as necessitated by threats to national security, and as a corollary to the evolution of theories or doctrines for the use of nuclear weapons. It is widely believed, however, that new weapon systems emerge not because of any military or security considerations but because technology by its own impetus often takes the lead over policy, creating weapons for which needs have to be invented and deployment theories have to be readjusted.[99]

The truth of the matter seems to be that at the end of World War II, when the United States found itself with an atomic bomb, it had no clearly defined policy as to what to do with it, or how to exploit its advantage, except possibly by giving it away. When this failed, however, the United States just let the bomb slide without doing anything to integrate it with policy. Not only was there a virtual shutdown of the facilities for producing the nuclear weapons, but practically nothing was done to safeguard the few existing bombs. The entire stockpile of existing weapons, as Professor George H. Quester said, "apparently was kept concentrated in a single location at Los Alamos, New Mexico, outside of Albuquerque, with the bombs left unassembled as a collection of components in the custody of the Atomic Energy Commission." Professor Quester also said, "under such circumstances, therefore, Soviet conventional bombers operating even on a one-way mission in early 1949 might have been able to cripple further production of bomb components, and perhaps to interfere with the assembly of those already produced. The possibility thus existed that the United States atomic advantage might be erased."[100] There was also the distinct possibility of the project being wrecked by sabotage. In other words, American policy at the time was a mixture of carelessness, inflated confidence and naive optimism.

One of the reasons for the failure of American policy in the handling of its atomic monopoly was the fact that so few people knew anything about the bomb before it was produced. Harry Truman, for example, was kept in the dark about the bomb all the time he was Vice-President and was not made privy to the secret until after the death of President Roosevelt.[101] The hiatus of solid thinking on the implications of the bomb was one of the reasons for the failure of policy in one of the most critical periods of the nation's history. The failure resulted in the furious and senseless race with the Soviet Union to be first in the world of nuclear weaponry. It was almost as though there was a giant scoreboard in the sky where points were chalked up for each advance in the nuclear race, so many points for being first with the fission weapons, additional ones for the hydrogen bomb, then a large number for the triumph of Sputnik, others for getting ahead with the independently tragetable re-entry vehicles, still more for the cruise missiles, the neutron bomb, spy satellites and so on ad infinitum. It's a winner-take-all game. Or is it? Can there be a winner? Is it a game? Will the winner be crowned with a laurel of thorns to prove that he is a bigger fool than his unworthy opponent?

Before proceeding with an examination of the arms race in more detail, it might be worthwhile to take a look at the SS-18 as a measure of Russian capabilities, and at the Poseidon submarine to get some idea of the destructive power available to the United States. The SS-18 is one of the most powerful land-based missiles in the Soviet system, and more than three hundred of them have been deployed against the United States. Some of the SS-18s carry 10 one-megaton warheads. Just one of these missiles could rain down on 10 targets in the United States the explosive power of 77 Hiroshima bombs on each target.[102] Imagine what the condition of the United States would be if 77 Hiroshima bombs hit each one of the largest metropolitan areas in the United States, demolishing New York, Los Angeles, Philadelphia, St. Louis, Baltimore, Newark, Detroit, San Francisco, Pittsburgh and Washington, D.C.

Of course, the United States has the same capability vis-a-vis the Soviet Union. Today, the United States has thirty-one Poseidon submarines and more are on the way. Some of these carry sixteen missiles, and each has ten warheads. The Poseidon warheads are smaller than one-megaton, to be sure, the equivalent of only three Hiroshimas each, but they would create enough destruction to level all the target cities in the Soviet Union even if more than one bomb had to be used on the four largest metropolitan areas. In his 1979 message on the State of the Union President Jimmy Carter said that one Poseidon submarine carried enough warheads "to destroy every large and medium-sized city in the Soviet Union."[103]

Are the statistics valid when Russian and American cities are compared with Hiroshima, and when the one-megaton bomb is equated with 77 Hiroshima bombs? To a certain extent, yes. The Hiroshima bomb was one of 13,000 tons of TNT, and by simple arithmetic it would take 77 such weapons to equal a one-megaton bomb, which is equivalent to 1,000,000 tons of TNT. The error in this calculation is very slight, not greater than one-third. Though the area covered by blast does not increase linearly with weapon yield, it comes close. In other words, while a 20-megaton bomb would do less than twice as much damage as a 10-megaton one, it comes close enough to be valid as a statistical tool. As to the target cities themselves, and looking at them from the point of view of population, Hiroshima is comparable in size to 80 percent or more of the target cities in Russia and the United States. While it would not compare with the five or six great metropolitan areas in Russia or the United States, it would be larger than the other cities in the Russian target system. And as for the United States, Hiroshima with its normal population of 740,000 would be nearly twice the size of Atlanta, Georgia, about the size of Boston, Massachusetts, and slightly larger than Dallas, Texas. For the large metropolitan areas of Russia or the United States, the enemy would simply use four or five additional bombs. In other words, what is meant by "assured-destruction" is 50 or 60 times as much damage as was meted out to Hiroshima in 1945. So from the point of view of "assured-destruction," it is that with a vengeance, if it ever comes to using the one-megaton weapons.

While the exact number of the nuclear warheads in the arsenals of the great powers probably is not known to any one person in the world today, enough it known to make it possible to have a high degree of confidence in the current estimates. For example, there was an official exchange of data between the United States and the Soviet Union on the strategic nuclear forces of both sides in the proposed SALT II treaty, and there is good reason to believe that these data are highly reliable. They are confirmed by other sources. The evidence is persuasive that the total number of nuclear warheads in existence today exceeds 40,000. The explosive force of these warheads ranges from a few kilotons, or even less, to 20 million tons, or more. There is in principle no upper limit to the explosive yield that may be attained today. In fact, atomic bombs of the Hiroshima type are now being used to trigger hydrogen bombs.

In 1980, a careful and completely objective study of nuclear weapons was made by the Secretary-General of the United Nations, and one of his conclusions was that

> The total strength of present nuclear arsenals may be equivalent to about one million Hiroshima bombs, i.e., some 13,000 million tons of TNT. It is often pointed out that this is equivalent to more than three tons for every man, woman and child on the earth. The arsenals of the United States and the Soviet Union contain most of these weapons, with the known remainder belonging to China, France and the United Kingdom.[104]

There are all kinds of bombs in the arsenals of the Superpowers today. Some of them could be called "baby" bombs of 50-kilotons each equal to about three Hiroshimas, to the "regular size" one-megaton bombs each equal to about 77 Hiroshimas, and the "Big Bertha" 20-metaton bomb, each equal to nearly 1,500 Hiroshimas.

The bombs that are capable of reaching the homelands of the two Super-powers by virtue of their launchers or delivery systems are referred to as "central strategic systems." The others are the ones designed for use in regional conflicts, say between the Soviet Union and Western Europe, or for tactical use on the actual battlefields of the world.

The speed and power of the nuclear weapons raise tactical and strategic questions that have not been encountered before in the human experience. The nuclear weapons release in one micro-second more power than was generated by all the conventional weapons in all the wars of history. In fact, in an all-out nuclear conflict these weapons would destroy more cities, burn more buildings and inflict more casualties than in all wars of history put together. No natural disaster has ever posed the threat to mankind inherent in these bombs, no volcano, no earthquake, no plague, nothing.

The nuclear balance of power between the United States and Russia today may be expressed in terms of the number of strategic warheads each side possesses. Using this yardstick, the United States has the edge, with 9,000 nuclear warheads against 6,000 for the Soviet Union. On the other hand, using the megatonnage yardstick, the Soviet Union has a distinct advantage. Whereas the total megatonnage of the U.S. weapons is on the order of 4,000, the corresponding figure for the U.S.S.R. would be 7,000. All this means, however, is that the Soviets have built missiles with larger yields than those of the United States. From the beginning, the United States has preferred the smaller missiles because the designers have greater confidence in the accuracy of their weaponry. The American concept is and always has been that with sufficient accuracy a smaller bomb would be equally as effective as a larger one. But when the word smaller is used in this connection it usually refers to a bomb that would be equivalent to some 25 Hiroshimas, not very small by the old standards of measurement.

The following Table is part of the Report prepared by the Secretary-General, and is a summary of the total number of warheads and the total megattonage available to the nuclear powers:

Table 1. Rough Estimates of Current Nuclear Arsenals[105]
(Expressed in thousands of warheads and total yield in megatonnage)

	Central Strategic		Other Systems		Total	
	War-heads	Mt.	War-heads	Mt.	War-heads	Mt.
US	9-11	3-4	16-22	1-4	25-33	4-8
USSR	6-7.5	5-8	5-8	2-3	11-15	7-11
UK (total)					200-1000	200-1000
China (total)					300	200-400
France (total					200	100

The 15,000 warheads in the "central strategic systems" of the USSR and the USA are equivalent to about 750,000 Hiroshimas to be divided between the two countries. However, Russia and the United States do not plan to achieve the common goal of "assured-destruction" in the same way. The Soviet Union has put its main reliance on its land-based missile system (ICBMs), a secondary reliance on its submarines (SLBMs) and almost no reliance on its manned bombers (Bisons and Bears). On the other hand, the United States has put its main reliance on its submarines (SLBMs) and split the difference between its long-range bombers (B-52s) and its land-based missile systems (ICBMs). Thus from the point of view of the distribution of the warheads (not the megatonnage) the United States and Russia are poles apart in their thinking.

A different picture would emerge if megatonnage were the basis for the comparison. For example, if there is sufficient warning of attack the manned bombers in the United States "would carry betweee one-fourth and one-third of all deliverable United States strategic nuclear weapons, adding up to perhaps half the total megatonnage, while the Soviet strategic nuclear payload is concentrated in its ICBMs."[106]

But concentrating on the distribution of the strategic warheads, the U.S. and the U.S.S.R. stand as follows on the delivery systems:

Leg of Triad	United States	Soviet Union
Land-based (ICBMs)	24%	75%
Submarine-based (SLBMs)	50%	20%
Air-based (Long-Range Bombers)	26%	5%
	100%	100% [107]

Thus it is apparent that the overkill potential is enornous on both sides. And it is growing. The rate of growth is astounding and, in view of the overkill factor, it is also surprising. From a basic requirement of not more than 200 to 400 one-megaton warheads to assure complete destruction of the war-making facilities in the United States or the Soviet Union, the two countries had 6,500 strategic nuclear warheads in 1960 and today have a minimum of 15,000. How many will they have five years from now? Why is it apparently impossible to check the descent into anarchy? Do both nations actually worship the false god of nuclear superiority? Is that the sole goal of the arms race?

Lest there be any question about the arms race being a senseless thing without an object, the United Nations report on nuclear weapons accepted the fact that both societies could be destroyed by an attack using a force of 400 megatons equivalent. In fact, the UN Report quite rightly said that "less than 400 equivalent megatons would suffice to cause 'unacceptable destruction' of either the USSR or the United States."[108] More than that. The Report said that in the United States (and to a lesser extent in the Soviet Union) the forces remaining after a surprise attack would be more than enough to achieve the goal of "assured destruction." This point is brought out in Table 3 of the UN Report, which emphasizes the fact that while thousands of megatons of bombs have been stockpiled in the Soviet Union and the United States, only a few hundred are required to bring about the destruction of the USSR or the US, or both.

Assumed Attacks Against the United States

Attack number	Total warheads & bombs required	Total megaton equiva-lents	Total Casual-ties (mill)	% Industry destroyed
1	300-400	144-166	40-60	25-35
2	400-500	244-266	50-70	35-45
3	500-600	344-366	60-80	45-60
4	700-800	544-566	70-90	60-65
	Assumed Attacks Against the USSR			
1	26	26	15	15
2	90	90	25	25
3	144	144	50	50
4	303	303	62	62 [109]

In addition to the 15,000 warheads in the "central strategic systems" of the United States and Russia, there are at least 21,000 in the other target systems designed for use against tactical targets on the battlefields of the world, including regional conflicts. This would not include the 700 warheads held in the arsenals of England, China and France. These are called tactical or Euro-system weapons. Though the yield of these weapons is smaller than the strategic weapons, it is nevertheless large, and the number of tactical warheads is actually larger than those in the strategic systems. There are some thirty thousand such warheads as opposed to fifteen thousand in the strategic systems.[110]

At the outset, one thing should be made clear about the tactical atomic bombs, and that is that calling them "tactical" nuclear weapons does not take away from the fact that they are nuclear weapons. The weapons themselves are all the same whether they are used for tactical or strategic purposes. It is largely a myth that some of the nuclear bombs are "clean" and others "dirty." The only difference is that the fission reaction gives off much larger amounts of radioactivity than the fusion reaction. Therefore, whether a bomb is clean or dirty depends on what fraction of its total energy is derived from fission. All bombs are dirty, the only difference is that some are dirtier than others.

The number of tactical nuclear weapons already deployed in the European zone is about seven thousand for the United States and three thousand for the Soviet Union.[111] The targets for these missiles are the enemy's field artillery, rocket and guided missile sites, airfields and command and communication centers. The average yield of weapons for battlefield targets might be of one to five kilotons and larger weapons, probably missiles with one hundred kiloton warheads, would be employed against the air bases and field com-communication centers.[112]

What would be the results of a tactical nuclear war fought out in Europe today along these lines? Europe would quickly become an atomic wasteland because basically the exchange would be a duel between the two opposing nuclear powers, and the fact that they used tactical atomic weapons instead of strategic ones would make very little difference. The weapon that devastated Hiroshima had a yield of only thirteen kilotons, and the "tactical" atomic bombs for use against the enemy's air bases and communication centers today have yields of one hundred kilotons each. This is equivalent to more than six Hiroshimas for every airfield and communication center, and for the people on the receiving end of the bombs it would be hard for them to remember that they were only tactical.

Would it be possible to fight a relatively clean war in Europe today with each side using only "tactical" atomic bombs against legitimate military objectives, a war that would not result in large numbers of civilian casualties, and one that would spare the cities, by and large? No, for several reasons. Tactical nuclear weapons are defined by the United Nations Report on the subject as being

> those nuclear weapons systems which, by virtue of their range and yield as well as the way they are incorporated in a military organization, have been designed or can be used for employment against military targets in a theatre of war. Such weapons are artillery shells, ground mobile rockets and missiles, air-launched bombs, rockets and missiles (with aircraft operating from carriers

> as well as land bases) and atomic demolition munitions (land mines). Naval forces of this kind comprise submarine-launched cruise or ballistic missiles, torpedoes and short-range, submarine-launched anti-submarine warfare rockets. Ground-based systems have ranges from about 15 kilometers (artillery) to several hundreds of kilometers (heavy missiles). Yield may vary from less than 01 to more than 100 kt.[113]

With weapons as diverse as these, and as varied in range and yield, it is next to impossible to think that a war involving troops in the hundreds of thousands could take place without the cities and urban populations becoming involved.

In the first place, large armies could not advance across the face of Europe in either direction without coming into close proximity to the ancient cities and capitals, and even if they could it is not reasonable to assume they would. What general would deliberately advance in such a way that his army would become an isolated target for nuclear attack? On the other hand, the generals would be drawn to the cities and ancient capitals because, if for no other reason, they are the transportation hubs controlling the railroad and highway networks without which the logistical support of large armies would not be possible. In World War II, while the public watched the great battles to see who won and lost the large cities, the generals of both armies fought for control of the railways. This was particularly true in the titanic struggle between Germany and Russia. In 1943 the Russians gained the victory when they denied the Germans the use of many north-south railroads, and in 1944 the Germans were doomed when the Russians cut the trunk lines from Germany and Poland into Russia. The same sort of struggle would take place today in an all-out east-west or west-east war.

Nevertheless, in the unlikely event of a great war breaking out in Europe today with both sides using only tactical atomic weapons, and using them only against legitimate military objectives, what would be the results for the civilian population of Europe? In such an event, the highly conservative UN Report predicted that approximately two hundred weapons of one hundred kilotons yields would be fired against strictly military objectives. The Report than went on to say,

> With this assumption, each of the 100 kt. weapons might kill or injure about 25,000 civilians, which would lead to a total of up to five million casualties. Thus the conclusion would be that the immediate effects of this nuclear war would be between five and six million civilian lives. This would hold only as long as the weapons were properly aimed, however. Each missile going astray and hitting an urban area instead of the intended target would add another quarter of a million to the total.[114]

Gruesome as these statistics are, they probably present an accurate picture of what a limited tactical war in Europe would look like. The civilian casualties in Europe would be equal to the total number of German, Japanese and Italian soldiers killed in the whole of World War II and would be five times as high as the battlefield losses for the United States, England and France in the same war.[115] In other words, there is no such thing as the tactical use of atomic weapons without the wholesale slaughter of people, particularly non-combatants.

The neutron bomb is no exception. This is a much publicized anti-tank weapon that, according to the theory, will emit neutrons that will penetrate tank armor and kill the crew without doing any other damage. However, in an article in *Foreign Affairs* (Winter 1982) Spurgeon M. Keeny challenged this idea. He said,

> It is frequently overlooked that the neutron bomb is in fact a nuclear weapon with significant yield. While it does emit some ten times as many neutrons as a comparable "ordinary" small nuclear weapon, it also kills by blast, heat and prompt radiation. For instance, one of the proposed neutron warheads for the Lance missile has a one-kiloton yield, which would produce the same levels of blast damage experienced at Hiroshima at a little less than one-half the distance from the point of detonation.[116]

The point of all this is that a nuclear weapon is a nuclear weapon. There is no meaningful difference between tactical and strategic nuclear weapons, and any idea that the use of tactical nuclear weapons in Europe could serve as a "firebreak" to the use of strategic nuclear weapons is almost absurd. The plain fact of the matter is that any use of nuclear weapons in a European conflict would almost certainly lead to their general use. The enormous civilian casualties that would result from any use of nuclear weapons make a tit for tat nuclear war in Europe something that is oout of the question.

The most provocative nuclear weapons in existence today are those that fall in the category of the use-them-or-lose-them weapons. This applies to practically all the tactical nuclear weapons and to all the silo-based nuclear weapons in the United States and Russia. It also applies to the manned bombers in both countries, but to a much lesser extent, because they can always fly out from under an attack, and because they can be recalled after being launched. However, submarines do not fit in this category at all, certainly not those at sea. Over twenty years ago, Admiral John T. Hayward argued the case for the Navy by stressing the fact that submarines were relatively invulnerable to surprise attack, and predicting that "in the years to come, any system that is vulnerable to surprise attack will fade from the scene."[117]

The last twenty years have made a false prophet of Admiral Hayward. The 1980 Report of the Secretary-General of the United Nations made note of the fact that between 1968 and 1980 "the total number of strategic nuclear warheads had increased from 4,500 to at least 9,300 for the United States and from 1,000 to at least 6,000 for the USSR."[118] At the same time, the Secretary-General noted that technological developments in the effectiveness of nuclear weapons had greatly increased between 1968 and 1980 "by the continued evolution of the nuclear explosion but mainly by improvements in the accuracy and flexibility of delivery." Thus, he said, "to retain in 1980 the same destructive capability, particularly against hard targets, as in 1968, fewer weapons would thus be needed. As has already been pointed out, however, nuclear weapons have greatly increased in numbers since 1968."[119]

In 1968, Defense Secretary Robert S. McNamara more or less admitted that the United States and Russia had both lost their minds in the senseless race to outdo each other in piling on nuclear weapons. He said:

> What we would much prefer to do is to come to a realistic and reasonably riskless agreement with the Soviet Union which would effectively prevent such an arms race. We both have strategic nuclear arsenals greatly in excess of a credible assured-destruction capability. These arsenals have reached that point of excess in each case for precisely the same reason: we each have reacted to the other's build-up with very conservative calculations. We have, that is, each built a greater arsenal than either of us needed for a second-strike capability, simply because both wanted to be able to cope with the worst plausible case.[120]

The worst plausible case to which McNamara referred was the case of the Soviet Union in the days of the atomic monopoly. The case continued over into the catch-up period, and today would be the case if either power succeeded in gaining a first-strike capability powerful enough to destroy the enemy's "counter-force" target system to the point of making him incapable of a "second-strike." So in the absence of common sense the race goes on and will continue on into infinity until common sense finally prevails on one side or the other. Other than that, the only way the race can end is to have one side go bankrupt, or alternatively to have one side push the button, in which case 200 million people may have to die in order to prove that man is still not a rational creature, but only the "missing link" between anthropoid apes and full-fledged human beings.

This goes back to the idea of how much is enough, to the force requirements to produce those results, and to the stance of the United States with respect to its nuclear retaliatory forces. Is it possible for the Soviet Union to strike first and avoid retaliation by taking out the whole of the system of land-based intercontinental ballistic missiles? The Communists could strike first of course, and in the process kill anywhere from two to 22 million Americans,[121] but the answer to the second question is no. That is to say, even if the Soviets succeeded in wiping out the 1,052 land-based missiles of the United States, an unlikely event, the United States would still have from 4,000 to 7,000 nuclear warheads on bombers and in submarines, which would be more than enough to achieve the goal of "assured destruction."[122] In fact, this could be achieved with from 50 to 150 bombs on target, and the fact that from 4,000 to 7,000 would be available means that there would be enough warheads left over to turn the Soviet Union into a nuclear wasteland. The only purpose that would be served by a large-scale attack on the land-based missile system of the United States would be to start a process of evolution in reverse, which could quickly take man back to the time when he was still walking around on all fours. "A republic of insects and grass," as depicted by Jonathan Schell, would be an optimistic way of looking at the future.[123]

Maxwell D. Taylor was so unimpressed with the essentiality of the land-based missile system that he argued that "the most effective way to attain a higher degree of invulnerability would be to remove from American soil the most inviting targets for a surprise attack, our land-based" inter-continental ballistic missiles. James Fallows has suggested that an even better solution might be "to leave the Minutemen where they are, since maintaining them costs very little, since they are probably not really 'vulnerable' anyway, and since their presence gives the Soviet Union another complication to consider."[124]

To a certain extent, the bombers and submarines complement each other. The bombers, given sufficient warning, could fly out from under an attack by intercontinental ballistic missiles, and they can carry the very heavy megatonnage weapons to the hard targets in the Soviet Union such as the missile silos. And the submarines, as the UN Report said, "could be utilized against larger and 'softer' targets, such as military bases, airfields and population centres."[125] The bombers can also carry cruise missiles that can be fired from a "stand-off" position as far as 1,500 miles away from the targets, and the submarines can carry missiles with warheads equivalent to from three to six Hiroshimas each. When the new Trident submarines come on line they will be armed with even more powerful and more accurate missiles.

The airplane-submarine combination brings up two altogether unexpected names: William (Billy) Mitchell and Guilio Douhet. Their fame as philosophers of air power has obscured the fact that they championed the submarine almost as much as the airplane. In his book on *Winged Defense*, published in 1925, Mitchell had almost as much to say about submarines as he did about airplanes. Mitchell's whole philosophy was that "the advent of air power *and* the use of submarines wrought a change in the methods of war." He was right on both points. Douhet had much the same thing to say in his book on *The Command of the Air*.[126]

Airplanes and submarines have come along about the way Mitchell and Douhet predicted they would, and even today in the midst of the space age their importance is growing. From the point of view of the distribution of the nuclear warheads, the United States is relying on its submarines to carry 52 percent of the load, and on its long-range bombers for 25 percent, with only 23 percent for the land-based missile system. In the case of the Soviet Union, the reliance on the submarines is 25 percent, with only four percent for the bombers and 71 percent for the land-based missiles.[127]

Under the circumstances of today, with the United States and Russia facing each other armed to the teeth with over 40,000 nuclear warheads, what is the greater danger, that war will come by accident or by design? If history be any guide, and there are exceptions to this, notably World War I, major wars are the result of planned aggression, as World War II was, not of accidents. However, the nuclear age may be different. Today, the possibility of a major war breaking out by accident is very great indeed. For example, within the last twenty months there have been no fewer than 147 occasions on which there have been computer malfunctions that have signaled a Soviet strategic attack. The signals were strong enough on at least four occasions to cause alert orders to be issued to the American retaliatory forces. Once, a false signal of a Soviet launch was flashed to earth by satellites that mistook the rising of the moon for a hostile attack. Any one of these accidents could have triggered World War III. In fact, they would have if the United States had been dedicated to the principle of "launch on warning."

Human error is just as likely to result in the outbreak of war in the nuclear age as computer malfunctioning. The closest the world has ever come to an all-out nuclear war was during the Cuban missile crisis of 1962. There, as Khrushchev said, the two most powerful nations in the world were "squared off against each other, each with its finger on the button. You'd have thought that war was inevitable."[128] It was almost an accident that it did not come off, and there is obviously no guarantee that future missile crises will end up in the same non-atomic way. In fact, the one great lesson of the Cuban missile crisis is that all-out nuclear war between Russia and the United States is not only possible, but probable, and not only because of computer malfunctioning, but because of human error. If this had arisen out of the Cuban missile crisis, it would have been like World War I, a war nobody wanted.

The "launch on warning" policy that would have triggered World War III on many occasions is a policy that both Russia and the United States have considered adopting because of the time factor in connection with nuclear attack. This was what was behind the Russian threat in 1983 to put missiles in Cuba or Nicaragua if the United States went ahead with its plan to deploy Pershing II missiles in Europe. The argument was that Russian targets would be only five or six minutes away from the Pershing launching sites, and that they would have to launch on warning or not launch at all, that is unless they put the United States in the same position with their missiles in Cuba or Nicaragua. Actually, this was not a very plausible argument because the United States is already within five or six minutes of an incoming nuclear missile attack. In fact, on one occasion "a mistake caused by a programming error flashed a warning of a Soviet submarine attack. According to the Pentagon, it took six minutes for U.S. command authorities to make a positive identification of the mistake; in a few minutes more, if there had been no mistake, a fusillade of Soviet submarine missiles would have struck our coastal cities."[129]

How well prepared is the United States to fight a nuclear war? To all outward appearances, the United

States is well prepared. Certainly, there is no lack of preparation from the point of view of the number of bombs in the stockpile, the different kinds of weapons systems, and the total megatonnage. There are also elaborate plans to cover every possible contingency from the lowest to the highest level of nuclear attack. In other words, all the options of the President are open. As Commander-in-Chief, he can respond as he sees fit to any kind of nuclear attack, be it a response in kind to a limited nuclear attack against military targets only, or a full-scale response taking in the whole of the Soviet Union, cities and all. These are the two extremes, of course, and in between them are various other choices. To facilitate the decision-making process there is always the little "black box" that is carried by a military aide to the President everywhere he goes. The "black box" is available to help the President make the right decision among the many options he has at the tip of his fingers.[130] In theory, once the President makes his decision, the submarine crews at sea, the pilots on their airfields, and the missile commanders in their underground silos, all get the message, and there is a fully coordinated and carefully calibrated nuclear response.

The whole system depends on the smooth functioning of an intricate and highly involved chain of command, and that in turn depends upon the system of communications that is very vulnerable to enemy attack. As to the way the chain of command works, a 1982 study depicts the process by saying that the current system,

> focuses on quickly informing the President of an attack and transmitting his decision to retaliate. Early warning of an attack would be transmitted by radio links and land lines to a series of command centers. North American Air Defense Command (NORAD), located deep inside Cheyenne Mountain in Colorado, would evaluate the incoming data. NORAD would then confer with the National Military Command Center (NMCC) in the Pentagon and the Strategic Air Command (SAC) headquarters in Omaha, Nebraska, to verify an attack and assess its characteristics. Finally, the Chairman of the Joint Chiefs of Staff, the Secretary of Defense and the President would be notified.[131]

This is where the little "black box" comes in. The communication system to support this process involves thirty switching stations and major radio facilities. The trouble is that not all of these can be expected to function properly, if at all, after an enemy attack. The anti-satellite capabilities of the Soviet union could be expected to take care of the "early warning" space vehicles of the United States and the ground components of the system (radio antennas, radar facilities and land-line switching centers) are vulnerable to nuclear attack as well as to sabotage. The estimate is that the Soviets could make it impossible for the American military communications system to work by the destruction of as few as fifty specific targets.[132]

It might not be necessary for the Soviets to destroy fifty specific targets to break the military chain of command of the United States in the event of nuclear war. The chances are that the system would break down of its own accord. High altitude nuclear explosions generate electric bursts known as electromagnetic pulses that can damage or disrupt power supplies, computers, radio stations, or even telephone systems at distances of a thousand miles or more from the point of burst. In addition, "high altitude nuclear explosions ionize the upper atmosphere, degrading or even blacking out radio communications in high frequencies."[133]

The impact of a surprise nuclear attack on the United States, particularly its impact on the military communications system, has been under intensive study since 1961, when a special committee highlighted the problem and concluded that the nation's ability to command and control its retaliatory capabilities would be in serious doubt after a nuclear attack. This is why there is an eighteen billion dollar item in the present budget for a continuing and expanding program for the improvement and survivability of the strategic communications system.[134] The 1961 study was brought up to date in 1982 when a blue-ribbon panel of experts sponsored by the Carnegie Endowment took a new look at the problem and found that the current system was still a "major weakness."[135] In fact, the Carnegie panel took cognizance of the fact that the strategic Command Control Communications and Intelligence (which the Pentagon calls C^3I) is still the "weakest link[136] in the system of America's defenses. The experts said that the survivability of C^3I "as a system sufficient to support our declared strategic policies is far more doubtful than the survivability of the forces."[137]

The implications of this flaw in the national defense structure are ominous, and particularly so in view of the fact that it is probably not correctable. What it means is that if the Soviets are successful in launching a surprise nuclear attack, C^3I would probably be instantly destroyed or rendered inoperative for a long period of time. In other words, Washington will be decapitated from the point of view of there being an orderly chain of command linking the President with the triad of nuclear forces, land, sea and air.

Is there a clear and present danger or only a remote possibility that "nuclear decapitation" will take place in the event of a nuclear attack? It would certainly be the height of folly to assume that the Soviets wuuld not have a go at this strategy. The price is very low, equal to less than one percent of the megatonnage available, and the results could be out of all proportion to the effort. In fact, the destruction of the military command system of the United States would be the single most effective thing the Russians could do to paralyze the United States. The only safe thing to do, therefore, is to assume that the decapitation of the United States would be the first item on the agenda of the Soviet planners.

Would the Soviets succeed in decapitating the United States if they launched a surprise nuclear attack? Obviously, the attack would be a large one, designed not only to achieve the decapitation objective, but also to disarm the retaliatory forces, and to wreck the social and industrial structure of the United States. The blue-ribbon panel of experts assembled by the Carnegie Endowment had a surprising answer to the question of the decapitating objective. The answer was no. The panel members said, "an attempt at a decapitating surprise attack would fail."[138] Why? The line of reasoning followed by the panel was that the "bolt from the blue" would not catch the United States by surprise as had been the case at Pearl Harbor and in the instance of the Cuban missile crisis, and accordingly the country would have taken all the necessary steps to make it impossible for a decapitating surprise attack to succeed. The panel members argued that the Soviets would not undertake a nuclear strike except "in the context of a grave crisis" between the United States and Russia, and that in that case the decapitating attempt would not succeed because the retaliatory forces would already have been given their orders, and precautions would have been taken "to ensure the safety of the President or his successors and their ability to make and transmit a decision that will be recognized as legitimate."[139]

What the panel members failed to take into consideration was that a "grave crisis" between Japan and the United States existed in 1941 when the Japanese took the country by surprise at Pearl Harbor. And also, when Khrushchev started putting his missiles in Cuba in 1962 there was a "grave crisis" between Russia and the United States over the spy plane and the Berlin Wall, and still the country got caught by surprise. In any case, the members of the panel did not take seriously the "abstract calculations" of the theoretical possibility of a surprise decapitating strike."[140] But the panel agreed that the present C^3I system would not provide "enduring support and direction for surviving forces *after* a large initial exchange."[141]

What this means is that the strategic command, control and communication system would disintegrate under the impact of a full-scale exchange of nuclear weapons, and that thereafter the Commander-in-Chief, if he was still alive, would not be able to know which American forces had actually survived, much less be in a position to give them orders as to what to do next. The war that would result would be a sort of guerrilla conflict, doubled in spades, with the nuclear decisions being made by isolated submarine crews, by young missile commanders in their underground silos and by surviving bomber crews beside their waiting aircraft.

This would not be a very fitting way, as Professor William H. McNeill said, for human history to come to an end, "with a bang of hydrogen nuclei and a whimper from irradiated humanity."[142] On the other hand, perhaps the end will come this way. Professor Leslie A. White might not have been wrong when he predicted that it would. He said that the

> belief and faith that civilization, won at such great cost in pain and labor, simply cannot go down in destruction because such an end would be too monstrous

> and senseless, is but a naive and anthropocentric whimper. The cosmos does little know nor will it long remember what man has done here on this tiny planet. The eventual extinction of the human race--for come it will sometime -- will not be the first time that a species has died out. Nor will it be an event of very great terrestrial significance.[143]

The voices of doom have been the loudest ones to be raised since the advent of the nuclear revolution in the 20th Century. The date of the revolution, by the way was not 1945 when the atomic bomb made its appearance on the scene in New Mexico and at Hiroshima and Nagasaki, but 1952 when the first full-scale thermonuclear explosion in history took place at the Eniwetok proving grounds in the Pacific. As Churchill said, "the atomic bomb, with all its terror, did not carry us outside the scope of human control or manageable events in thought or action, in peace or war"; on the other hand, he also said that with the hydrogen bomb, "the entire foundation of human affairs was revolutionized."[144] Raymond Aron expressed a similar thought when he said, "the step from the atomic to the thermonuclear bomb was at least as radical as that from so-called conventional chemical explosives to atomic (or fission) bombs; some, in fact, date the *qualitative* revolution from the advent of the H-bomb rather than the A-bomb."[145]

Even though 1952 may be the correct date for the nuclear revolution, by 1945 or even earlier it was already apparent to some thoughtful observers that a crisis had been reached in the duel between the offense and the defense in warfare, and that the offense had outstripped the defense so far that it might never be able to catch up. General Arnold gave expression to this when he said that destruction had become too cheap, and accordingly predicted that the future belonged to the power that could solve the problem of defense. In other words, it was the view of these far-sighted thinkers that the empire of the world might come in sight if the problem of the defense could be solved.

This kind of thinking is in harmony with an idea of history that was put forward by Major General J.F.C. Fuller in 1945. He did this in a book called *Armament and History*, and the general idea was called the Constant Tactical Factor. The main point was that war never stands still, but swings from one extreme to another, as the offense asserts its superiority over the defense and as the defense rebounds to assert itself over the offense. The only certainty about war is the certainty of change. General Fuller concluded by saying, "it is this rhythm which has always had the profoundest influence in shaping history."[146]

History abounds in examples of how the rhythm of warfare has influenced the course of history. As long as the castles and fortified towns could be defended successfully, they were the basic political units in Europe, but when the development of gunpowder and cannon made this impossible the age of feudalism came to an end, and in place of the castles and fortified cities the modern territorial states became the basic political units, each with its impregnable walls of sovereignty. However, the defense came to the fore again in the First World War when the bullet on the defensive proved to be superior to the bullet on the offensive, and trench and siege warfare resulted, much like the condition in the Middle Ages when the castles and fortified towns were the basic political units. On the other hand when World War II came around, the tank and the airplane reversed the balance of World War I and restored authority to the offense. Then with the advent of the nuclear weapons and giant rockets, defense was all but forgotten, and talk began to circulate about One World or None, the wooden walls of sovereignty obviously being inadequate to protect against the new offensive weapons. The question is whether the state will cease to be the basic unit of society or whether the pendulum will swing back again and restore the authority of defense.

While the question was still up in the air two very distinguished scientists, Jerome B. Wiesner and Herbert F. York, pronounced judgment on the issue, and it was a resounding no. "Both sides in the arms race," they said, "are thus confronted by this dilemma of steadily increasing military power and steadily decreasing national security. It is our considered professional judgment that this dilemma has no technical solution. If the great powers continue to look for solutions in the area of science and technology only, the result will be to worsen the situation. The clearly predictable course of the arms race is a steady open spiral downward into oblivion."[147]

This is reminiscent of another very profound prediction on the part of a distinguished scientist with respect to the invention of the airplane. Fifty-seven days before the Wright brothers domonstrated that man could fly in a heavier-than-air, powered machine, the acknowledged dean of science Simon Newcomb handed down his judgment on the idea of flying. The cautious man said:

> There are many problems which have fascinated mankind ever since civilization began which we have made little or no advance in solving. The only satisfaction we can feel is that we have shown their solution to be impossible. The mathematician of today admits that he can neither square the circle, duplicate the cube, or trisect the angle. May not our mechanicians, in like manner, be ultimately forced to admit that aerial flight is one of the great class of problems with which man can never cope and give up all attempts to grapple with it?[148]

In this era of denial, doubt, and the failure of arms control after nearly fifty years of trying, man is about ready to throw in the sponge. George F. Kennan did this on May 19, 1981, when in a brilliant essay on the human condition in the nuclear age, he said:

> Over all these years the competition in the development of nuclear weaponry has proceeded steadily, relentlessly, without the faintest regard for all these warning voices. We have gone on piling weapon upon weapon, missile upon missile, new levels of destruction upon old ones. We have done this helplessly, almost involuntarily, like the victims of some sort of hypnotism, like men in a dream, like lemmings heading for the sea, like the children of Hamlin marching blindly along behind their Pied Piper. And the result is that today we have achieved,

> we and the Russians together, in the creation of these devices and their means of delivery, levels of redundancy of such grotesque dimensions as to defy rational understanding.[149]

It may be that Kennan was right in heaping ridicule on the results that have been obtained after 40 years of trying to gain peace and security through arms control, beginning with the abortive Baruch Plan of 1946. This has been a blind alley, and it should have been evident from the beginning that it would be. Warfare cannot be excised by arms control. Arms are only the symptoms of the disease, not the disease itself, and the control or lack of control of arms does not necessarily lead to either war or peace. Kennan is also right about the accumulation of the wrong kind of weapons, piling offensive weapons upon offensive weapons, missiles upon missiles, and new levels of destruction upon old ones. This does not lead to peace and security. It is also a blind-alley approach to the problem of national goals. It is also the exact opposite of the approach to peace and security that General H.H. Arnold and other thoughtful strategists urged upon the country 40 years ago when they warned that destruction had become too cheap, and that the country which solved the problem of defense would be in a position to rule the world.

That being so, one of the worst mistakes the country may have made in the post-war world was to allow itself to get involved in the arms control talks that led to the signing of a treaty in 1972 (SALT-I) which among other things, prohibited either country from the "development, testing, or deployment of ABM (anti-ballistic missile) components which are sea-based, space-based, or mobile land-based."[150] However, each side was allowed to build missile defenses for two cities or sites (such as the Russian defense of Moscow). This was later reduced to one, and today the United States has none.[151] This was a fundamental mistake, a "first step" in the wrong direction, and as it turned out a fatal "first step" from which the present sorry plight of the world may well have proceeded. Eleven years later in 1983 the United States began to take the first timid steps to correct the original mistake. It is never too late to lock the barn door, of course, but 11 years after the animal has escaped or been stolen is a little late to begin thinking about locking the barn door.

In 1983, America's outstanding physicist and father of the H-bomb, Edward Teller, stood up and let himself be counted as being opposed to the policy of "mutually assured destruction" instead of "mutually assured survival." Just as the Wright brothers ignored Simon Newcomb's false pronouncement about the folly of trying to unlock the secret of controlled human flight, so Teller would ignore the scientists like Wiesner and York who spoke ex-cathedra on the fact that science and technology would not be able to find a way out of the dilemma of the nuclear age.

Teller called his new policy "mutually assured survival," and in doing so took a swipe at the defense posture of the United States when he said that "the one practical certainty is that we will not remedy our present weakness by the procedures which have prevailed in the Pentagon for the past few decades." On the other hand, he said, "I am sure that very small sized nuclear explosives (for example, a nuclear bomb equivalent to about 50 tons of TNT) sent up to meet incoming missiles in the atmosphere can succeed in disabling them there without detonating the missiles. The effects of these little explosions will hardly be noticeable on the ground. However, this will require an assured warning from eyes in the sky." Teller concluded by repating the assertion that the eyes in the sky were the key element in the control of outer space. He said:

> This cannot be accomplished without a survivable observation system to deliver warning of war preparations, particularly the launching of destructive missiles. A limited presence in space with a variety of means for observation satellite survival is certainly a prime requirement for defense. Establishing such a system should be possible if necessary funds and even more necessary ingenuity are applied.[152]

What Teller did not say was that Bernard Baruch was probably right on June 14, 1946, when he said, "we are here to make a choice between the quick and the dead. That is our business. Behind the black portent of the new atomic age lies a hope which, seized upon with faith, can work our salvation. If we fail, then we have damned every man to be the slave of Fear. Let us not deceive ourselves: We must elect World Peace or World Destruction."[153]

FOOTNOTES--CHAPTER XII

[1]Halford J. Mackinder, *Democratic Ideals and Reality* (New York: W.W. Norton & Company, 1962), 27.

[2]Ecclesiastes, i, 8.

[3]Henry F. Woods, *American Savings* (New York: Duell, Sloan and Pearce, 1945), 1948.

[4]Robert Frank Futrell, *Ideas, Concepts, Doctrine: A History of Basic Thinking in the United States Air Force 1907-1964* (Alabama: Air University, Maxwell Air Force Base, 1974), 431.

[5]Walter A. McDougall, "Technology and Statecraft in the Space Age" *The American Historical Review*, Oct. 1982, v. 87, n. 4, p. 1021.

[6]Nikita Khrushchev, *Khrushchev Remembers* (Boston: Little Brown and Company, 1970), 362.

[7]*Ibid.*

[8]*Ibid.*, 517.

[9]Walter A. McDougall, "Technology and Statecraft" in the Space Age,"*The American Historical Review*, Oct. 1982, v. 87, n. 4, p. 1014.

[10]*Ibid.*, 1015.

[11]Michael Mandelbaum, *The Nuclear Revolution* (New York: Cambridge University Press, 1981), 96.

[12]James R. Shepley and Clay Blair, Jr., *The Hydrogen Bomb* (New York: David McKay Company, 1954), 5.

[13]George H. Quester, *Nuclear Diplomacy* (New York: The Dunellen Company, Inc., 1970, 91-92.

[14]Nikita Khrushchev, *Khrushchev Remembers* (Boston: Little, Brown and Company, 1970), 482.

[15]*Ibid.*

[16]Louis J. Halle, *The Cold War as History* (New York: Harper & Row, 1967), 297.

[17]Leslie H. Gelb with Richard Betts, *The Irony of Vietnam: The System Worked* (Washington: The Brookings Institute, 1979), 57.

[18]Neil Sheehan, *The Pentagon Papers* (New York: Bantam Books, 1971).

[19]*Ibid.*

[20]*Ibid.*, 30.

[21]Leslie H. Gelb with Richard K. Betts, *The Irony of Vietnam: The System Worked* (Washington: The Brookings Institute, 1979), 333.

[22]Louis J. Halle, *The Cold War as History* (New York: Harper & Row, 1967), 262.

[23]*Ibid.*, 296.

[24]*Ibid.*

[25]*Ibid.*, 297

[26]Robert Frank Futrell, *Ideas, Concepts, Docrtine: A History of Basic Thinking in the United States Air Force 1907-1964* (Alabama: Air University, Maxwell Air Force Base, 1974), 213.

[27]*Ibid.*

[28]Michael A. Guhin, *John Foster Dulles* (New York: Columbia University Press, 1965), 229-30.

[29]*Ibid.*, 229.

[30]John Foster Dulles, "Policy for Security and Peace," *Foreign Affairs*, April 1954, 354-55.

[31]*Ibid.*, 229.

[32]Robert Frank Futrell, *Ideas, Concepts, Doctrine: A History of Basic Thinking in the United States Air Force 1907-1964* (Alabama: Air University, Maxwell Air Force Base, 1974), 308.

[33]Louis J. Halle, *The Cold War as History* (New York: Harper & Row, 1967), 284.

[34]George H. Quester, *Nuclear Diplomacy* (New York: The Dunellen Company, Inc., 1970), 91-92.

[35]Louis J. Halle, *The Cold War as History* (New York: Harper & Row, 1967), 283.

[36]Nikita Khrushchev, *Khrushchev Remembers* (Boston: Little, Brown and Company, 1970), 407-08.

[37]*Ibid.*, 408.

[38]*Ibid.*, 405.

[39]*Ibid.*

[40]*Ibid.*, 515-16.

[41]*Ibid.*, 516.

[42]*Ibid.*, 435.

[43]*Ibid.*, 434.

[44]George H. Quester, *Nuclear Diplomacy* (New York: The Dunellen Company, 1970), 124.

[45]*Ibid.*

[46]Nikita Khrushchev, *Khrushchev Remembers* (Boston: Little Brown and Company, 1970), 435.

[47]Adam B. Ulam, *The Rivals* (New York: The Viking Press, 1971), 259.

[48]Robert Frank Futrell, *Ideas, Concepts, Doctrine: A History of Basic Thinking in the United States Air Force 1907-1964* (Alabama: Air University, Maxwell Air Force Base, 1974), 232.

[49]Louis J. Halle, *The Cold War as History* (New York: Harper & Row, 1967), 279.

[50]Bernard Brodie, *Strategy in the Missile Age* (Princeton: Princeton University Press, 1965), 155.

[51]Winston S. Churchill, *Closing the Ring* (Boston: Houghton Mifflin Company, 1951), 519-20.

[52] Giulio Douhet, *The Command of the Air* (New York: Coward-McCann, 1942, tr. Dino Ferrari).

[53] *Ibid.*, 58.

[54] Robert Frank Futrell, *Ideas, Concepts, Doctrine: A History of Basic Thinking in the United States Air Force 1907-1964* (Alabama: Air University, Maxwell Air Force Base, 1975), 89.

[55] *Ibid.*

[56] *Ibid.*, 223.

[57] *Ibid.*, 225.

[58] *Ibid.*, 304.

[59] *Ibid.*

[60] *Ibid.*

[61] Maxwell D. Taylor, *The Uncertain Trumpet* (New York: Harper & Brothers, 1960), 35.

[62] *Ibid.*, 37.

[63] *Ibid.*, 64.

[64] *Ibid.*

[65] *Ibid.*, 45.

[66] *Ibid.*, 46.

[67] *Ibid.*, 66.

[68] *Ibid.*, 118.

[69] *Ibid.*

[70] *Ibid.*, 123

[71] *Ibid.*, 148.

[72] Robert Frank Futrell, *Ideas, Concepts, Doctrine: A History of Basic Thinking in the United States Air Force 1907-1964* (Alabama: Air University, Maxwell Air Force Base, 1974), 310.

[73] Peter de Mendelssohn, *Design for Aggression* (New York: Harper & Brothers, 1946), 220.

[74] Nikita Khrushchev, *Khrushchev Remembers* (Boston: Little, Brown and Company, 1970), 494.

[75] George H. Quester, *Nuclear Diplomacy* (New York: The Dunellen Company, 1970), 234.

[76] *Ibid.*, 239.

[77] *Ibid.*, 244.

[78] Chalmers M. Roberts, *The Nuclear Wars* (New York: Praeger Publishers, 1970), 60.

[79] Nikita Khrushchev, *Khrushchev Remembers* (Boston: Little, Brown and Company, 1970), 496.

[80] Chalmers M. Roberts, *The Nuclear Wars* (New York: Praeger Publishers, 1970), 59.

[81] Nikita Khrushchev, *Khrushchev Remembers* (Boston: Little, Brown and Company, 1970), 500.

[82] *Ibid.*, 517.

[83] *Ibid.*

[84] Elie Able, *The Missile Crisis* (New York; J.B. Lippincott Company, 1966), 51.

[85] *Ibid.*, 43.

[86] Joseph M. Jones, *The Fifteen Weeks* (New York: The Viking Press, 1955), 272.

[87] K. E. Meyer and Tad Szulc, *The Cuban Invasion: The Chronicle of a Disaster* (New York, 1962), 72.

[88] Nikita Khrushchev, *Khrushchev Remembers* (Boston: Little, Brown and Company, 1970), 500.

[89] *Ibid.*

[90] Robert Frank Futrell, *Ideas, Concepts, Doctrine: A History of Basic Thinking in the United States Air Force 1907-1964* (Alabama: Air University, Maxwell Air Force Base, 1974), 281

[91]Robert S. McNamara, *The Essence of Security* (New York: Harper & Row, 1968), x.

[92]*Ibid.*, 52-53.

[93]The United States Strategic Bombing Survey, *The Effects of Atomic Bombs on Hiroshima and Nagasaki.* (Washington: Government Printing Office, 1946), 4.

[94]Bernard Brodie, *Strategy in the Missile Age* (Princeton: Princeton University Press, 1965), 157.

[95]Michael Mandelbaum, *The Nuclear Question* (New York: Cambridge University Press, 1981), 85.

[96]Report of the Secretary General of the United Natins, *Nuclear Weapons* (Brookline: Autumn Press, Inc., 1980), 10.

[97]Michael Mandelbaum, *The Nuclear Question* (New York: Cambridge University Press, 1981), 48.

[98]Bernard Brodie, *Strategy in the Missile Age* (Princeton: Princeton University Press, 1965), 157.

[99]Report of the Secretary-General of the United Nations, *Nuclear Weapons* (Brookline: Autumn Press, Inc., 1980), 28.

[100]George H. Quester, *Nuclear Diplomacy* (New York: The Dunellen Company, Inc., 1970), 33.

[101]Harry S. Truman, *Year of Decisions* (New York: Doubleday & Company, 1955), 10.

[102]Edward M. Kennedy, *Freeze!* (New York: Bantam Books, 1982), 11

[103]James Fallows, *National Defense* (New York: Random House, 1982), 144-45.

[104]Report of the Secretary-General of the United Nations, *Nuclear Weapons* (Brookline: Autumn Press, Inc., 1980), 8.

[105]*Ibid.*, 27.

[106]*Ibid.*, 15.

[107]Edward M. Kennedy, *Freeze!* (New York: Bantam Books, 1982). 14.

[108]Report of the Secretary-General of the United Nations, *Nuclear Weapons* (Brookline: Autumn Press, Inc., 180), 88.

[109]*Ibid.*

[110]*Ibid.*, 27.

[111]*Ibid.*, 23.

[112]*Ibid.*, 77.

[113]*Ibid.*, 21.

[114]*Ibid.*, 79.

[115]Quincy Wright, *A Study of War* (Chicago: The University of Chicago Press, 1967), 1542.

[116]Spurgeon M. Keeny, Jr. and Wolfgang K.H. Panofsky, "MAD VERSUS NUTS," *Foreign Affairs* (Winter 1981/82, vol. 60, n. 2), 297.

[117]Robert Frank Futrell, *Ideas, Concepts, Doctrine: A History of Basic Thinking in the United States Air Force 1907-1964* (Alabama: Air University, Maxwell Air Force Base, 1974), 310.

[118]Report of the Secretary-General of the United Nations, *Nuclear Weapons* (Brookline: Autumn Press, Inc., 1980), 6.

[119]*Ibid.*, 16.

[120]Robert S. McNamara, *The Essence of Security* (New York: Harper & Row, 1968), 61-62.

[121]Stanley R. Sloan and Robert C. Gray, *Nuclear Strategy and Arms Control* (New York: Foreign Policy Association, 1982), 45.

[122]*Ibid.*, 46.

[123]Jonathan Schell, *The Fate of the Earth* (New York: Alfred A. Knopf, 1982), 2.

[124]James Fallows, *National Defense* (New York: Random House, 1982), 166.

[125]Report of the Secretary-General of the United Nations, *Nuclear Weapons* (Brookline: Autumn Press, Inc., 1980), 15.

[126]William Mitchell, *Winged Defense* (New York: G. P. Putnam's Sons, 1925), 110.

[127]Stanley R. Sloan and Robert C. Gray, *Nuclear Strategy and Arms Control* (New York: Foreign Policy Association, 1982), 47.

[128]Nikita Khrushchev, *Khrushchev Remembers* (Boston: Little, Brown and Company, 1970), 500.

[129]Edward M. Kennedy, *Freeze!* (New York: Bantom Books, 1982), 131.

[130]Carnegie Endowment for International Peace, *Nuclear Strategy Issue of the 1980s* (New York: 30 Rockefeller Plaza, 1982), 95.

[131]*Ibid.*, 94-95.

[132]Edward M. Kennedy, *Freeze!* (New York: Bantam Boosk, 1982), 81.

[133]Carnegie Endowment For International Peace, *Nuclear Strategy Issues of the 1980s* (New York: 30 Rockefeller Plaza, 1982), 97.

[134]*Ibid.*, 84.

[135]*Ibid.*, 16.

[136]*Ibid.*, 85.

[137]*Ibid.*

[138]*Ibid.*, 15.

[139]*Ibid.*

[140]*Ibid.*

[141]*Ibid.*, 17.

[142] William H. McNeil, *The Rise of the West* (Chicago: The University of Chicago Press, 1963), 807.

[143] Leslie A. White, *The Science of Culture* (New York: Grove Press, Inc., 1949), 391.

[144] Michael Mandelbaum, *The Nuclear Revolution* (New York: Cambridge University Press, 1981), 3.

[145] Raymond Aron, *The Great Debate* (New York: Doubleday & Company, 1965), 7.

[146] Major General J.F.C. Fuller, *Armament and History* (New York: Charles Scribner's Sons, 1945), 21.

[147] Robert Frank Futrell, *Ideas, Concepts, Doctrine: A History of Basic Thinking in the United States Air Force 1907-1964* (Alabama: Air University, Maxwell Air Force Base, 1974), 427.

[148] William Fielding Ogburn, *The Social Effects of Aviation* (new York: Houghton Mifflin Company, 1946), 53-54.

[149] Edward M. Kennedy, *Freeze!* (New York: Bantam Books, 1982), 231.

[150] The Harvard Nuclear Study Group, *Living with Nuclear Weapons* (New York: Bantam Books, 1983), 94.

[151] Michael Mandelbaum, *The Nuclear Future* (Ithaca: Cornell University Press, 1983), 45.

[152] Edward Teller, *The High Frontier Study* (Skokie: Defense Science 2000), 1983.

[153] Chalmers M. Roberts, *The Nuclear Years* (New York: Praeger Publishers, 1970), 123

INDEX

About the Author

James Trapier Lowe was a pioneer in the field of Air Intelligence, and in 1945 the Legion of Merit was bestowed on him for his contribution to the "theory of strategic bombing." He has taught at Georgetown University, at the George Washington University, and at the University of Miami's Center for Advanced International Studies. He was a Carnegie Fellow in International Law in 1935, and in 1956 was awarded Georgetown's Vicennial Medal. As a theoretician of world power, Lowe published a book on *Geopolitics and War* in 1981 bringing up to date Halford J. Mackinder's theory of the Heartland. The present volume is a biography of an idea, *A Philosophy of Air Power*, more properly called "Aerospace Power." This idea goes back to the beginning of the human experience-- resulting first in the conquest of the air, and then of space.